Brief Edition

Computers
Understanding Technology
Seventh Edition

Lisa A. Bucki | Faithe Wempen | Floyd Fuller | Brian Larson

St. Paul

Vice President, Content and Digital Solutions: Christine Hurney
Director of Content Development, Computer Technology: Cheryl Drivdahl
Developmental Editors: Katrina Lee and Stephanie Schempp
Contributing Writers: Jan Marrelli, Brienna R. McWade, and Nancy Muir
Technical Review Consultant: Debbie Abshier, Abshier House
Director of Production: Timothy W. Larson
Production Editors: Carrie Rogers and Melora Pappas
Senior Design and Production Specialist: Valerie A. King
Photo Researcher and Illustrator: Hespenheide Design
Copy Editors: Eric Braem; Susan Freese, Communicáto, Ltd.; Renee LaPlume, LaPlume Communication; Suzanne Clinton
Indexer: Terry Casey
Vice President, Director of Digital Products: Chuck Bratton
Digital Projects Manager: Tom Modl
Digital Solutions Manager: Gerry Yumul
Senior Director of Digital Products and Onboarding: Christopher Johnson
Supervisor of Digital Products and Onboarding: Ryan Isdahl
Vice President, Marketing: Lara Weber McLellan
Marketing and Communications Manager: Selena Hicks

Care has been taken to verify the accuracy of information presented in this book. However, the authors, editors, and publisher cannot accept responsibility for web, email, newsgroup, or chat room subject matter or content, or for consequences from the application of the information in this book, and make no warranty, expressed or implied, with respect to its content.

Trademarks: Microsoft is a trademark or registered trademark of Microsoft Corporation in the United States and/or other countries. Some of the product names and company names included in this book have been used for identification purposes only and may be trademarks or registered trade names of their respective manufacturers and sellers. The authors, editors, and publisher disclaim any affiliation, association, or connection with, or sponsorship or endorsement by, such owners.

Paradigm Education Solutions is independent from Microsoft Corporation and not affiliated with Microsoft in any manner.

Cover Illustration Credit: © iStock/enjoynz
Interior Image Credits: Following the index.

We have made every effort to trace the ownership of all copyrighted material and to secure permission from copyright holders. In the event of any question arising as to the use of any material, we will be pleased to make the necessary corrections in future printings.

ISBN 978-0-76388-759-9 (print)
ISBN 978-0-76388-760-5 (digital)

© 2020 by Paradigm Publishing, LLC
875 Montreal Way
St. Paul, MN 55102
Email: CustomerService@ParadigmEducation.com
Website: ParadigmEducation.com

All rights reserved. No part of this publication may be adapted, reproduced, stored in a retrieval system, or transmitted in any form or by any means, electronic, mechanical, photocopying, recording, or otherwise, without prior written permission from the publisher.

Printed in the United States of America

29 28 27 26 25 24 23 22 21 20   1 2 3 4 5 6 7 8 9 10

Contents

Preface ix

Chapter 1
Touring Our Digital World 1

- 1.1 **Being Immersed in Digital Technology** . 2
  - Digital Information 3
  - Computerized Devices versus Computers 4
- 1.2 **Discovering the Computer Advantage** 5
  - Speed 6
  - Accuracy 7
    - Practical Tech: Tapping the Crowd's Opinion 7
  - Versatility 8
  - Storage 9
  - Communications 9
- 1.3 **Understanding How Computers Work** ... 12
  - Data and Information 12
  - The Information Processing Cycle 13
- 1.4 **Differentiating between Computers and Computer Systems** 14
- 1.5 **Understanding Hardware and Software** . 16
  - Hardware Overview 16
    - Tech Ethics: Close the Gap 16
  - Software Overview 19
- 1.6 **Identifying Categories of Computers** ... 21
  - Smart Devices (Embedded Computers) and the Internet of Things 22
    - Cutting Edge: Previewing the Latest at CES ... 22
  - Voice-Controlled Smart Speakers 24
  - Wearable Devices 25
    - Practical Tech: Creating Your Own Automation . 25
    - Cutting Edge: A Computer in Your Eye .. 26
  - Smartphones, Tablets, and Other Mobile Devices 27
    - Hotspot: An Inside Look at Your Small Intestine . 27
    - Practical Tech: Dealing with Older Devices ... 28
  - Personal Computers and Workstations 29

- Servers (Midrange Servers and Minicomputers) 31
- Mainframe Computers 32
- Supercomputers 33
  - Cutting Edge: Beyond Super 33
- Chapter Summary 34
- Key Terms 35
- Chapter Exercises 37

Chapter 2
Sizing Up Computer and Device Hardware 41

- 2.1 **Working with Input Technology** 42
  - Keyboards 42
    - Practical Tech: Are Some of Your Keys Not Working? 43
  - Touchscreens 45
    - Practical Tech: Touchscreen Gestures .. 45
  - Mice and Other Pointing Devices 46
  - Joysticks and Other Gaming Hardware 47
    - Practical Tech: Pointing Device Actions . 47
  - Graphics Tablets and Styluses 48
  - Scanners 48
  - Bar Code Readers 50
    - Hotspot: QR Codes 50
  - Digital Cameras and Video Devices 51
    - Hotspot: Photo Storage in the Cloud ... 52
  - Audio Input Devices 53
    - Cutting Edge: Biometric Security 53
  - Sensory and Location Input Devices 54
    - Tech Ethics: The Chip Debate 54
- 2.2 **Understanding How Computers Process Data** 55
  - Bits and Bytes 55
  - Encoding Schemes 56
- 2.3 **Identifying Components in the System Unit** 57
  - Power Supply 57
  - Motherboard 58
  - Buses 59

	Expansion Slots and Expansion Boards	59
	Ports	60
	Storage Bays	63
2.4	**Understanding the CPU**	**63**
	Internal Components	64
	Speed and Processing Capability	65
	Hotspot: Coming Soon: Transistors on Steroids!	65
	CPU and System Cooling	67
2.5	**Understanding Memory**	**68**
	RAM Basics	68
	Cutting Edge: Nanotechnology	68
	Static versus Dynamic RAM	69
	Practical Tech: Upgrading Your RAM	69
	RAM Speed and Performance	70
	RAM Storage Capacity	70
	ROM and Flash Memory	71
	Cutting Edge: Goodbye to Flash Memory?	71
2.6	**Exploring Visual and Audio Output**	**72**
	Displays	72
	Cutting Edge: New Applications for OLED	74
	Projectors	76
	Virtual Reality and Augmented Reality Hardware	76
	Speakers and Headphones	78
	Practical Tech: Connecting Speakers to Your Computer	78
2.7	**Differentiating between Types of Printers**	**79**
	Inkjet Printers	79
	Laser Printers	81
	Multifunction Devices and Fax Machines	83
	Thermal Printers	83
	Plotters	84
	Special-Purpose Printers	84
	Chapter Summary	85
	Key Terms	87
	Chapter Exercises	90

Chapter 3

Working with System Software and File Storage 95

3.1	**Defining System Software**	**96**
3.2	**Understanding the System BIOS**	**97**
3.3	**Understanding What an Operating System Does**	**99**
	Starting Up the Computer	100
	Providing a User Interface	101
	Practical Tech: Command Line: What Is It Good For?	101
	Running Applications	103
	Configuring and Controlling Devices	103
	Managing Files	103
3.4	**Looking at Operating Systems by Computer Type**	**104**
	Personal Computer Operating Systems	104
	Practical Tech: Resetting Your Windows PC	104
	Hotspot: OS Simulators	106
	Mobile Operating Systems	107
	Cutting Edge: Fuchsia OS	107
	Server Operating Systems	109
3.5	**Exploring Types of Utility Programs**	**110**
	Antivirus Software	111
	Firewalls	111
	Diagnostic Utilities	112
	Uninstallers	112
	Disk Scanners	112
	Disk Defragmenters	113
	File Compression Utilities	113
	Backup Utilities	113
	Spam Blockers	113
	Antispyware	114
3.6	**Programming Translation Software**	**114**
3.7	**Working with File Storage Systems**	**115**
	How File Storage Works	115
	File Types and Extensions	117
	Files and Shortcuts	118
	Basic File Management Skills	118

Contents

3.8 Understanding File Storage Devices and Media ... **119**
- Hard Disk Drives ... 120
- Solid-State Drives and USB Flash Drives ... 122
  - Practical Tech: How Fast Is Your SSHD? ... 122
- Network and Cloud Drives ... 122
  - Cutting Edge: Hybrid Drives ... 122
- Optical Storage Devices and Media ... 123
  - Tech Ethics: The Stored Communications Act ... 126

3.9 Understanding Large-Scale Storage ... **127**
- Local versus Network Storage ... 127
- RAID ... 128
- Chapter Summary ... 130
- Key Terms ... 132
- Chapter Exercises ... 134

Chapter 4

Using Applications to Tackle Tasks ... 137

4.1 Distinguishing between Types of Application Software ... **138**
- Individual, Group, or Enterprise Use ... 138
- Desktop Applications versus Mobile Apps ... 138
- Software Sales and Licensing ... 138
  - Hotspot: The App-ification of the Software Industry ... 139
  - Practical Tech: Purchasing and Downloading Software Online ... 141
- Application Download Files ... 142
  - Tech Ethics: Paying for Shareware ... 142

4.2 Using Business Productivity Software ... **143**
- Word-Processing Software ... 144
- Desktop Publishing Software ... 146
- Spreadsheet Software ... 147
- Database Management Software ... 149
- Presentation Software ... 151
- Software Suites ... 153
- Personal Information Management Software ... 153
  - Practical Tech: Syncing between Phones and Computers ... 153
- Project Management Software ... 154
- Accounting Software ... 155
- Note-Taking Software ... 155

4.3 Selecting Personal Productivity and Lifestyle Software ... **156**
- Lifestyle and Hobby Software ... 156
- Legal Software ... 156
- Personal Finance and Tax Preparation Software ... 157
- Educational and Reference Software ... 157
  - Cutting Edge: To Catch a Plagiarist ... 158
  - Tech Ethics: Should You Cite Wikis in Academic Papers? ... 159

4.4 Exploring Graphics and Multimedia Software ... **160**
- Painting and Drawing Software ... 160
- Photo-Editing Software ... 161
- 3-D Modeling and CAD Software ... 161
  - Practical Tech: Graphics File Formats ... 162
- Animation Software ... 163
- Audio-Editing Software ... 163
- Video-Editing Software ... 165
- Web Authoring Software ... 166
  - Cutting Edge: Mobile-First Design ... 166

4.5 Understanding Gaming Software ... **167**
- Types of Gaming Software ... 167
- Game Ratings ... 169
  - Cutting Edge: Gaming-Optimized PCs ... 169

4.6 Using Communications Software ... **170**
- Email Applications ... 170
- Web Browser Applications ... 172
- Text-Based Messaging Applications ... 172
- Voice over Internet Protocol Software ... 173
- Web Conferencing Software ... 173
- Groupware ... 173
- Social Media Apps ... 174
  - Hotspot: Instant Apps ... 174
- Chapter Summary ... 175
- Key Terms ... 178
- Chapter Exercises ... 180

Chapter 5
Plugging In to the Internet and All Its Resources 183

5.1 Exploring Our World's Network: The Internet **184**
  Communications 184
  Telecommuting, Collaboration, and the Gig Economy 185
  Practical Tech: Saving Money on Mobile Service 185
  Entertainment and Social Connections 186
  Electronic Commerce 186
  Research and Reference 187
  Distance Learning 188
  Practical Tech: Free Online Learning Opportunities 188

5.2 Connecting to the Internet **189**
  Hardware and Software Requirements 190
  Types of Internet Connections 191
  Hotspot: Understanding Airplane Mode 194
  Practical Tech: Shopping for an ISP 195

5.3 Navigating the Internet **197**
  Web Browsers 197
  IP Addresses and URLs 198
  Practical Tech: Is Your Computer System IPv6 Ready? 198
  The Path of a URL 199
  Packets 201

5.4 Understanding Web Page Markup Languages **202**
  HTML and CSS 202
  XML 203
  Website Publishing Basics 203

5.5 Viewing Web Pages **204**
  Basic Browsing Actions 204
  Audio, Video, and Animation Elements 205
  Ads 206
  Hotspot: Wireless, USA 207
  Private Internets 208
  Cutting Edge: Introducing Web 3.0, the Semantic Web 208

5.6 Searching for Information on the Internet **210**
  Search Engine Choices 210
  Default Search Engine 211
  Advanced Search Techniques 211
  Searching with an Intelligent Personal Assistant 212
  Bookmarking and Favorites 213

5.7 Using Other Internet Resources and Services **213**
  Electronic Mail 214
  Practical Tech: Mail on Mobile 214
  Social Media, Sharing, and Networking 216
  Chat and Instant Messaging 216
  Message Boards 218
  Blogs 218
  Online Shopping 219
  News and Weather 220
  Business, Career, and Finance 220
  Tech Ethics: Keeping Clean in Company Communications 220
  Government and Portals 221
  FTP and Online Collaboration 221
  Peer-to-Peer File Sharing 223
  Voice over Internet Protocol 223
  Audio, Video, and Podcasts 224
  Cutting Edge: Accelerating Innovation 224
  Audio Books and Ebooks 226
  Hotspot: Replacing TV with Telephone? 226
  Health and Science 227
  Online Reference Tools 227
  Gaming and Gambling 228
  Distance Learning Platforms 228
  Web Demonstrations, Presentations, and Meetings 229

5.8 Respecting the Internet Community **230**
  Netiquette 230
  Practical Tech: The Internet is Forever 230
  Moderated Environments 231
  Net Neutrality 231
  Privacy Issues 232
  Copyright Infringement 232

Contents

- Chapter Summary ... 232
- Key Terms ... 235
- Chapter Exercises ... 237

Acronyms and Abbreviations ... 241

Glossary and Index ... 247

Image Credits ... 270

Preface

For billions of people worldwide, computers, handheld devices such as mobile phones and tablets, and the internet help them perform integral and essential life activities. In the home, we use computers to communicate quickly with family and friends, manage our finances, control appliances, enjoy music and games, shop for products and services, and much more. In the workplace, computers enable us to become more efficient, productive, and creative employees, and make it possible for companies to connect almost instantly with suppliers and partners everywhere, including the other side of the world. Mobile devices help keep us engaged and productive wherever we are, providing essential features such as messaging and navigation, productivity software, and games and online media for entertainment on the go.

Computers: Understanding Technology, Seventh Edition, will help prepare you for the modern workplace, in which most careers require basic computer skills. Many jobs require more advanced skills with a variety of software and hardware, giving employees with the desired skills and knowledge an advantage. This program will help you start developing the tech skills you need for job survival and career growth, as well as introduce you to a wide variety of technology career possibilities.

Program Overview

Computers: Understanding Technology, Seventh Edition, introduces basic concepts in computer and information technology. Instructors and experts participating in surveys, reviews, and focus groups have contributed to ongoing improvements that have emerged in the various editions of this book. Throughout the various editions, we have incorporated their feedback to create an innovative, effective computer concepts program intended to meet and exceed your needs and expectations. In this newest edition, we updated figures, references, and examples throughout the program resources to align with the latest versions of Microsoft and Apple operating systems (Windows 10 and macOS Mojave) and Microsoft Office 365. We reorganized some of the content to make for more focused chapters, and we also updated facts and figures throughout to represent the most current information available.

This program is offered in two versions to match the needs of the most common computer concepts courses. The brief version consists of five chapters covering the digital world, hardware, software, the internet, and computer applications. The comprehensive version consists of 13 chapters, beginning with the identical five chapters and adding eight chapters covering networking and communication, the cloud, maintaining and managing devices, security and ethics, technology in business, programming, big data, and the future of computing.

Program Methodology and Structure

Computers: Understanding Technology, Seventh Edition, is a competency-based, objective-driven program that provides concept-level feedback. The program is divided into chapters, each focusing on a single goal. The chapters are divided into numbered sections, each focusing on a corresponding numbered learning objective. Each section covers a number of key concepts (also called *key terms*), which are boldfaced for easy reference. Careful study of the key concepts in a section supports mastery of the related learning objective, and mastery of the learning objectives in a chapter contributes to accomplishment of the chapter goal.

Chapter Features

The following visual guide shows how you can use this program's comprehensive content and enriching features to reinforce and expand on the concepts covered.

Clear Goals and Objectives

Chapter Goal provides a concise statement of the objectives covered in the chapter.

> **Chapter Goal**
> To learn how the internet works and how you can use it properly to perform a variety of life-enhancing activities

Learning Objectives are numbered and align with the major sections of the chapter.

> **Learning Objectives**
> 5.1 Describe the overall types of activities made possible by the internet.
> 5.2 Explain how to connect to the internet, including needed hardware and software and different types of connections.

Numbered section headings divide a chapter into major sections that align with the chapter learning objectives.

> **5.1 Exploring Our World's Network: The Internet**
> The internet is the largest computer network in the world. Its design closely resembles a client/server model, with network groups acting as clients and internet service providers (ISPs) acting as servers. (You will read more about networks and network

Rich Visuals, Engaging Features, and Helpful Online Connections

Infographics, figures, tables, and photos add interesting facts and visual emphasis to support your understanding of important concepts.

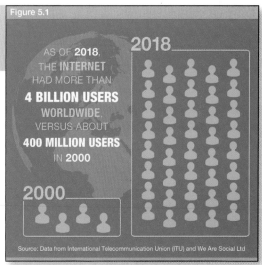

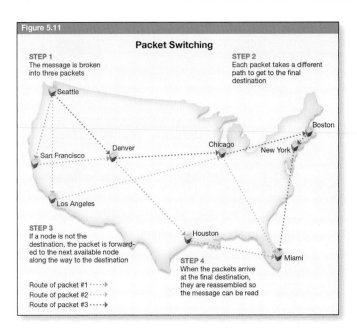

Table 5.2 Common Domain Suffixes Used in URLs

Domain Suffix	Type of Institution or Organization	Example
.com	commercial organization	Ford, Intel
.edu	educational institution	Harvard University, Washington University
.gov	government agency	NASA, IRS
.int	international treaty organization, internet database	NATO
.mil	military agency	US Navy
.net	administrative site for the internet or ISPs	EarthLink
.org	nonprofit or private organization or society	Red Cross

Practical Tech features provide real-world examples and advice about the best technologies or methods you can use to accomplish tasks or achieve objectives.

Practical TECH

The Internet Is Forever

The fact that Facebook and Twitter feeds seem to scroll off and Snapchat snaps seem to disappear immediately may tempt you to act out or reveal too much information online. You should remember, though, that the servers hosting your data may archive it for a long time, and other users may capture and keep your material. For example, it's possible for a recipient to make a screenshot of a Snapchat snap and post it on another service, such as Facebook or Twitter. In short, don't put anything online that you wouldn't want a future employer or anyone else to see. When posting "selfies," avoid photos that show you nude,

Hotspot features focus on wireless technology and its interesting twists, perspectives, and uses related to communications and community-building issues.

Hotspot

Understanding Airplane Mode

Making a call using a mobile phone during a commercial airline flight is prohibited in the US by the Federal Aviation Administration (FAA). However, the FAA does allow passengers to use some personal electronic devices (PEDs) during all phases of a flight, including takeoff and landing. Most major airlines offer Wi-Fi service, so now you can use your ebook reader or tablet throughout a flight. (Because of the sizes of laptop computers and certain devices, you must still stow them during takeoff and landing.) As of early 2014, additional rule changes with regard to the use of cell phones were also under consideration. service in the device's settings. (Depending on the device and manufacturer, you may see another name for airplane mode, such as *flight mode, offline mode,* or *stand-alone mode*.) Using airplane mode prevents the device from sending and receiving calls, text messages, and other forms of data. Enabling airplane mode may also turn off additional signaling features that could interfere

Cutting Edge features showcase compelling new technologies.

Cutting Edge

Accelerating Innovation

The development of advanced, high-speed internet applications and technologies is being facilitated by the Internet2 research platform. Internet2 is a consortium of more than 200 universities that are working in partnership with industry and government. It enables large US research universities to collaborate and share huge amounts of complex scientific information at amazing speeds. The goal is to someday transfer these capabilities to the broader Internet community.

Internet2 provides a testing ground for universities to work together and develop advanced Internet technologies, such as telemedicine, digital libraries, and virtual laboratories. An example of this collaboration was the Informedia Digital Video Library (IDVL) project. IDVL uses a combination of speech recognition, image understanding, and natural language processing technology to automatically transcribe, partition, and index video segments. Doing so enables intelligent searching and navigation, along with selective retrieval of information.

Internet2 operates the Internet2 Network backbone, which is capable of supporting speeds of more than 100 gigabits per second (Gbps). To learn more about Internet2, visit the Internet2 website at https://CUT7.ParadigmEducation.com/Internet2.

Tech Ethics features highlight ethical issues and situations in IT. Throughout the book, these features provide an ongoing discussion of ethics in the profession.

Tech Ethics

Keeping Clean in Company Communications

Electronic communications in the workplace often involve legal privacy issues. Companies generally keep backups of communication in the event of a data loss, but some industries require that all forms of digital communication be archived for legal reasons. For example, the financial industry requires that all email and instant message content be kept for several years. Doing so is necessary to ensure it will be available as evidence in the event of an investigation by the Securities and Exchange Commission or another legal body.

Employees should keep the content of email and instant messages in the workplace professional and courteous, because their employers will have digital records of who said what and when.

Bold blue links connect you to websites with relevant information.

1. **Obtain your domain name.** You can check to see if a particular domain name (for example, *mydomainname.com*) is available by using name registration services such as GoDaddy (at https://CUT7.ParadigmEducation.com/GoDaddy) and Network Solutions (at https://CUT7.ParadigmEducation.com/NetworkSolutions). If the name you want is available, you can register it by paying a fee. Registration services also enable you to renew your domain name, and in some cases, they allow you to

Preface

Comprehensive Reinforcement

A variety of study aids, exercises, and assessments offer ample opportunities to check your understanding of the chapter content.

Chapter Summary features highlight the most important concepts for each major section of the chapter.

> **Chapter Summary**
>
> **5.1 Exploring Our World's Network: The Internet**
> The internet is used for communication, working remotely (**telecommuting**) and collaboration. Computers can be used for social connections and entertainment purposes, including playing games, listening to music, and viewing movies and videos. Electronic commerce (e-commerce) refers to the internet exchange of business infor-

Key Terms features list the key concepts you should know after completing each major section of the chapter.

> **Key Terms**
>
> *Numbers indicate the pages where terms are first cited with their full definition in the chapter. An alphabetized list of key terms with definitions is included in the end-of-book glossary.*
>
> **5.1 Exploring Our World's Network: The Internet**
> telecommuting, 185
> World Wide Web (web), 187
> distance learning, 188
> learning management system (LMS), 188

Chapter Exercises features offer opportunities to apply your knowledge through the following categories of questions and prompts.
- Tech to Come: Brainstorming New Uses
- Tech Literacy: Internet Research and Writing
- Tech Issues: Team Problem-Solving
- Tech Timeline: Predicting Next Steps
- Ethical Dilemmas: Group Discussion and Debate

The Cirrus Solution
Elevating student success and instructor efficiency

Powered by Paradigm, Cirrus is the next-generation learning solution for developing skills in Microsoft Office and knowledge of computer concepts. Cirrus seamlessly delivers complete course content in a cloud-based learning environment that puts students on the fast track to success. Students can access their content from any device anywhere, through a live internet connection. Cirrus is platform independent, ensuring that students get the same learning experience whether they are using PCs, Macs, or Chromebook computers.

Cirrus provides access to all student resources, delivered in a series of **scheduled assignments** that report to a grade book to track progress and achievement.

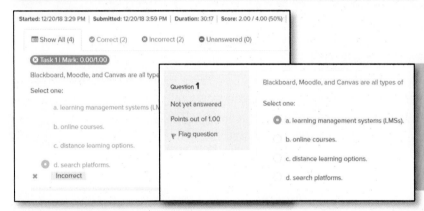

Objective quizzes (multiple-choice, true/false, matching, and completion quizzes) provide immediate feedback and an entry into the grade book.

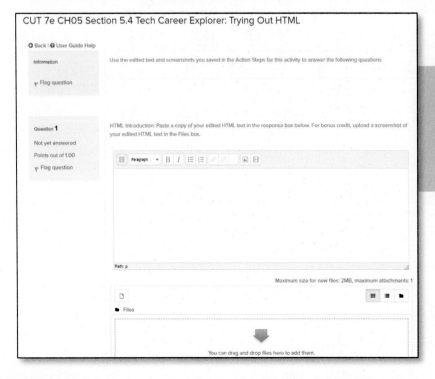

Essay quizzes provide response boxes with a text editor, and in some cases file upload options, to collect student work for instructor review and evaluation.

Dynamic Training

Cirrus online courses for *Computers: Understanding Technology*, Seventh Edition, include interactive assignments to guide and enrich student learning. Each activity ends with a quiz to check understanding and mastery.

Watch and Learn lessons offer opportunities to view section content in a video presentation, read the content, and check comprehension of the key concepts.

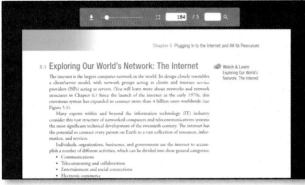

Video activities highlight interesting tech topics, directly related to section content.

Practice activities engage students in interactions to reinforce and test their understanding of key diagrams in the section content.

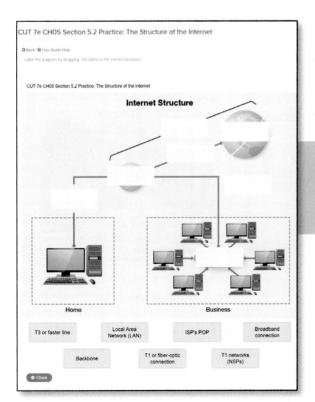

Tracking Down Tech activities introduce each chapter by challenging students to get off the computer and out of the study lab—to explore on campus and beyond to learn about technology while completing a scavenger hunt.

Tech Career Explorer activities help students explore a wide variety of tech career options and opportunities. Action steps range from completing interactive tutorials to researching and writing.

Hands On activities provide step-by-step instructions for specific tech tasks that are directly related to the section content.

Article activities expand section content with relevant and interesting topics. Bonus exercises challenge students to dig deeper into article topics, using open-ended questions to prompt critical thinking.

Review and Assessment

Interactive chapter review and assessment activities in the Cirrus environment offer students with all types of learning styles ample opportunity to reinforce learning and check understanding.
- Chapter Summary
- Chapter Glossary
- Game
- Flash Cards
- Terms Check: Matching
- Knowledge Check: Multiple-Choice
- Key Principles: Completion
- Tech Illustrated: Figure Labeling
- Chapter Exercises
- Chapter Exam

Student eBook

The Student eBook makes *Computers: Understanding Technology*, Seventh Edition, content available from any device (desktop, laptop, tablet, or smartphone) anywhere. The ebook is accessed through the Cirrus online course.

Instructor eResources

Accessed through Cirrus and visible only to instructors, the Instructor eResources for *Computers: Understanding Technology*, Seventh Edition, include the following support:
- Answer keys and rubrics for evaluating responses to chapter activities and exercises that require manual grading.
- Lesson blueprints with teaching hints, lecture tips, and discussion questions.
- Syllabus suggestions and course planning resources.
- Exam item banks in RTF format for instructors who are not using Cirrus. Instructors can use the banks of over 2,000 multiple-choice and true/false exam items to create customized tests. The banks are organized by learning objective. Each key concept in the textbook has a related quiz item and exam item—so the quizzes can be used for formative assessment and the exams for summative assessment. Approximately 10 percent of the questions include graphics based on key figures and tables from the textbook. Cirrus includes prebuilt quizzes for each numbered section (learning objective), prebuilt exams for each chapter, and a bank of exam items for creating customized assessments.

Acknowledgments

Creating and publishing a book requires the dedicated efforts of many people. Throughout this project, we authors have had the pleasure and privilege of working closely with the highly skilled and quality-focused professionals at Paradigm Education Solutions.

We offer our sincere appreciation to Vice President, Content and Digital Solutions, Christine Hurney for her continued support of this program. Director of Content Development, Computer Technology, Cheryl Drivdahl provided invaluable leadership in guiding the complicated update process. Digital Projects Manager Tom Modl deserves special recognition for the enhancements he and his team brought to the interactive and online components of this total learning package. We thank Developmental Editors Katrina Lee and Stephanie Schempp, as well as Production Editor Melora Pappas, for their hard work in vetting and coordinating changes to the text, images, and instructional content for this program, as well as Copy Editor Eric Braem for his thorough attention to accuracy and consistency.

Consultants and Reviewers

We thank the instructors and other professionals who provided valuable feedback about the enhancements made for this edition. As instructors who teach introductory computer courses, and as practicing professionals who are knowledgeable about the latest computer technologies, they brought a real-world perspective to the project.

Gayle Flentge
Mineral Area College
Park Hills, MO

Lana LaBruyere
Mineral Area College
Park Hills, MO

Laura J. Larimer, MBA
Education Technology Consultant
Indianapolis, IN

Walter Pauli
Community College of Allegheny
County—Boyce Campus
Monroeville, PA

Chris Pichereau
Independent IT Consultant
Tucson, AZ

Darlene Putnam
Thomas Nelson Community College
Hampton, VA

William R. Stanek
Lecturer & Author
Seattle, WA

William R. Stanek, Jr.
SecOps, Milliman, Inc.
Seattle, WA

Chapter 1

Touring Our Digital World

Chapter Goal

To explore the expanding computer and technology field and learn how computers work and impact our lives

Learning Objectives

- **1.1** Give examples of digital technologies.
- **1.2** Discuss the advantages of using computers.
- **1.3** Briefly explain how computing works.
- **1.4** Differentiate between computers and computer systems.
- **1.5** Identify the hardware and software that make up a computer system.
- **1.6** Describe the categories of computers.

Online Resources
The online course includes additional training and assessment resources.

Tracking Down Tech
Watching Technology Add Up in Your Life

1.1 Being Immersed in Digital Technology

Computers and other "smart" digital devices permeate our daily lives. High-definition TVs (HDTVs) display amazingly clear, colorful images of sports events, reality shows, and other popular programs. Smart TVs can connect to the internet, enabling you to download movies and other content on demand. Smartphones and tablets equipped with a plethora of useful apps simplify and speed up your daily routines. Appliances with smart technology built in help you make coffee at a set time, control home heating and cooling costs, manage home lighting, and more. The typical vehicle is likely run by an onboard computer and may include such features as global positioning system (GPS) navigation, hands-free calling via a smartphone and Bluetooth, hard disk storage for digital music, digital music streaming from satellite radio stations or portable devices, and even a Wi-Fi hotspot or built-in cellular capability. Generally speaking, *smart devices* are connected, interactive electronics used to share information and perform daily activities.

Watch & Learn
Being Immersed in Digital Technology

Video
A Digital Lifestyle

Many appliances now include digital technology that enables you to control them using your smartphone or tablet.

Smartphones have become so common in everyday life that most people would be lost without them. Smartphones can run a variety of useful apps, including navigation apps that help users get where they're going.

Manufacturers increasingly use computerized robots to build products and manage shipping. Businesses rely on electronic mail (email) to communicate internally and externally, as well as a growing number of other information technology tools. **Information technology (IT)** is the use of computer, electronics, and telecommunications equipment and technologies to gather, process, store, and transfer data. For example, even a small business such as your dentist or hair stylist likely tracks appointments using a computerized system that may also include the capability to send you an appointment reminder via text message. The next several years could see the rapid growth of technology in the medical field, as providers strive to make medical care easier to deliver and more cost effective. Around the world, we spend trillions of dollars on all kinds of information technology every year (see Figure 1.1).

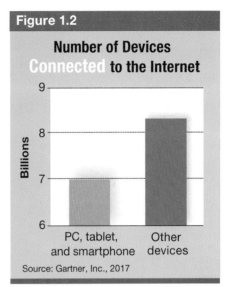

In 2018, approximately 89 percent of US residents used the internet from home, school, or another location, according to the Pew Research Center. In 2017, the number of computers, tablets, smartphones, and other internet-connected devices exceeded 15 billion units (see Figure 1.2). In today's world, very few people go even a single day without interacting with some type of digital technology. What about you? How digital is your life?

Like most people, you probably realize that you interact with electronic devices every day. The extent to which computers and digital technology drive daily life has led historians to characterize today's world as a "digital world." You likely know that the term *digital* has something to do with computers. Here's a closer look at what the term really means and why it's important.

Digital Information

The term **digital** describes an electronic signal that's processed, sent, and stored in discrete (separate) parts called *bits*. In contrast, an analog electronic signal is continuous, but the voltage, frequency, and current vary to represent information in a wave form. Other forms of analog signals include sound waves, light waves (which produce variations in intensity and color), and more. For example, the continuous sweep of the second hand on an old clock face provides an analog representation of time passing, while the numbers on a digital clock face present discrete, digital indications of the time.

On and *off* electrical states represent the discrete parts (bits) of a digital signal. Those states correspond to the digits 1 (on) and 0 (off). In a computer or digital device, this system of 1s and 0s corresponding to on and off electrical currents represents all information. Thus, computers use digital information, and computer technology in general is considered *digital technology*. You will read more about this fundamental IT concept in later chapters, but for now, remember that *digital* refers to information represented by numbers.

Video
Analog versus Digital

Video
Digitizing Information

This smart plug, one of iDevices' home automation products, lets you use your smartphone or tablet to control any appliance you plug into the device.

Computerized Devices versus Computers

Digital processing occurs within miniature electrical transistors etched onto a tiny square of silicon or another material called a *chip*, a *microprocessor*, or an *integrated circuit* (*IC*). Digital cameras, smartphones and basic-feature mobile phones, electronic appliances, and computers all contain electronic chips. However, the chips within computerized devices differ considerably depending on the purpose or functionality of the device. Generally, computers require more from chips in terms of power and capability, a distinction that separates electronic devices into two broad groups: special-purpose (or embedded) computers and general-purpose computers (also simply called *computers*).

The chip comprising an *embedded computer* (or *embedded system*) performs just a few specific actions. For example, the embedded chip in a digital camera automatically controls the speed of the lens so that the right amount of light enters through the lens. The embedded chip in a bar code scanner reads the bar code on a clothing tag, converts it to digital information, and sends it to the computer or system that uses the digital information to identify the item and its price. The tiny chip in a digital thermometer helps determine the body temperature of a patient at a medical clinic. Embedded chips in home appliances enable you to program the appliances for a delayed start and other automatic operations, or you can set up home automation via smart plugs and switches that enable you to control appliances and lighting.

The microprocessor chips in a *general-purpose computer*, on the other hand, provide numerous powerful capabilities and work with programs that enable the system to perform a range of complex processes and calculations. For example, you can use a computer with a word-processing program installed to create, edit, print, and save various kinds of documents, including letters, memos, reports, and brochures. A **computer**, therefore, is defined as an electronic device that

- has one or more chips, memory, and storage that together enable it to operate under the control of a set of instructions called a **program**;
- accepts data that a user supplies;
- manipulates the data according to the programmed instructions;
- produces the results (information) from the data manipulation; and
- stores the results for future use.

Hands On
Finding Out What Processor Your Computer Has

Numerous companies design and/or manufacture microprocessors. Some, such as longtime chip pioneer Intel, make multiple types of processors for many types of computers, including desktop and larger computers, mobile laptop computers and tablets, smartphones, and embedded systems. Others focus on particular types of processors. For example, Qualcomm designs and licenses processing technology primarily for mobile and wireless devices, and has been moving into embedded technologies for markets such as automotive and consumer electronics. Still other chip companies excel in a particular niche or two, such as the graphics processors pioneered by NVIDIA, as well as its mobile processors.

Computer processors, such as Intel Core processors, may have multiple processing cores and enable both desktop and mobile computers to handle powerful business and creative activities while conserving power.

1.2 Discovering the Computer Advantage

Watch & Learn
Discovering the Computer Advantage

Before the early 1980s, the average person had never seen a computer, let alone used one. The limited number of computers that existed were relatively large, bulky systems confined to secure computer centers in corporate and government facilities. Referred to as *mainframes*, these computers required intense maintenance, including special climate-controlled conditions and several full-time operators for each machine. Because of the expense and difficulty involved in operating early mainframes, most were used only by computer programmers and scientists for complex operations, such as handling large amounts of business data and compiling and analyzing study data. Other than

Early mainframe computers like this one were large, bulky, and difficult to operate. Many of them used large reels of magnetic tape for storage.

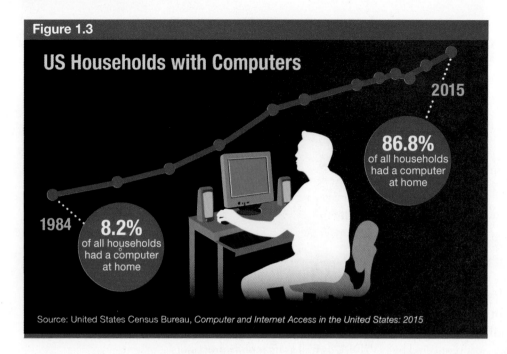

Figure 1.3 US Households with Computers
1984: 8.2% of all households had a computer at home
2015: 86.8% of all households had a computer at home
Source: United States Census Bureau, *Computer and Internet Access in the United States: 2015*

a few researchers and technicians who had security clearances, most employees were prohibited from entering the areas housing these computers.

Beginning in the early 1980s, the computer world changed dramatically with the introduction of the **microcomputer**. This new type of computer was also called the *personal computer* (or *PC*), because it was designed for easy operation by an individual user. Compared to mainframes, personal computers were more affordable, significantly smaller, and much easier to use. Within a few years, many workplaces relied on PCs for handling a variety of tasks. Today, you can find personal computers alongside standard appliances in homes and schools (see Figure 1.3).

Today's computers come in a variety of shapes and sizes and differ significantly in computing capability, speed, and price. For example, a powerful computer capable of processing millions of scientific data points in a few minutes may cost millions of dollars, while a small office desktop computer or tablet used for creating correspondence and budget forecasts typically costs less than $1,000. Whatever the size, power, or cost, all computers offer advantages over manual technologies in the following areas:

- speed
- accuracy
- versatility
- storage capabilities
- communications capabilities

Video Computer History I
Video Computer History II
Video Computer History III
Video Computer History IV

Speed

Today's computers operate with lightning-like speed, and processing speeds will continue to increase as computer manufacturers introduce new and improved models. Contemporary PCs can execute billions of program instructions in one second. Some of the most powerful computers (such as supercomputers) can execute quadrillions of instructions per second. A rate this fast is needed to process the huge amounts of data involved in forecasting weather, mapping how blood flows through the human body, constructing scientific models that re-create the Big Bang, conducting national cybersecurity operations to thwart cyberattacks, and more.

Accuracy

People sometimes blame human errors and mistakes on a computer. In truth, computers offer extreme accuracy when they are equipped with error-free programs and are processing accurate user-entered data. Computer professionals like to use the phrase *garbage in, garbage out (GIGO)*, which means that if a program has logical or other errors and/or the data entered for processing has errors, then the resulting output can only be inaccurate.

Let's say you're shopping and the store owner uses a computer and program to track the sale, calculate the sales tax due, and print your receipt. If the program has an error, such as the wrong sales tax rate, or the store owner types the wrong price, the receipt will show the wrong amount due. Using high-quality, error-free computer programs and double-checking the accuracy of the data you enter will help ensure you won't get garbage out.

Computers, such as those found on communications, imaging, and weather satellites, can process information with speed and accuracy. For example, the National Aeronautics and Space Administration (NASA), the US Geological Survey (USGS), and Google Earth rely on satellites to take digital images of the Earth from space and then organize and beam that information back for analysis and mapping.

Practical TECH

Tapping the Crowd's Opinion

You don't always need to pay high prices for computer programs. You can often find good free or low-cost programs and apps for your computer or mobile device at a variety of online stores and download websites. However, a free or low-cost program isn't a bargain if it doesn't work as expected, has errors, is difficult to use, or otherwise wastes your time. Fortunately, most online stores and websites ask users to rate how well programs performed and then display the results. Some also offer a "Comments" feature that lets users explain their ratings. Before you download a program, check out the ratings and comments from other users. These will give you a sense of how well the program works and whether it will perform as advertised.

Versatility

Computers and mobile devices provide amazing versatility. With the right programs or apps installed, a computer or device can perform a variety of personal, business, and scientific applications. Families use computers for communications, budgeting, online shopping, completing homework assignments, watching movies, playing games, and listening to music. Banks use computers to conduct money transfers and account withdrawals. Retailers use computers to process sales transactions and check on product availability. Manufacturers can manage entire production, warehousing, and selling processes with computerized systems and smart devices. Schools use computers for keeping records, conducting distance-learning courses, scheduling events, and analyzing budgets. Universities, government agencies, hospitals, and scientific organizations conduct life-enhancing research using computers.

One of the most ambitious computer-based scientific research programs of all time was the Human Genome Project (HGP). This project represented an international effort to sequence the 3 billion DNA (deoxyribonucleic acid) pairs of letters in the human genome, which is the genetic material that makes up every person. Scientists from all over the world can now access the genome database and use the information to research ways of improving human health and fighting disease. They can also share that information with others, as the US government's National Human Genome Research Institute does in its website (see Figure 1.4). Other projects around the globe are applying computing power to map the complex structures of the human brain.

Members of a family may use computers for a variety of purposes: budgeting and paying bills, corresponding with friends and relatives, shopping online, completing homework assignments, viewing and printing photos, playing games, viewing video content, and listening to music.

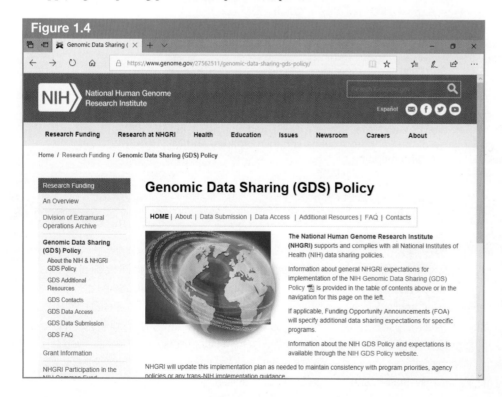

Figure 1.4

The National Human Genome Research Institute (NHGRI) funds genetic and genomic research, studies the related ethics, and provides educational resources to the public and to health professionals. The NHGRI uses computers to gather and process the data from its various research projects, as well as to present that information in a variety of digital formats that members of the public can access via computer.

Storage

Program and data storage revolutionized early computing by making computers incredibly flexible. Once you have installed a program, it's stored in the computer. This means that you and other users can access it again and again to process different data. For example, you can install the spreadsheet program Microsoft Excel and then access it repeatedly to track budget items and to explore possible outcomes if income and expenses change.

A flash drive can hold a large amount of data in a small physical space. It is also highly portable, allowing you to move information easily from one computer to another.

Computers can store huge amounts of data in comparably tiny physical spaces. For example, one compact disc (CD) can store about 109,000 pages of magazine text. Internal storage devices and many external storage devices offer capacities many times larger than a CD.

Communications

Most modern computers and devices contain equipment and programs that enable them to communicate with other computers through cable connections, fiber-optic cables, satellites, and other wireless connections. Connecting these devices allows users to share programs, data, information, messages, and equipment (such as printers). The structure in which computers are linked using special programs and equipment is called a *network* (see Figure 1.5). Wireless networks enable users to communicate and share data over wireless connections by connecting devices such as tablets, smartphones, and wireless-capable printers.

A network can be relatively small or quite large. A local area network (LAN) operates within a relatively small geographical area, such as a home, office building,

Figure 1.5

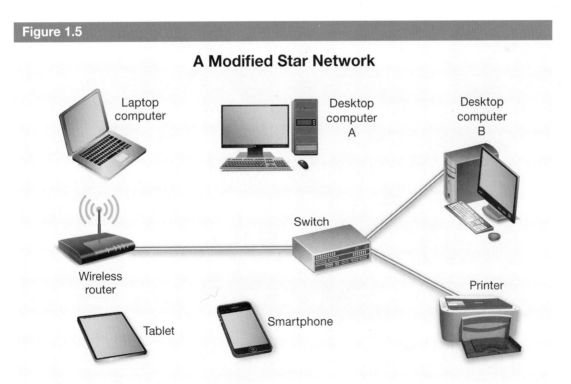

A Modified Star Network

Networks commonly use a star topology, in which computers and devices are connected through a central switch. In this modified star network, the wireless router that's connected to the switch enables computers and devices to connect to the network wirelessly.

factory, or school campus. A wide area network (WAN) links many LANs and might connect a company's manufacturing plants located throughout the United States.

Constant, quick connections, along with other computer technologies, have helped boost productivity for many manufacturers. For example, some manufacturers reduce production time by networking production equipment to share information with other planning and production systems in the organization.

The Internet: A Super Network The network you are most likely familiar with is the internet, which is the world's largest network (and often considered a WAN). The **internet** (also called the *net*) is a worldwide network made up of large and small networks linked by communications hardware, software, telephone, cable, fiber-optic, wireless, and satellite systems for the purposes of communicating and sharing information (see Figure 1.6). Network service providers (NSPs) provide access to internet service providers (ISPs), who sell various types of internet access to consumers. The ISPs use different connection types to deliver the access.

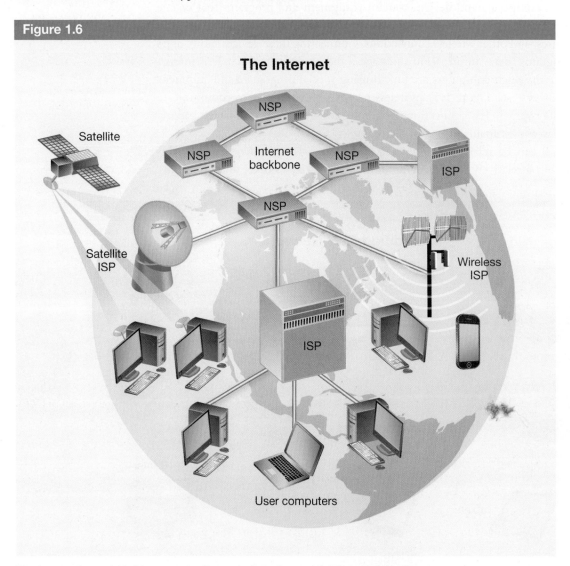

Figure 1.6

The internet is a worldwide network of large and small networks that are linked by communications hardware and software for the purposes of communicating and sharing information.

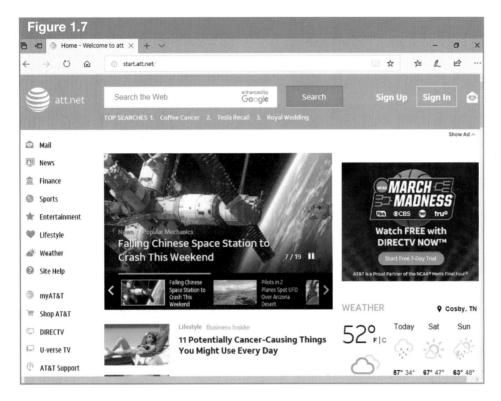

Figure 1.7 In addition to providing internet access, an ISP such as AT&T offers daily weather reports, news stories, other types of information, and access to user email and account information on its website (shown here).

As of 2018, more than 4 billion people worldwide used the internet for various purposes:
- sending and receiving email
- finding information, such as weather forecasts, maps, stock quotes, news reports, airline schedules, and news and feature articles
- buying and selling products and services
- taking online courses
- using social media
- accessing entertainment, such as online games, movies, and music
- downloading software and apps, or using them online

In addition to offering internet access, many ISPs provide a portal that displays daily weather reports, stock quotes, news, and other types of information (see Figure 1.7, for example). Users can leave the ISP's portal to access other websites of interest.

The World Wide Web A widely used part of the internet is the World Wide Web (WWW or *web*), a global system within the internet that enables users to move from one linked website or web page to another. A website is a collection of web pages stored on a web server. A web page is an electronic document that may contain text, images, sound, video, and links to other web pages and websites. Some websites and web pages have interactive features, such as online shopping carts, pop-up descriptions, and self-guided slide shows. Other websites offer cloud-based services—such as online file storage—that you'll read more about in Chapter 7.

When using the web, you can navigate directly to a particular page or browse to a location by following hyperlinks. When you aren't sure where to find the information you need, you can use a search engine, which is a website or service that enables you to enter search terms in a search box. The search engine uses those terms to locate and retrieve a list of websites and web pages that might contain the requested information.

Tech Career Explorer
Taking a Look at Web Jobs

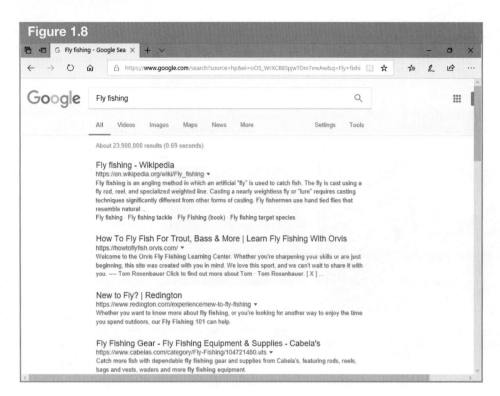

Figure 1.8 A search engine, such as Google, enables the user to search for, locate, and retrieve information available on the World Wide Web.

For example, if you type "Fly fishing" into a search box, the search engine will display a list of links to web sources that contain information about those terms, as shown in Figure 1.8.

1.3 Understanding How Computers Work

Understanding the broad steps involved in information processing is key to recognizing the significance of computer technology. A computer accepts the data a user enters, processes the data according to instructions provided by the program, and then outputs the data in a useful form: information. The rest of this section explores how the recurring series of events that make up the *information processing cycle* change data into information.

Watch & Learn
Understanding How Computers Work

Data and Information

Data consists of raw, unorganized facts and figures. By itself, a piece of data may be meaningless. For example, the fact that an employee has worked 40 hours in one week doesn't provide enough meaning to enable the payroll department staff to calculate his or her total pay and issue a paycheck. Entering additional data—such as the employee's pay rate, number of allowances, and number of deductions—and then processing the data enables payroll staff to generate useful information, including paychecks, earnings statements, and payroll reports.

Data that's been processed (organized, arranged, or calculated) in a way that converts it into a useful form is called **information**. Once a computer has created information, it can display the information on screen or print it on paper. The computer also can store information for future use, such as processing monthly or quarterly payroll reports.

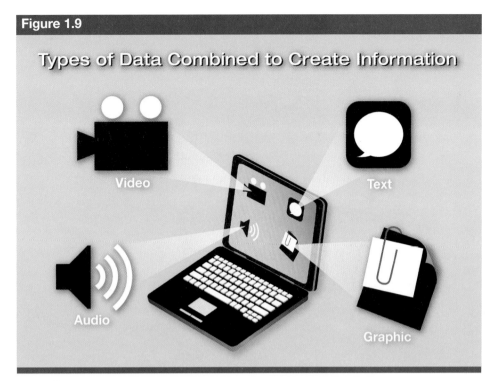

Figure 1.9 Types of Data Combined to Create Information

Practice
Types of Data Combined to Create Information

Text, graphic, audio, and video data can all be combined to create a compelling message or presentation.

Data that's been entered into a computer can be of one of the following types (see Figure 1.9), or a combination:

- **Text data** consists of alphabetic letters, numbers, and special characters. It might also consist of the formulas you enter in a spreadsheet program. These data typically enable the computer to produce output such as letters, email messages, reports, and sales and profit projections.
- **Graphic data** consists of still images, including digital photographs, charts, drawings, and 3-D images. Graphic data may also include various types of maps and mapped data, such as an online road map, a topographical map, a property survey map, or maps with various types of weather data.
- **Audio data** refers to sound, such as voice sound effects, and music. For example, you can use a microphone to record the narration for an onscreen presentation, which the computer stores in digitized form. Or you can download music from an online service and play it back through your computer's speakers or transfer it to another playback device.
- **Video data** refers to moving pictures and images, such as webcam data from a videoconference, a video clip, or a full-length movie. For example, you might use a video camera, smartphone, or tablet to record video of a new product demonstration (demo). After transferring the video to your computer, you could play it back onscreen, share it with potential customers, or include it in a larger project for promoting the new product.

The Information Processing Cycle

Information processing (also called *data processing*) means using a computer to convert data into useful information. Processing data into information involves four basic functions: input, processing, output, and storage. During processing, the computer may perform these four functions sequentially, but not always.

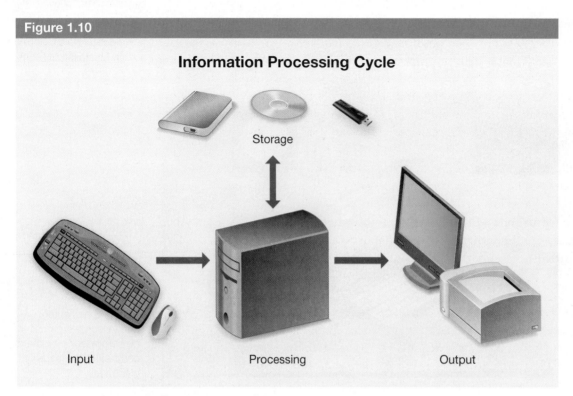

Figure 1.10

During the information processing cycle, data is entered into a computer, processed, sent as output, and then stored (if required for future use).

Figure 1.10 illustrates how these four functions form the **information processing cycle**.

The term used to identify the first step in the information processing cycle—**input**—is also used to describe physically entering data into a computer or device. After you enter data using an input device such as a keyboard, mouse, or touchscreen, the computer handles the data according to the programmed instructions. (Keep in mind that your inputs will likely also include selecting commands and options in the program being used, as well as audio and video from other devices such as webcams and microphones.) **Processing** occurs in the computer's electrical circuits and creates information called **output**. Next, the computer sends the output to one or more output devices, such as a monitor or printer. Usually, you will also direct the computer to send the information to **storage** (such as a hard disk, flash drive, external hard disk, or cloud storage area) for future use. The information a computer processes can be output in a variety of forms:

- written or textual form, such as research reports and letters
- numerical form, such as a spreadsheet analysis of a company's finances
- verbal or audio form, such as recorded voice and music
- visual form, such as digital photos, drawings, and videos

1.4 Differentiating between Computers and Computer Systems

Watch & Learn
Differentiating between Computers and Computer Systems

Technically, the term *computer* identifies only the system unit: the part of a desktop computer system that processes data and stores information. A **computer system**, however, includes the system unit along with input devices, output devices, and external storage devices.

The user's individual needs and preferences will determine which additional devices he or she needs. For example, an engineer in a structural design firm needs a powerful system unit that's capable of running sophisticated design software, along with one or more large monitors, a plotter or wide format printer, and a multifunction device that provides printing, scanning, copying, and faxing capabilities. Most mobile computers and devices already include the built-in input and output devices that most users are likely to want, such as speakers, cameras, and touchscreens. But if you are shopping for your first desktop PC, you will probably purchase an entire system, including the system unit, keyboard, mouse, monitor, external storage devices, and perhaps a printer. Figure 1.11 shows a variety of input, processing, output, and storage devices that may be included in a PC system.

The rest of this book uses the terms *computer* and *PC* to refer to a computer system that includes all of the devices necessary to input programs and data, process the data, output the results, and store the results for future use. Chapter 2 goes into detail about computer system components, including input, output, and storage devices.

You also can think about a computer system in terms of how many people can use it simultaneously. A **single-user computer system**, as the name implies, can accommodate one user at a time. This term applies to PC systems found in homes and small businesses and organizations. A **multiuser computer system** can accommodate many users at a time over a network. Large businesses and organizations typically use such systems to enable several managers and employees to simultaneously access, use, and update information stored in a central storage location. For example, a payroll clerk can access and view an employee's payroll record from one computer connected to the system while a shipping clerk tracks a customer's shipment from a computer in the warehouse. In addition, the users can interact with each other easily and quickly.

Article
Major Advances in the History of Computing

Practice
Computer System Components

Figure 1.11

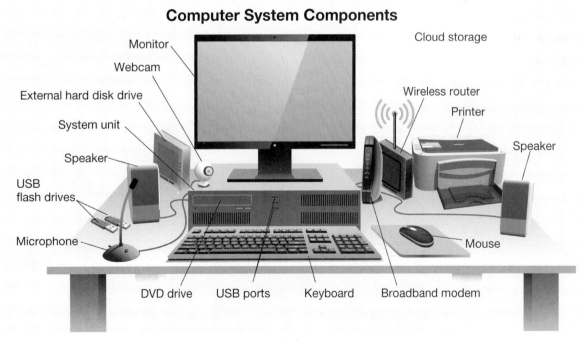

Computer System Components

At a minimum, a desktop computer system consists of more than just the system unit, which contains the hard disk drive, video processor, central processing unit, motherboard, and power supply. It also includes a monitor, keyboard, and mouse. Users can add a variety of other input, output, communications, and storage devices.

This book provides a lot of discussion about general-purpose, single-user computers—personal computers—that enable users to complete a variety of computing tasks. These are the computers you will most likely work with in your home, in your school's computer lab, and on the job. However, this book also covers large-scale, multiuser systems and their applications; reading about these computers will give you a sense of the scope of computing and how it affects many aspects of society.

1.5 Understanding Hardware and Software

Watch & Learn
Understanding Hardware and Software

A computer system includes both hardware and software. The combination of the two that makes up a particular computer system depends on the user's requirements. Given the number of hardware devices and software programs available in the marketplace, you can customize a system for nearly any purpose—from typical business use, to graphic design, to home media and entertainment, to gaming. Manufacturers typically offer a system unit, monitor, and mouse and keyboard package; the choice of the printer and other hardware devices is left up to the buyer. The majority of desktop and laptop computers come with the operating system software preinstalled, plus some basic software such as a word-processing program.

Hardware Overview

Hardware includes all the physical components that make up the system unit plus the other devices connected to it, such as a keyboard or monitor, as discussed earlier in this chapter. You can also call a connected device a **peripheral device**, because in many cases it is outside of (or peripheral to) the computer. Examples include a keyboard, mouse, webcam, and printer. Some peripheral devices, such as a monitor and internal hard disk drive, are essential components of a personal computer system. Hardware devices can be grouped into the following categories:

- system unit
- input devices
- output devices
- storage devices
- communications devices

Tech Ethics

Close the Gap

Close the Gap, which supports efforts by the United Nations (UN), is a worldwide nonprofit organization that aims to improve access to computers and communication technology in developing countries to help them improve educational and economic performance. The word *gap* in the organization's name refers to what's called the *digital divide*: the gap in opportunities between people who have access to technology and those who do not. By 2018, Close the Gap had supported almost 4,900 projects around the world and delivered hundreds of thousands refurbished computers and components (donated by its corporate partners) to educational, medical, entrepreneurial, and social projects in developing countries. Close the Gap also recycles donated equipment that has reached the end of its life.

Considering the volume of electronics waste that's being generated worldwide, Close the Gap creates win–win outcomes. Corporations have a place to send their old equipment, and recipients receive assistance in bridging the digital divide.

Has your family ever donated an old computer or mobile phone to a charity or needy person?

(Chapters 2 and 3 go into detail about the system unit, input devices, output devices, and storage devices, and Chapters 5 and 6 discuss communications and networking hardware and setup.)

The System Unit The system unit is a relatively small, metal and plastic cabinet that houses the electronic components involved in processing data. The cabinet holds the main circuit board (called the *motherboard*), which includes slots and sockets for installing and connecting other electronic components. Once installed on the motherboard, the components can communicate with each other to change data into information.

The main components of the motherboard are the central processing unit (CPU) (also called the *microprocessor* or *processor*) and internal memory. The CPU consists of one or more electronic chips that read, interpret, and execute the instructions that operate the computer and perform specific computing tasks. When executing a program, the processor temporarily stores in the computer's memory the program's instructions and the data needed by the instructions in the computer's memory. **Main memory** (also called *primary storage* or *random access memory* [*RAM*]) consists of banks of electronic chips that provide temporary storage for instructions and data during processing.

Article
The Continuing Evolution of Personal Computing Technology

Input Devices An input device is a form of hardware that enables the user to enter program instructions, data, and commands when using a program on the computer. The program determines the input devices needed. Common input devices are the keyboard, mouse, microphone, graphics tablet, touchscreen, and a stylus or similar pen-style device, as well as audio and video input devices such as microphones and various types of cameras.

Output Devices An output device is a form of hardware that makes information available to the user. Popular output devices include display screens (monitors),

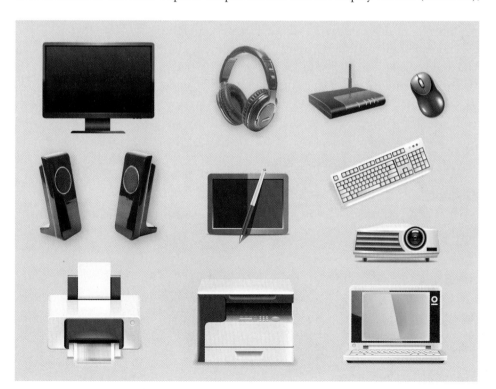

Just for the fun of it, try naming all the devices you see in this image, and categorizing each one as a system unit, an input device, an output device, a storage device, or a communications device. Compare results with a classmate to see how well you know your devices.

printers, TV screens, and speakers. Some types of output devices (such as printers) produce output in *hard copy* (tangible, or physically fixed) form, such as on paper or plastic. Other output devices (such as monitors) produce output in *soft copy* (intangible, or changeable) form, which can be viewed but not physically handled.

Storage Devices In contrast to memory, which stores instructions and data temporarily during processing, a **storage device** (often called a *storage medium* or *secondary storage*) provides more permanent storage of programs, data, and information. After you have stored data and information, you can retrieve, modify, display, import, export, copy, share, or print it.

Although the terms *storage device* and *storage medium* are often used interchangeably, they actually have slightly different meanings. In fact, a storage device records programs, data, and/or information to a storage medium and then retrieves them from the storage medium. For example, a DVD-R drive (optical storage device) writes data to a DVD-R disc (storage medium), and the drive can also later retrieve data from the disc.

Connecting an external hard drive will add large amounts of storage to your system.

External hard drives provide a cost-effective way to expand storage beyond the internal storage in the system unit. These drives provide as much space as or more than some internal hard disks; they can plug directly into a USB port on the system and require little setup. Some external hard disks can also connect to your home or small network to enable easy sharing between users.

Communications Devices A **communications device** makes it possible for multiple computers to exchange instructions, data, and information. One common communications device is a **modem** (*mo*dulator plus *dem*odulator): an electronic device that converts computer-readable information into a form that can be transmitted and received over a communications system. In the early days of the internet (and even before that), many modems enabled communications through standard telephone landlines. Today, most modems enable connections through newer and faster internet infrastructures (which you will read about in Chapter 5). In addition, many computer systems and mobile devices include hardware for connecting to a wired or wireless network and for transferring information over short distances using technologies such as Bluetooth, which Chapter 6 covers in greater detail.

Communications devices for wireless networking enable users to exchange information between computers.

Software Overview

Software (also called *programs*) consists of the instructions that direct operation of the computer system and enable users to perform specific tasks, such as word processing. The two main types of software are system software and application software. Chapters 3 and 4 provide a more in-depth look at each of these types of software beyond the brief introduction you'll read next.

System Software System software tells the computer how to function and includes operating system software and utility software. The operating system contains instructions for starting the computer and coordinates the activities of all hardware devices and other software. Most PCs use a version of the Microsoft Windows operating system (see Figure 1.12). Apple computers come with the macOS operating system (shown in Figure 1.13), although they can also be set up to run Windows. Other users

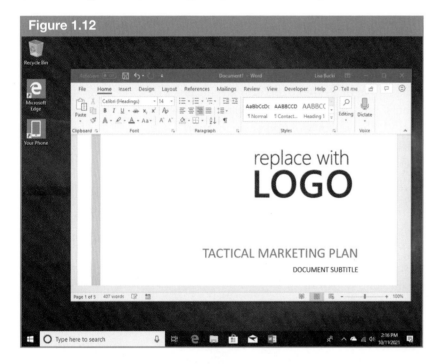

The Windows operating system runs computers from a wide variety of manufacturers.

Apple's computers are powered by the macOS operating system.

may prefer the open source Linux operating system, available in various versions called *distributions*. Both macOS and Linux share origins from the Unix operating system, originally developed decades ago. Utility software consists of programs that perform administrative tasks, such as checking the computer's components to determine whether each is working properly, managing disk drives and printers, and looking for computer viruses.

New mobile devices are effectively small computers, and they also have operating systems. The latest mobile operating systems include iOS for Apple iPhones and iPads (see Figure 1.14) and Android for devices from other manufacturers. Although Android™ was released by Google, it's considered *open source*; this means that hardware manufacturers are free to tailor it to the capabilities of specific devices. For that reason, Android is used to run everything from smartphones to Kindle Fire ebook readers. Google's Chrome OS open source operating system entered the market more recently. Developed to provide a platform for basic web browsing on inexpensive, lightweight laptops called *Chromebooks*, Chrome OS may soon feature a wider variety of applications.

Application Software Programs, also called *applications*, are installed on your computer and run along with the operating system or an app to perform specific tasks, such as creating word-processing documents, preparing spreadsheets, searching databases, developing reports, and presenting slide shows. You can find an application for doing nearly any personal, school, or business task. On mobile devices, applications are known as *apps* because they are often streamlined and have focused capabilities. Some of the programs installed on the Windows Start menu are known as *apps*.

Cloud Apps and Services Unlike programs and apps that install on your computer, cloud apps and services are delivered online. You generally access these apps and services through web browser software installed on your system or device. The cloud app or service itself may not download to your system, although it may download "helper" components that enable your web browser to work with the online software.

Most cloud-based apps and systems require either a free account or a paid subscription. For example, with your free Microsoft or Google account, you get access to cloud apps for word processing, spreadsheet calculations, email, and more, as well as online file storage. Paid subscription services give you access to more sophisticated programs and services, such as cloud-based business accounting systems. Some subscription services use a hybrid approach. For example, with a Microsoft Office 365 business-level subscription, users can download and install the Office software and also use cloud-based features for team collaboration and account management. Similarly, Adobe Creative Cloud for teams combines the ability for users to download and install graphics creation software, with the ability to use online features that keep file versions in sync and to access the Adobe Stock image service. Chapter 7 explores cloud apps in more detail.

Figure 1.14

iPhones and iPads run iOS, Apple's operating system for its mobile devices.

1.6 Identifying Categories of Computers

Rapid advances in computer technology often blur the differences among types of computers, and industry professionals sometimes disagree on how computers should be categorized. Typically, the categories are based on differences in usage, size, speed, processing capabilities, and price, but there's a lot of overlap even among very different types of computers. You can think of computers as falling into the categories listed here and described in more detail in the following sections:

- smart devices (embedded computers)
- voice-controlled smart speakers
- wearable devices
- smartphones, tablets, and other mobile devices
- personal computers and workstations (desktops, laptops, and convertibles)
- midrange servers and minicomputers
- large servers and mainframe computers
- supercomputers

Table 1.1 summarizes the basic differences among these categories of computers.

Watch & Learn
Identifying Categories of Computers

Article
Enterprise versus Personal Computing

Table 1.1 Categories of Computers

Category	Size	Number of Users Accommodated	Approximate Price Range
Smart devices (embedded computers)	Generally small, contained within another product	As many as are using the host appliance or item	$ Often included as part of the price of the appliance, vehicle, or other item in which the device is embedded; otherwise, about $50 to hundreds of dollars
Voice-controlled smart speakers	Fit on a shelf or corner of a desk or table	Any user in the speaker's range within the home or other location	$ $49 to $400 or more
Wearable devices	Fit on the wrist, head, or elsewhere	A single user, although these devices can often share information with other devices	$ $100 to $1,500 or more
Smartphones, tablets, and other mobile devices	Fit in the hand(s) and/or a pocket	A single user or as part of a network	$ $100 to several hundred dollars
Personal computers and workstations (desktops, laptops, and convertibles)	Fit on a desk, in a briefcase, on a lap, or on another workspace	A single user or as part of a network	$$ A few hundred to a few thousand dollars
Servers (midrange servers and minicomputers)	Fit in a large cabinet or small room	Dozens or hundreds of users concurrently	$$$ A few thousand to hundreds of thousands of dollars
Mainframe computers	With needed equipment, occupy part of a room or a full room	Hundreds or thousands of users concurrently	$$$$ Several thousand to millions of dollars
Supercomputers	With needed equipment, occupy a full room	Thousands of users concurrently	$$$$$ Several million to more than one billion dollars

Smart Devices (Embedded Computers) and the Internet of Things

Technology seems to be entering a new era of convergence, in which different technologies are combined into a single package. The result is a product that features all the benefits of the individual technologies plus additional benefits derived from merging the component technologies. Perfect examples of this convergence are found in embedded computers, which are used to make other consumer products "smart."

Previously, the term **smart device** referred strictly to a smartphone. Today, however, the term refers more broadly to any type of appliance or device that has embedded technology. Devices with embedded technology that can send and receive data wirelessly or via the internet make up the **Internet of Things (IoT)**. Even cars and trucks now feature central computers (as shown in Figure 1.15). In addition, numerous sensors within vehicles provide data on the performance and status of various systems. For instance, sensors might report the air pressure from your tires to the computer, which turns on a tire pressure warning light when the pressure gets low. In-dash systems provide GPS navigation and location information for your travel needs, such as listings of nearby restaurants, gas stations, and attractions. You may also be able to connect a music player, such as an iPod, using a USB cable and control playback from the car's audio system. If the system includes a hard disk drive, you can copy digital music files to it from a USB flash drive and then play it back through the system. Some in-dash computers can connect or pair with your smartphone wirelessly over Bluetooth; this feature enables you to make hands-free calls and stream music from a service such as Pandora or Spotify.

Like the smart washer and dryer mentioned earlier in the chapter, many smart devices bring additional convenience, capabilities, and energy savings to appliances and other tools in the home or office.

> Video
> What Is the Internet of Things?

Figure 1.15

Number of **lines** of computer code in an average high-end car's onboard programming — 100 Million

By 2030, up to **15%** of vehicles on the road will be **fully autonomous**

Source: Paul Gao, Hans-Werner Kaas, Detlev Mohr, and Dominik Wee, "Disruptive Trends That Will Transform the Auto Industry," McKinsey & Company, 2016; Bob O'Donnell, "Your Average Car Is a Lot More Code-Driven than You Think," *USA Today*, 2016

Cutting Edge

Previewing the Latest at CES

Every January, the Consumer Electronics Show (CES) takes place in Las Vegas, Nevada. Technology companies and other types of businesses converge there to introduce new products and emerging technologies. Some of the products that are listed in Table 1.2 were introduced at the CES event.

At the 2018 CES, Whirlpool introdshowcased its Yummly app and how it integrates with Whirlpool's line of smart kitchen appliances. Yummly can use image recognition to look at ingredients you have on hand and suggest recipes to use those items, reducing food waste. From there, Yummly can send recipe cooking instructions to the smart range or microwave.

Chapter 1 Touring Our Digital World

According to the research firm Gartner, as of 2017, 8.4 billion smart, connected IoT devices were in use worldwide, surpassing the number of people on the planet.

Table 1.2 lists a sampling of the types of smart devices that have brought technology to products that had changed little for decades. You can control many of these devices with your smartphone (using either an app or text messaging); doing so allows you

Table 1.2 Examples of Smart Devices

Categories	Example Products	Example Functions
Locks	Goji Smart Lock Kwikset Kevo Touch-to-Open Smart Lock Schlage Sense Smart Deadbolt Yale Real Living Touchscreen Deadbolt	Enable using methods other than a key to open the door. (For example, the lock might detect your smartphone or a key fob as you approach, so you finish unlocking by touching the lock or entering a code on a touchscreen.) Some enable creation of multiple unique codes to allow access by multiple users; others have an alarm that will signal potentially illegal entry attempts.
Large appliances	Samsung Smart Home LG SmartThinQ Whirlpool app	Allow control of large appliances (such as washers and dryers) remotely; also perform other functions.
Small appliances	Crock-Pot Smart Slow Cooker (enabled by WeMo)	Can calculate cooking time and process and send reminders. Enable you to turn the appliance on and off, change the temperature, and change time settings from anywhere.
Security and smoke detectors	iSmartAlarm Home8 Nest Protect smoke and carbon monoxide alarm Ring Video Doorbells, Security Cameras, and Security Systems	Include motion sensors, contact sensors for doors and windows, and alarms; also provide the ability to arm, disarm, and monitor security devices from a remote location, often without a monthly service fee.
Thermostats	Nest Learning Thermostat Honeywell Wi-Fi Smart Thermostat	Learn the home's heating and cooling habits to program themselves; may include extra devices, such as a humidity sensor.
Lighting and smart outlets and switches	Belkin WeMo Smart Light Switch Connected by TCP Wireless LED Lighting Kit Insteon LED Bulbs	Enable remote control of lighting and plugged-in devices, such as turning on/off, dimming, and changing LED lighting colors.
Garage door controllers and other devices	Garageio LiftMaster MyQ Garage LG Smart TV	Allow performing a wide variety of functions, including using your smartphone to control your garage door. Allow internet access through your TV for streaming video from services such as Hulu and YouTube.

to monitor their operation, turn them on and off, and perform other tasks either locally or from a remote location. These emerging technologies form the basis for a concept called the **smart home** (or *connected home*) where numerous networked home devices provide specific, usually automated functions, and share data and media.

Voice-Controlled Smart Speakers

A voice-controlled **smart speaker** (such as the Amazon Echo, Google Home, or Apple HomePod) offers hands-free interaction with the internet as well as technology in a smart home. With their sophisticated speech recognition capabilities, these speakers have made it practical for people to get information by verbally asking for it or even to control other smart devices. A smart speaker device is typically a small box or cylinder with an integrated speaker and microphone. It can be placed anywhere in the user's environment and can connect to the internet wirelessly.

Each smart speaker runs its own **intelligent personal assistant** software that uses natural language processing to interpret the voice questions and requests picked up by the smart speaker's microphone and then translate those questions and requests into a response or action. Leading intelligent personal assistants include Alexa (Amazon Echo), Google Assistant (Google Home), and Siri (Apple HomePod). Some of these intelligent personal assistants also work on other devices, such as your smartphone or tablet.

You get the intelligent personal assistant's attention by saying a key phrase (usually its name) and then asking it a question. The software parses your verbal question, turns it into a web query, gets the information from the web, and then delivers the information to you verbally. For example, if you say, "Hey, Google, will it rain today?" the Google Home device recognizes your question, figures out that you want to know about the weather in your location, sends your query to a weather website, gets the information from the site, and a few seconds later, you hear the answer to your question—such as, "Yes, it will likely rain this afternoon"—through the device's speaker. You can give verbal commands to buy things online, stream music, control other smart devices, and more. The questions, requests, and commands that a particular intelligent personal assistant understands are known as its *skills*.

An intelligent personal assistant device works entirely on verbal input and output.

Other types of smart home control systems are also available. For example, some appliance manufacturers offer a *smart home hub* for controlling compatible devices. These and other smart devices may use specific low-power wireless communications standards such as Zigbee and Z-Wave. These standards are used to create personal area networks, a topic covered in Chapter 6.

Wearable Devices

Wearable computers continue to emerge and specialize to serve specific functions beyond what's offered by a simple pair of eyeglasses or wristwatch. As the name implies, a **wearable computer** is a computer device that's worn somewhere on the body—most often on the head or wrist.

Wrist-based smartbands and watches typically gather data about your physical activities. Examples include the Jawbone UP, and various models from FitBit and other makers. You can transfer the data the device collects to a smartphone or computer over a wireless, USB, or other direct connection. Using an app or website, you can review metrics such as steps taken, miles run, route followed, level of activity, and more. The use of these self-tracking tools and the personal data they provide is part of a movement called the *quantified self*.

A smartwatch provides you with access to mobile computing capabilities and information through a wireless internet connection or connection to a smartphone. Products include the Apple Watch, Fitbit Surge, Samsung Galaxy Gear, and Pebble.

A smartwatch, such as this Apple Watch, typically interfaces with your smartphone. By wearing this device, you can manage calls, text messages, and reminders such as appointment alerts from your wrist.

Practical TECH

Creating Your Own Automation

A hot technology called Arduino enables you to create your own smart automation projects. According to Adruino's creators, it is intended for use by "artists, designers, hobbyists and anyone interested in creating interactive objects or environments." The Arduino microcontroller board can receive input from sensors and other means, and it can send output to control lights, motors, and actuators (devices that run mechanisms or systems). The Arduino platform includes a programming language and a programming (development) environment, and it's open source, which means anyone can freely modify the source code. To get started with Arduino, you buy the board and accessories or a kit, download and install the free software, review the generous quantity of examples at the company's website (available at https://CUT7.ParadigmEducation.com/Arduino), and then start coding and building. You can also incorporate Arduino-compatible components from other sources.

The avid Arduino community has developed projects ranging from an aquarium control system to interactive visual art installations consisting of LED lights that respond to viewers' movements. You can join other enthusiasts and learn more at Arduino workshops and events held all over the world. For instance, an annual Arduino Day features 24 hours of official and unofficial events in locations worldwide. What objects or systems would you like to automate at home?

Wrist-based devices are growing in popularity. The research firm Gartner predicted that sales of smartbands, smartwatches, sports watches, and other wearables would grow to more than half a billion units per year by 2021. Vandrico, a wearable technology experts firm, has established The Wearables Database to track and analyze the growing number of wearable devices that are available (see Figure 1.16). You can visit the firm's website at https://CUT7.ParadigmEducation.com/Vandrico to see the latest data, such as the number of devices available, as well as average costs, typical uses, and number of devices worn at particular locations.

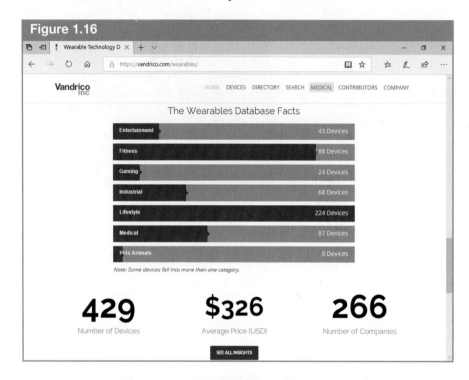

Figure 1.16

Vandrico's wearable technology database tracks the number of available wearable devices and provides information about average cost and range of purposes.

Cutting Edge

A Computer in Your Eye

Have you ever thought about what kind of superpower you would want to have, if such a thing were possible? If bionic eyesight tops your list, then you will be glad to know that engineers at the University of Washington (UW) in Seattle are working on it. A research team at the university is developing an electronic contact lens that could have a major impact on both the medical and computer industries.

The lens is still in its early stages, but the UW team has developed a prototype with an electronic circuit and red light-emitting diodes (LEDs). Although the prototype isn't yet functional, its creation represents a breakthrough for the engineers. They hope to eventually outfit each lens with several types of circuits and antennas, hundreds of LEDs, and a wireless signal. Lens wearers would use it like a regular computer, but the images would appear in front of the eye, rather than on a traditional screen.

What about the lens's medical capabilities? The lens would use the biomarkers present on the surface of the eye to monitor how the body's systems are functioning and then transmit that information directly to doctors. This technology could lead to making quicker and more accurate diagnoses and greatly reduce the need to take blood samples.

Verily, a division of Alphabet (the parent holding company that evolved from and owns Google) is working on a similar contact lens that's designed specifically to monitor blood glucose. Wearing this lens would eliminate the need for diabetics to prick their fingers to test blood glucose levels.

While the design of the bionic lens is far from completion, UW researchers are moving forward. They are currently studying how to power and mass produce the device.

Smartphones, Tablets, and Other Mobile Devices

The term *handheld computer* used to refer to dedicated small computers, such as the Pocket PC and Apple Newton. The user typically operated these devices by tapping on the screen with a stylus. These types of devices have mostly become obsolete. Today, a *handheld computer* comes in one of two general forms: smartphones and tablets. (Some other smart devices—including advanced portable music players, such as the iPod Touch, and some mobile gaming devices—may also be thought of as handheld computers.) You can hold one of these devices in your hands and use it for almost the same functions as a full-featured computer.

A handheld computer is small enough to fit in the palm of a user's hand.

A **smartphone** is a cell phone that can make and receive calls and text messages on a cellular network; it can also connect to the internet via the cellular network or a wireless network and perform numerous computing functions. Smartphones measure up to 6 inches or so diagonally and can fit in your hand. A **tablet** (sometimes called a *tablet computer* or *tablet PC*) is larger than a smartphone and generally used on your lap or on a table. Tablets can also connect wirelessly to the internet and perform computing activities. Only some models of tablets can connect to the cellular network, and to do so, they must be covered by a data plan from a cell phone carrier.

You operate a smartphone or tablet primarily through its touchscreen by using a variety of gestures, such as tapping and swiping. These devices also typically include a microphone and a combination still/video camera. Smartphones and tablets use a special class of mobile processor (or mobile CPU) that's typically designed to save power and facilitate longer battery life. Some include additional chips that provide location services (via GPS) and other capabilities. These devices also come with a special mobile operating system installed.

Hands On
Checking Out Your Mobile OS

Hotspot

An Inside Look at Your Small Intestine

While the medical community has been using endoscopy to diagnose and treat problems in the esophagus and stomach, and colonoscopy to see the large intestine, the small intestine has remained out of view until recently. Feeding patients barium and then making X-rays of the small intestine has provided only limited diagnostic and treatment value. Now, physicians can use *capsule endoscopy* to get a direct view of the small intestine. A patient swallows a capsule containing up to two video camera chips, as well as a light, battery, and radio transmitter. Over the course of about eight hours, the capsule takes images as it passes through the esophagus, stomach, and small intestine, transmitting the images to a receiver worn by the patient. The physician later transfers the images from the receiver for review and diagnosis. Capsule endoscopy now provides a more patient-friendly way of diagnosing problems such as gastrointestinal bleeding and Chrohn's disease.

A smartphone with a diagonal screen measurement greater than 5.25 inches, sometimes called a *phablet*, combines mobile calling with easier web browsing and multimedia viewing made possible by the larger screens.

To add functions to mobile devices, you can download and install apps that perform specific activities—for instance, sending and receiving email, taking and editing photos and videos, using social media sites (such as Facebook and Twitter), video chatting, online banking, and gaming (see Figure 1.17). These devices use flash storage (the same type of storage found in portable USB flash drives) rather than a hard disk drive. The amount of storage determines how many apps you can install, how many photos you can take, and so on. You may need or want to synchronize (sync) your mobile device with your main computer from time to time to perform updates, back up content, organize contact lists, and so on. For some tablet models, you can get an add-on keyboard, which enables you to use the tablet more like a laptop computer. Other popular peripheral devices you may use with a smartphone or tablet include speaker docks, external battery chargers, and enhanced cameras.

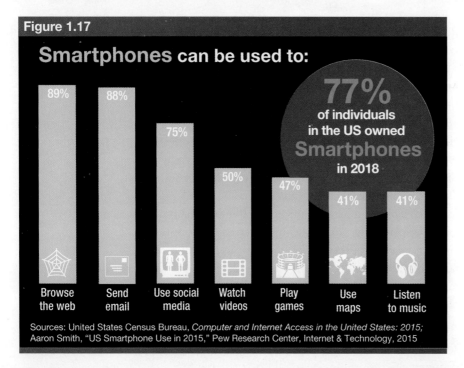

Figure 1.17

Smartphones can be used to: Browse the web 89%, Send email 88%, Use social media 75%, Watch videos 50%, Play games 47%, Use maps 41%, Listen to music 41%. 77% of individuals in the US owned Smartphones in 2018.

Sources: United States Census Bureau, *Computer and Internet Access in the United States: 2015*; Aaron Smith, "US Smartphone Use in 2015," Pew Research Center, Internet & Technology, 2015

Practical TECH

Dealing with Older Devices

Small mobile devices evolve so quickly that some users upgrade them as quickly as they can afford to. Most major data services carriers, such as Verizon and AT&T, provide trade-in and recycling programs. These programs may save you money on a new phone and ensure that your device won't wind up in a landfill when you leave it behind. If your carrier doesn't offer such a program or you would like to find out what your phone is worth, you can check prices at a general auction website, such as eBay. Better yet, you can go to a specialized website such as Gazelle (available at https://CUT7.Paradigm Education.com/Gazelle), which purchases and resells used mobile devices. Other websites offer the opportunity to donate an older phone or recycle a very old device that no longer has any value. No matter how you decide to dispose of your outdated device, be sure to first completely erase your personal data.

As the name suggests, a desktop computer fits on top of a work area. A tower-type system unit, like the one shown here, can be placed on the floor.

Personal Computers and Workstations

A **personal computer (PC)**—often referred to as simply a *computer*—is a self-contained computer that can perform input, processing, output, and storage functions. A PC must have at least one input device, one storage device, and one output device, as well as a processor and memory. Smartphone and tablet devices technically meet these criteria. However, PCs are generally larger, use a keyboard and mouse or a touchpad for input, and can run full-blown applications for business and other purposes. PCs might have multiple processors (or *microprocessors* or *CPUs*). (Think of the CPU as the "brain" of the computer.) The three major groups of PCs are desktop computers, portable (mobile) computers, and workstations.

Tech Career Explorer
Put Your Sales Skills to the Tech

Tech Career Explorer
Step Up to the Service Side of Tech

Article
DIY with Raspberry Pi

Desktop Computers A **desktop computer** is a PC that's designed to allow the system unit, input devices, output devices, and other connected devices to fit on top of, beside, or under a desk or table. This type of computer may be used in a home, a home office, a library, or a corporate setting. The monitor and speakers may have wired connections to the main system unit. Peripheral devices such as a mouse, keyboard, and printer may have wired or wireless connections. Desktop computers may have a wired or wireless connection to the local network and/or the internet.

Desktop computers offer greater expansion and upgrade capabilities than mobile computers (described in the next section). Standard desktop computers are made so that you can "crack the case" (open the system unit) and add or replace internal hardware on your own. They include expansion slots that enable upgrading the RAM and adding features such as a graphics card for a second monitor. *All-in-one desktop computers* combine the system unit and the monitor. They save space on your desk surface, but typically need to be upgraded by a computer repair technician when upgrades are desired.

Portable (Mobile) Computers The mobility of today's workforce and the need for internet access anytime, anywhere have increased the demand for mobile computers. A *portable computer* (sometimes called a *mobile computer*) is a PC that's small enough to be moved around easily.

There are several types of portable computers. As its name suggests, a **laptop computer** can fit comfortably on your lap. As this type of computer decreased in size, many users began using a new term for it: **notebook computer**. Another type of

A laptop computer is a PC designed for people who frequently move about in their work.

portable computer, called a *netbook*, emerged as a low-cost alternative to a full laptop computer. A *netbook* is usually smaller and less powerful than a laptop; it's intended primarily for web browsing and emailing over the internet. Each of these types of portable computers has a flat screen, and the screen and keyboard fold together like a clamshell for easy transport.

Because many people are more comfortable entering data on an external keyboard than a touchscreen, *convertible models* (sometimes called *2-in-1 computers*) attempt to bridge the gap between a tablet and a laptop. A convertible computer has a touchscreen like a tablet, but it also has a keyboard that either attaches and detaches (as on the Microsoft Surface) or folds and snaps in behind the screen (as on Dell and Lenovo models). This feature lets you use either the keyboard or the touchscreen.

A high-performance class of small laptop computer called an Ultrabook emerged in 2012. Intel set the standard for this new class of portable computers. An Ultrabook must use low-power Intel Core processors, have solid-state drives for storage (rather than a hard disk drive), and feature a more durable chassis (body). While this lightweight machine tends to have a long battery life, it also may lack features present in other types of laptops, such as a DVD-R drive and certain ports.

Workstations **Workstations** resemble desktop PCs but provide more processing power and capabilities. A workstation is a high-performance, single-user computer with advanced input, output, and storage components. A workstation can be networked with other workstations and with larger computers. Previously, workstations used a different operating system and different hardware than a standard desktop computer. Today, however, workstations generally use a standard Windows or Mac operating system.

Workstations are typically used by companies for complex applications that require considerable computing power and high-quality graphics resolution. Examples of these applications include computer-aided design (CAD), computer-aided manufacturing (CAM), desktop publishing, video editing, and software development.

Workstations provide the enhanced power and graphics capabilities required for some applications such as programming, data analysis, computer-aided design, and desktop publishing and graphics.

Workstations generally come with large, high-resolution graphics displays and/or multiple displays, built-in network capability, and high-density storage devices. Like PCs, workstations can serve as single-user computers but more often connect to a network and have access to larger computers (often, servers and mainframes).

Servers (Midrange Servers and Minicomputers)

In a large organization, networked computers typically connect to a larger and more powerful computer called a *server* (sometimes referred to as the *host computer*). Although the size and capacity of network servers vary considerably, most are midrange rather than large mainframe computers (discussed later in the next section). Dell and Hewlett-Packard Enterprise are among the leading manufacturers of servers.

A **server** (formerly known as a *midrange server or minicomputer*) is a powerful computer that's capable of accommodating numerous client computers at the same time. Midrange servers are widely used in networks to provide users with computing capability and resources available through the network. Those resources generally include internet content and software that's critical to business operations, such as an accounting system, manufacturing planning software, customer relationship management (CRM) software, and more. Because servers must respond to the needs of many users, system reliability and access have become key considerations in planning and maintaining these computers.

In terms of physical hardware, a modern server often consists of multiple different server and storage units in storage racks. These so-called rack servers consume a lot of power and generate a lot of heat; to ensure optimum operation, they are often located in a separate room with enhanced power and cooling capabilities. Blade servers also mount in racks, but they require less space and use less power.

A server is a powerful computer that can accommodate (serve) multiple users in a network. Many servers use hardware that's configured in a rack like this—a setup known as a *rack server*.

Mainframe Computers

In a small-business environment, a personal computer can be used as a server. Large businesses, which need more powerful servers, may use mainframe computers. Bigger, more powerful, and more expensive than a midrange server, a **mainframe computer** can accommodate hundreds or thousands of network users performing different computing tasks. A mainframe's internal storage can handle hundreds of millions of characters. Mainframe applications are often large and complex. These computers are useful for dealing with large, ever-changing collections of data that can be accessed by many users simultaneously. A mainframe computer can also function as a network server.

Government agencies, banks, universities, and insurance companies use mainframes to handle millions of transactions each day. The latest generation of mainframe computers from IBM is called IBM Z and operates on a variety of advanced software platforms.

Article
Enterprise Computing Trends

Article
A History of Computers Timeline

A mainframe computer is a large, powerful, expensive computer system that can accommodate multiple users at the same time. The tremendous processing capability of mainframes differentiates them from server systems.

Supercomputers

Supercomputers are the current "Goliaths" of the computer industry. A **supercomputer** is an extremely fast, powerful, and expensive computer. Many supercomputers are capable of performing quadrillions of calculations in a single second. Performing the same number of calculations on a handheld calculator would take 2 million years!

Supercomputer designers achieve stunning calculation speeds by joining thousands of separate microprocessors. Many supercomputers provide enough disk storage capacity for dozens of petabytes of data. (One petabyte is the equivalent of 1 quadrillion alphabet letters, numbers, or special characters.)

In a move to expand supercomputing into the realm of the unimaginable, Oak Ridge National Laboratory has developed Summit, the most powerful supercomputer in the world. With high-bandwidth technology by IBM, sizable memory, and more than 27,000 deep-learning optimized NVIDIA GPUs (graphics processing units), Summit can theoretically process data at a rate of 200 petaflops. (A petaflop is equivalent to 1 quadrillion calculations per second.) Summit will be used for high-end scientific and research projects, such as simulating stellar explosions. Summit went online in mid-2018 and became available for researchers in early 2019.

Supercomputers are among the world's fastest, most powerful, and most expensive computers. They are capable of processing huge amounts of data quickly and accommodating thousands of users at the same time. The Summit supercomputer at Oak Ridge National Laboratory in Tennessee is the most powerful supercomputer in the United States.

Cutting Edge

Beyond Super

Quantum computers exceed the capabilities of supercomputers. So far, though, they haven't been used much for practical applications because of their high cost and the difficulties of operating them. Quantum computers will be used to perform computations beyond what supercomputers can currently tackle, for example, evaluating trillions of amino acid combinations to find a particular protein for diagnosis and drug development. Chapter 13 introduces this exciting and emerging generation of computers.

Chapter Summary

1.1 Being Immersed in Digital Technology

Information technology (IT) is the use of computer, electronics, and telecommunications equipment and technologies to gather, process, store, and transfer data. We live in a **digital** world, in which computer technology increasingly powers the devices that are part our daily lives. Those devices include smart TVs and appliances, smartphones, and embedded computers in everything from wearable devices to automobiles. Technologies such as embedded chips, computers, networks, and the internet and World Wide Web enable us to communicate globally. No digital device has had a greater impact on our lives than the computer. A **computer** is an electronic device that operates under the control of programmed instructions, called a *program*, stored in its memory.

1.2 Discovering the Computer Advantage

Computers offer advantages in the areas of speed, accuracy, versatility, storage, and communications. As a result, they are widely used in homes, schools, businesses, and other organizations for communicating, managing finances, analyzing data, planning, researching, and hundreds of other purposes. The **internet** and the World Wide Web, in which networks of computers around the world are linked, continue to serve an important function in all areas of human activity.

1.3 Understanding How Computers Work

Data is raw, unorganized content, such as facts and figures. Data entered into a computer consists of one or more of the following types: **text data**, **graphic data**, **audio data**, and **video data**. **Information** is data that's been processed (manipulated, organized, or arranged) in a way that's converted it into a useful form. Using a computer to convert data into useful information is called **information processing** (or *data processing*). The **information processing cycle** involves four stages: **input**, **processing**, **output**, and **storage**. It accepts data (input) that's manipulated (processed) according to the instructions and output as information, which may be placed in storage for future use.

1.4 Differentiating between Computers and Computer Systems

The term *computer* identifies only the system unit—the part of a desktop computer system that processes data into information. A **computer system** includes the system unit along with input, output, storage, and communications devices.

1.5 Understanding Hardware and Software

There are two general types of computer system components: hardware and software. **Hardware** includes all of the physical components of the computer and the **peripheral devices** connected to it. The main hardware components are the system unit, input devices, output devices, **storage devices**, and **communications devices**. A **modem** is an electronic device that converts computer-readable information into a form that can be transmitted and received over a communications system.

 Software consists of the programs that contain instructions that direct the operation of the computer system and programs that enable users to perform specific tasks. There are two main types of software: system software, consisting of the operating system and utility software, and application software, or *apps* on a mobile device. Mobile devices also use separate mobile operating systems. Cloud apps and services are delivered online. Users typically access them through a web browser (or a downloaded app) and must have a free account or subscription.

When executing a program, the processor temporarily stores the program's instructions and the data needed by the instructions in the computer's memory. Main memory provides temporary storage for instructions and data during processing.

1.6 Identifying Categories of Computers

A **smart device** is an everyday appliance or piece of equipment that has capabilities provided by an embedded computer; examples include the computing technologies being introduced in vehicles and home appliances. Devices with embedded technology that can send and receive data wirelessly or via the internet make up the **Internet of Things (IoT)**. Voice-controlled **smart speakers** run **intelligent personal assistant** software to interpret verbal requests and commands, even running other smart devices as part of a **smart home** system.

Similarly, a **wearable computer**, which is typically worn on the wrist, performs specific computing functions; examples include tracking the number of steps taken or miles run and helping the user respond to alerts and messages from a smartphone. **Tablets** and **smartphones** are handheld computers; these internet-connected devices are operated via a built-in touchscreen. A **personal computer (PC)** is a self-contained computer capable of input, processing, output, and storage; types of PCs are **desktop computers**, portable computers (which are also called *mobile computers* and include **laptop computers** and **notebook computers**), and **workstations**.

A **server** can accommodate numerous of client computers or terminals (users) at the same time. A **mainframe computer** is capable of accommodating hundreds to thousands of network users performing different tasks. A **supercomputer** exceeds the power and capabilities of a mainframe; the fastest supercomputer in the United States is capable of processing quadrillions of calculations per second.

Key Terms

Numbers indicate the pages where terms are first cited with their full definition in the chapter. An alphabetized list of key terms with definitions is included in the end-of-book glossary.

Chapter Glossary

Flash Cards

1.1 Being Immersed in Digital Technology
information technology (IT), 3
digital, 3
computer, 4
program, 4

1.2 Discovering the Computer Advantage
microcomputer, 6
internet, 10

1.3 Understanding How Computers Work
data, 12
information, 12
text data, 13
graphic data, 13
audio data, 13
video data, 13
information processing, 13
information processing cycle, 14
input, 14
processing, 14
output, 14
storage, 14

1.4 Differentiating between Computers and Computer Systems

computer system, 14
single-user computer system, 15
multiuser computer system, 15

1.5 Understanding Hardware and Software

hardware, 16
peripheral device, 16
main memory, 17
storage device, 18
communications device, 18
modem, 18
software, 19

1.6 Identifying Categories of Computers

smart device, 22
Internet of Things (IoT), 22
smart home, 24
smart speaker, 24
intelligent personal assistant, 24
wearable computer, 25
smartphone, 27
tablet, 27
personal computer (PC), 29
desktop computer, 29
laptop computer, 29
notebook computer, 29
workstation, 30
server, 31
mainframe computer, 32
supercomputer, 33

Chapter Exercises

Complete the following exercises to assess your understanding of the material covered in this chapter.

Tech to Come: Brainstorming New Uses

In groups or individually, contemplate the following questions and develop as many answers as you can.

1. Futurists predict that computers will be everywhere. For example, some bridges have computers and sensors that alert city planners when parts of these structures are weak or overly stressed and in need of repair. What other objects should have the same type of warning or notice capability built into it?

2. Many of us already use wearable computers. What workplace dilemmas, problems, or limitations might be solved if we wore computers that are capable of collecting and analyzing personal data while we are on the job?

3. Computer literacy is extremely important for workers, because the use of computers has become commonplace in many occupations. What are some examples of how computers are used in your field of study or future career?

Tech Literacy: Internet Research and Writing

Conduct internet research to find the information described, and then develop appropriate written responses based on your research. Be sure to document your sources using the following format, which is recommended by the MLA Handbook, *Eighth Edition.*

Use a period after each part of the citation except the publisher or sponsor, which should be followed by a comma. If you cannot find some of the information described below, include whatever is available.

- Name of author, compiler, director, editor, narrator, performer, or translator, if known
- Title of work (in italic type if work is independent; in roman type and enclosed in quotation marks if work is part of larger work, such as an article within a journal)
- Title of overall website (in italic font), if different from title of work (if website is not titled, use a genre label, such as *Home Page, Introduction,* or *Online posting,* in roman type)
- Version or edition
- Publisher or sponsor of website if known; if not known, use *N.p.*
- Date of publication (day, month, and year, as available); if nothing is available, use *n.d.*
- Medium of publication (*Web.*)
- Date you accessed the site (day, month, and year)
- URL, in angle brackets <> (optional; MLA does not require this)

- Terms Check
- Knowledge Check
- Key Principles
- Tech Illustrated
  Star Network
- Tech Illustrated
  Information
  Processing Cycle
- Chapter Exercises
- Chapter Exam

For example: David E. Sanger and Nicole Perlroth. "Cyberattack Disrupts Printing of Major Newspapers." *The New York Times*. 30 December 2018. Web. 1 January 2019. <https://www.nytimes.com/2018/12/30/business/media/los-angeles-times-cyberattack.html>.

1. Suppose that you have been offered a free personal computer system of your choice and can select the input, output, and storage devices you want. Create a list of the ways you will use your new computer. Then research various computer systems and components advertised in magazines and on the internet. Create a computer system that will meet your needs, and write a paragraph explaining why you selected a particular system or combination of components.

2. The internet provides easy access to a wealth of information and is considered to be a timesaver for busy people. Prepare a written report that explains what aspects of your life have been simplified or improved by using the internet. Include predictions for additional internet capabilities you expect to use in the next five years.

3. Many projects are under way to expand the use of wearable computers in the workplace, in military applications, and for personal use. Using your school library and other sources of information, research the uses of wearable computers. Based on your findings, write an article that describes ways in which wearable computers can enhance our daily lives.

4. Two websites gather stats for people browsing the web. Go to **https://CUT7.ParadigmEducation.com/StatCounter** and **https://CUT7.ParadigmEducation.com/NetMarketShare**, and use the tools provided to find statistics on what web browsers are installed for users browsing the web. Write a brief article summarizing your findings, including one or more charts, if possible.

Tech Issues: Team Problem-Solving

In groups, develop possible solutions to the issues presented.

1. The students in today's classrooms are more diverse and have a wider range of abilities than students from previous decades. In fact, some experts claim that the learning rate in classrooms today is twice as fast that of years past. Imagine how computers will help instructors teach so many different types of students. Consider both traditional and distance learning courses and programs.

2. Artificially intelligent robots are likely to have many uses in the future. What new applications of this technology seem possible in the areas of manufacturing, health care, and home maintenance?

3. Since computers were first introduced, there has been considerable debate about their effects on employment. For example, some people argue that computers have replaced many workers and are therefore harmful to society. Others argue that computers increase productivity and that the computer industry has created many new high-paying jobs. In your group, discuss both sides of this issue: Overall, have computers had a good effect or a bad effect on employment?

Tech Timeline: Predicting Next Steps

The following timeline outlines some of the major events in the development of computing. Research this topic to predict at least three important milestones that could occur between now and the year 2040, and then add your predictions to the timeline.

1941 Konrad Zuse completes work on the world's first programmable digital computer, called the Z3.

1942 Dr. John Atanasoff and Clifford Berry complete work on the *Atanasoff–Berry Computer* (or *ABC*), a forerunner of today's modern computers.

1958 Jack Kilby, an engineer at Texas Instruments, and Robert Noyce, an electrical engineer at Fairchild Semiconductor, simultaneously and independently invent the integrated circuit, thereby laying the foundation for fast computers and large-capacity memory.

1981 IBM enters the personal computer field by introducing the IBM PC.

1993 World Wide Web technology and programming is officially proclaimed to be in the public domain and available to all.

2004 Wireless computer devices—including keyboards, mice, and wireless home networks—become widely accepted among users.

2006 Five million subscribers connect to BlackBerry Internet Service (BIS) for work and personal communications.

2007 The first iPhone is released by Apple.

2010 The first iPad is released by Apple.

2014 Apple introduces the Apple Watch, calling it a "comprehensive health and fitness device."

2019 The Summit supercomputer becomes available for use by researchers.

Ethical Dilemmas: Group Discussion and Debate

As a class or within an assigned group, discuss the following ethical dilemma.

The term *plagiarism* refers to the unauthorized and illegal use of another person's writing or creative work. For example, a student may copy text from a magazine article or web page and submit the report without crediting the author of the material.

In your group, discuss the issue of plagiarism for class assignments and reports. What types of plagiarism are you aware of? Are there tools that instructors can use to detect and protect against plagiarism?

Chapter 2

Sizing Up Computer and Device Hardware

Chapter Goal

To understand and identify computer input, processing, and output devices

Learning Objectives

2.1 Identify types of input devices.

2.2 Explain how computers represent and process data with bits and bytes.

2.3 Describe the components of a system unit.

2.4 Explain how a CPU functions.

2.5 Differentiate between types of memory.

2.6 Differentiate between types of display technologies and understand digital audio.

2.7 Describe several printer technologies and their advantages and disadvantages.

Online Resources
The online course includes additional training and assessment resources.

Tracking Down Tech
Identifying Input and Output Devices

2.1 Working with Input Technology

As you read in Chapter 1, the four steps of the information processing cycle are input, processing, output, and storage. We will begin this chapter by looking at the first step in that process: input.

Input is any data or instruction entered into a computer. You can enter input in a variety of ways, such as typing on a keyboard, clicking with a mouse, dragging and tapping on a touchscreen, and speaking into a microphone.

An **input device** is any hardware component that enables you to perform one or more input operations. The types of input devices that are available depend on the type of computer. For a desktop personal computer (PC), for example, the most common input devices are a keyboard and a mouse. For a laptop PC, a touchpad is the most common pointing device, replacing the mouse. On a tablet PC, the screen is touch sensitive and functions as an input device. Figure 2.1 shows a variety of common input devices.

Input devices (and output devices, too) can either come with the computer or be purchased separately. A device that comes with a computer system is sometimes referred to as an Original Equipment Manufacturer (OEM) device.

Watch & Learn
Working with Input Technology

Keyboards

The most common input device is the **keyboard**, which is used to enter text characters (letters, numbers, and symbols). Some computers, such as laptops, have the keyboard built in; other computers accept a keyboard as an external device. An external keyboard can be plugged into the computer or connected wirelessly through a radio frequency (RF) technology such as Bluetooth.

The most common keyboard is alphanumeric and has a QWERTY layout. The term **alphanumeric keyboard** means that the keyboard has both letters (*alpha-*) and numbers (*-numeric*), and the term **QWERTY layout** refers to the first six letters in

Video
Keyboard Layout

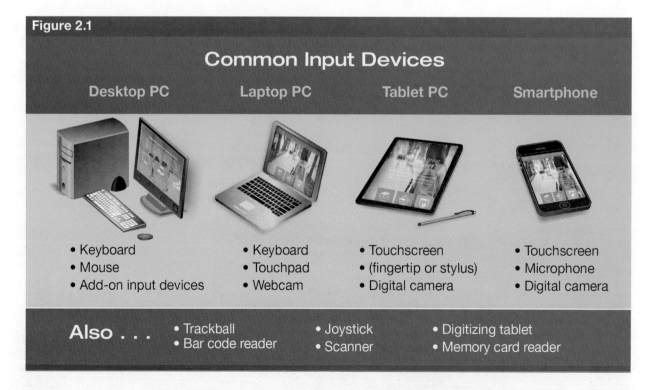

Figure 2.1

Common Input Devices

Desktop PC	Laptop PC	Tablet PC	Smartphone
• Keyboard • Mouse • Add-on input devices	• Keyboard • Touchpad • Webcam	• Touchscreen (fingertip or stylus) • Digital camera	• Touchscreen • Microphone • Digital camera

Also . . .
- Trackball
- Bar code reader
- Joystick
- Scanner
- Digitizing tablet
- Memory card reader

the top row of letters on the keyboard. Figure 2.2 shows a typical external keyboard with a QWERTY layout. Notice that in addition to the letters and numbers, there are several other types of keys. Function keys (F1 through F12) run across the top of the keyboard; these keys are assigned specific purposes in different operating systems and applications. Other special-purpose keys, such as Ctrl and Alt, are located to the right and left of the space bar. To the immediate right of those keys are cursor control keys (arrow keys), which control the movement of the cursor. Along the right side of the

Figure 2.2

An Alphanumeric Computer Keyboard

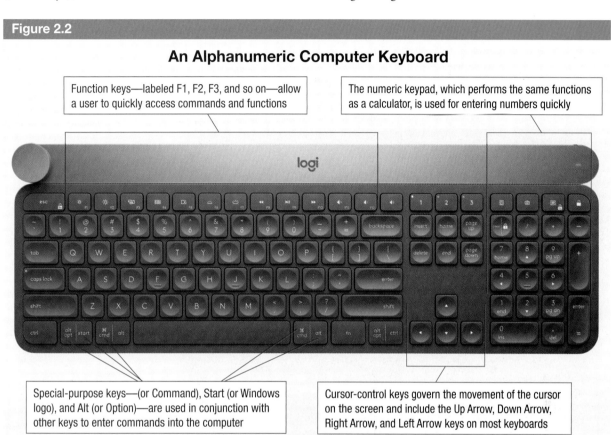

Function keys—labeled F1, F2, F3, and so on—allow a user to quickly access commands and functions

The numeric keypad, which performs the same functions as a calculator, is used for entering numbers quickly

Special-purpose keys—(or Command), Start (or Windows logo), and Alt (or Option)—are used in conjunction with other keys to enter commands into the computer

Cursor-control keys govern the movement of the cursor on the screen and include the Up Arrow, Down Arrow, Right Arrow, and Left Arrow keys on most keyboards

An alphanumeric computer keyboard is organized into several groups of related keys. Some alphanumeric keyboards, including the one shown here, are designed for use with both Windows and Mac computers.

Practical TECH

Are Some of Your Keys Not Working?

A keyboard is a grid of uncompleted circuits. When you press a key, a connection is made inside the keyboard, completing the circuit and sending an electrical pulse to the computer. The computer interprets the electrical pulse and inputs the corresponding character in the software you are using.

A keyboard gets dirty through normal use, and various types of debris collect under the keys. The debris may prevent the circuit from being completed, which will cause the keys not to work. You can clean underneath the keys using compressed air or (if it's an external keyboard) by turning the keyboard upside down and shaking it. To clean the surface of the keys, use a paper towel or soft cloth and a small amount of an all-purpose household cleaning product. Don't spray or pour the cleaning liquid directly onto the keyboard, because moisture can cause the electronics to short-circuit. Instead, spray or pour the liquid onto the cloth and wipe the cloth across the keyboard.

keyboard is a numeric keypad. The positions of some of the additional keys may vary between keyboards made by different manufacturers, and not all keyboards have all of these keys.

Not all computers require having a full alphanumeric keyboard; some work most efficiently with a special-function keyboard. For example, at a fast-food restaurant, employees take customers' orders by pressing buttons on a customized grid or touchscreen panel that has a separate button for each item on the menu. Computerized voting booths also use a special-function keyboard or touchscreen; in this case, the layout that has been customized for both the voting software and the candidates running for office.

A device with a touchscreen, such as a tablet PC or smartphone, may have a **virtual keyboard**, which is a software-generated, simulated keyboard (see Figure 2.3). To use a virtual keyboard, you "type" by clicking or tapping the virtual keys on the screen. Virtual keyboards are common on mobile devices.

Figure 2.3

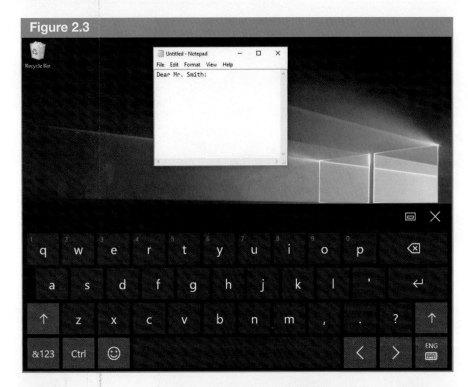

A virtual keyboard enables you to type without having a physical keyboard. You can use a pointing device (such as a mouse) to click the keys on the screen, or on a touchscreen, you can tap the virtual keys with your finger.

A touchscreen enables the user to interact with the computer by tapping and sliding his or her fingers across the surface.

Touchscreens

Many portable computing devices have touchscreens as input devices. A **touchscreen** is a touch-sensitive display that is produced by laying a transparent grid of sensors over the screen of a monitor. You can interact with the computer by tapping and sliding your fingers across the screen. Tablet PCs typically use touchscreens, and so do smartphones. Many self-service kiosks—such as shopping mall directories, bridal registries at department stores, and check-in stations at airline terminals—also use touchscreens.

Video
Touchscreens

Video
How Do Touchscreens Work?

Practical TECH

Touchscreen Gestures

Touchscreens are different from other types of input devices, and you will need some special skills to use this form of device. Here are a few important terms you should know:

- **Tap** *(or* **touch***)*. Tap the screen with your finger, pressing and quickly releasing on the same spot. You tap to make a selection or issue a command.

- **Pinch.** Touch two fingers to the screen in different spots and then drag the fingers together. You pinch to zoom out (and display a larger area).

- **Stretch** *(or* **unpinch***)*. Touch two fingers to the screen in adjacent spots and then drag the fingers farther apart. You stretch to zoom in (and display a smaller area).

- **Drag** *(or* **slide** *or* **swipe***)*. Touch one finger to the screen and then slide it along the surface. You drag to perform a variety of functions, depending on the software and context. For example, dragging can open menu bars, exit applications, scroll the display, or move items around on the screen.

- **Rotate.** Touch two fingers on the desired object or area and then drag them in a circular motion.

Mice and Other Pointing Devices

Most operating systems represent program features and commands visually on the screen by incorporating a graphical user interface (GUI, pronounced "gooey"), which contains buttons, drop-down menus, and icons. The easiest way to interact with a GUI is with a **pointing device**—that is, an input device that moves an onscreen pointer (usually, an arrow). Pointing devices typically include buttons for making selections. The left button is usually the primary button, and the right button is generally assigned to special functions, such as displaying a pop-up menu of choices.

The most common type of pointing device is a **mouse**, a handheld device that you move across a flat surface (like a desk) to move the onscreen pointer. An alternative is a **trackball**, which is a stationary device with a ball that you roll with your fingers to move the pointer. Like a keyboard, a mouse or trackball connects to your computer using either a cord or wireless technology such as Bluetooth.

A **touchpad** (also called a *track pad*) is a small, rectangular pad that's sensitive to pressure and motion. Many portable computers have built-in touchpads. The onscreen pointer moves as you slide your finger across the surface of the touchpad. Tapping the pad is the equivalent of pressing the primary button on a mouse. Most touchpads also have nearby buttons that work like the buttons on a mouse.

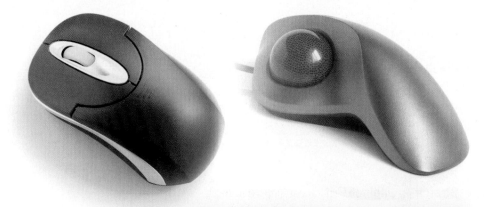

Both a mouse (left) and a trackball (right) connect to your computer and allow you to control an onscreen pointer.

Sliding your finger across the touchpad on a laptop computer is like moving a mouse across your desk. In this photo, below the touchpad are two buttons that function like left and right mouse buttons.

Chapter 2 Sizing Up Computer and Device Hardware

47

A joystick can be tilted in any direction to move an object on the screen.

A game controller pad offers the same buttons and positions as gaming consoles such as the Xbox and PlayStation. The game pad can be held in both hands, and the main buttons are pushed by the thumbs.

Joysticks and Other Gaming Hardware

People who play video games on their PCs are sometimes dissatisfied with the in-game performance when using ordinary keyboards and pointing devices for input. They may prefer to use specialized input devices, such as joysticks, steering wheels, and game controller pads.

A **joystick** is an upright stick mounted vertically to a base. You can tilt the stick in any direction to move a pointer or other object on the screen. Joysticks are popular controllers for flight simulator games. A **steering wheel** simulates a vehicle's steering wheel and is often used for driving games. A **game controller pad** (or *game pad* or *controller*) offers the same button types and positions as do the controllers for dedicated gaming consoles such as the Xbox and PlayStation. People who come to PC gaming after spending time playing console games may prefer to use the game pad's familiar layout rather than having to learn keyboard and mouse commands for their favorite games.

Practical TECH

Pointing Device Actions

Moving the onscreen pointer is likely the action you will perform most often with a pointing device. But after you have positioned the pointer at the item you want to work with, you must take one of these actions:

- **Click.** Press and release the left mouse button once without moving the pointer. You usually click to select something; for example, clicking may open a menu or select an icon.
- **Double-click.** Press and release the left mouse button twice in quick succession without moving the mouse pointer. You usually double-click to activate something; for example, if you double-click the icon for a document, the document will open.
- **Right-click.** Press and release the right mouse button once without moving the pointer. You usually right-click to open a shortcut menu, from which you can select a command (by left-clicking it). If your mouse doesn't have a right mouse button (for example, if it is an older Mac mouse with just a single button), you can right-click by pressing the Ctrl key (or the Command key on a Mac) and holding it down while you click the mouse button.
- **Drag.** Press and hold the left mouse button and then move the mouse. You usually drag to move whatever you have selected from one place to another onscreen.

Engineers, drafters, and artists use a graphics tablet to create precise, detailed drawings.

Graphics Tablets and Styluses

Artists, engineers, and others who need precise control over input can use a graphics tablet to simulate drawing on paper. A **graphics tablet** is a flat grid of intersecting wires that form uncompleted digital circuits. Each two-wire intersection has a unique numeric value. When you touch a stylus to a spot on the surface of the tablet, the circuit is completed and the computer receives a digital code that corresponds to the location of the circuit on the grid. A **stylus** resembles an inkless pen and is used much like a normal pen, except that instead of leaving ink marks directly on the tablet, it causes marks to be made in the computer software.

Tech Career Explorer
Computer-Aided Design Drafter

Scanners

A **scanner** is a light-sensing device that detects and captures text and images from a printed page. It works by shining a bright light on the page and measuring the amount of light that bounces back from each spot. Lighter areas reflect the scanner's light, and darker areas absorb the scanner's light—so the darker the area, the less amount of light is bounced back. The strength of the captured light is digitized (converted to digits) and the **digitized information** is sent to the computer for recording. Figure 2.4 outlines the scanning process. Color scanners measure the amounts of red, green, and blue individually. The sensor that records the light strength is a **charge-coupled device (CCD)**.

The scanned material is stored as a **bitmap**, which is a matrix of rows and columns of pixels. A **pixel** is the smallest picture element that a monitor can display (an individual dot on the screen) and is represented by one or more bits of binary data. The more bits that are used to represent each pixel, the more different colors that can be recorded. The measurement of bits used to store information about a pixel is known as *color depth*. (This term will be discussed more later in this chapter.) For example, a 1-bit image shows each pixel as black (represented by a 1) or white (represented by a 0). A 4-bit image differentiates among 16 shades of gray, because there are 16 possible combinations of 1 and 0 in a four-digit string (0000 through 1111). A color scanner splits the number of bits among the three colors; for example, a scan with a 24-bit color depth uses 8 bits for red, 8 for green, and 8 for blue. Modern scanners typically use 30 to 48 bits per pixel. Most of those bits describe the pixel's color and shade, but some are used for error correction.

Chapter 2 Sizing Up Computer and Device Hardware

Resolution is the number of pixels captured per inch of the original page; this value is measured in **dots per inch (dpi)**. A scanner with 1,200 dpi resolution can capture 1,200 pixels per inch in each direction (horizontally and vertically). Scanners capture images at higher resolutions than most people need for displaying them, as computer screens typically display only 96 dpi. When scanning materials for use in a printout, though, you need higher resolutions and must pay more attention to dpi. The drawback of scanning at a high resolution is that the resulting image file is many times larger than that of scan at a lower resolution.

A flatbed scanner is used to scan flat items, such as pages of a document and printed photos.

When scanning text, **optical character recognition (OCR)** software is used. This specialized software converts the bitmap images of text to actual text that you can work with in a word-processing or other text-editing program.

Different types of scanners operate differently. A **flatbed scanner** has a large, flat, glass-covered surface, similar to a copy machine. You place the item to be scanned face

Practice
The Scanning Process

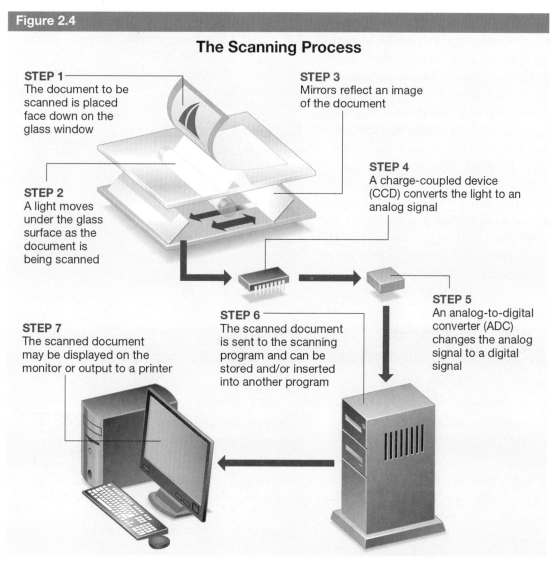

Figure 2.4

The Scanning Process

STEP 1 The document to be scanned is placed face down on the glass window

STEP 2 A light moves under the glass surface as the document is being scanned

STEP 3 Mirrors reflect an image of the document

STEP 4 A charge-coupled device (CCD) converts the light to an analog signal

STEP 5 An analog-to-digital converter (ADC) changes the analog signal to a digital signal

STEP 6 The scanned document is sent to the scanning program and can be stored and/or inserted into another program

STEP 7 The scanned document may be displayed on the monitor or output to a printer

A scanner captures text and graphic images and converts them into a format the computer can understand and display.

down on the glass, close the cover, and issue the command to begin scanning; a bar inside the scanner moves a light and a sensor across the glass, capturing the image. Specialty scanners are also available. For example, medical offices often use a small scanner with a slot for insurance cards; a card is inserted in the slot, scanned, and then ejected.

Traditional scanners capture only flat, two-dimensional pages and images, but a **3-D scanner** captures complete size and shape information about an object. The collected data can then be used in a 3-D modeling application, from which it can be printed on a 3-D printer or altered to produce a modified copy of the object.

Bar Code Readers

Nearly every product you buy today is marked with a **bar code**, which is a set of black bars of varying widths and spacing. Manufacturers assign each product a unique bar code called a **universal product code (UPC)** and then stores use the bar code to look up the item's price. Warehouses generate bar codes using label printers, creating stickers they can use to tag inventory. Shipping companies generate bar codes to track packages as they travel to their destinations. Hospitals use bar-coded wristbands to verify patients' identities before treating them and giving them medication.

A **bar code reader** captures the pattern in a bar code and sends it to the computer, where special software translates the bar pattern into meaningful data (such as price). A bar code reader is similar to a scanner in that it shines a light on the code and measures the amount of light bounced back from each spot to determine which areas are light and which are dark. Unlike a scanner, though, a bar code reader is designed to read codes on three-dimensional (3-D) objects, not just flat pieces of paper. Bar code readers are also not nearly as technologically sophisticated as scanners or smartphone cameras. They typically record only 1 bit (black or white) for each pixel that's scanned.

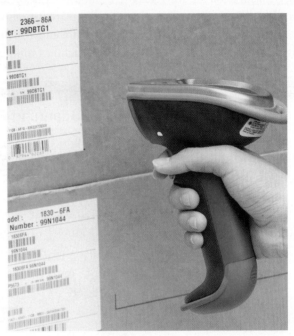

A bar code reader reads information in printed bar codes and sends it to a computer.

Hotspot

QR Codes

On everything from billboards to store windows, you are likely to see mysterious-looking squares of black-and-white splotches. These are QR (quick response) codes—two-dimensional bar codes that advertisers and organizations use to direct you to information about products and services. A smartphone uses QR code reader software, along with a built-in digital camera, to read a QR code and direct the device's browser to a website that is linked to that code. A QR code in a product ad might link to a website that provides a coupon or other sales promotion, along with product information. Try it for yourself now if you have a smartphone. Scan the code shown here, and see where it takes you. Then do a web search for QR code generators, and make your own QR code.

Digital Cameras and Video Devices

A **digital camera** captures images in a digital format that a computer can use and display. Technologically, a digital camera is similar to a scanner: it captures the image with a charge-coupled device and uses multiple bits to describe each pixel of the image in terms of its red, green, and blue values. The main difference between the devices is that a digital camera takes pictures of 3-D objects, not just flat pieces of paper. Digital cameras are built into almost all cell phones sold today and can also be purchased as standalone devices. Figure 2.5 shows the process of capturing an image with a stand-alone digital camera and then transferring the image to a computer.

Like a scanned image, a digital photo is represented by pixels, and its color depth is measured as the number of bits used per pixel (for example, 24 bits). The resolution of the camera is the number of pixels in the highest resolution it's capable of producing; for example, a camera might have a maximum resolution of 4,096 horizontal pixels and 4,096 vertical pixels. Digital camera resolution is described in megapixels (a **megapixel** is 1 million pixels). To determine the number of megapixels, multiply the

> **Practice**
> Capturing an Image with a Digital Camera

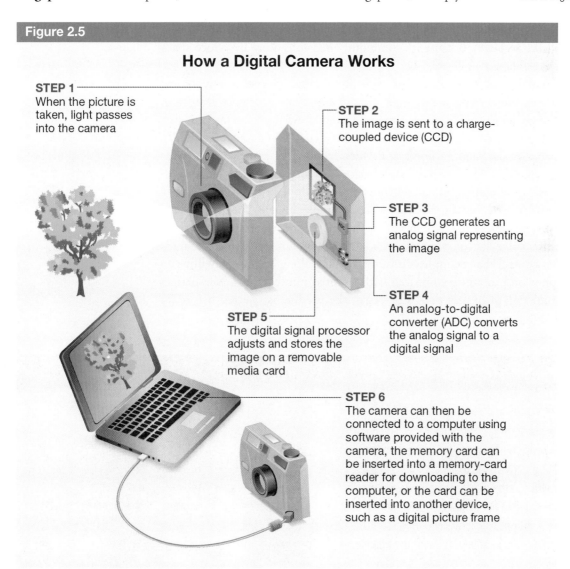

Figure 2.5

How a Digital Camera Works

STEP 1 — When the picture is taken, light passes into the camera

STEP 2 — The image is sent to a charge-coupled device (CCD)

STEP 3 — The CCD generates an analog signal representing the image

STEP 4 — An analog-to-digital converter (ADC) converts the analog signal to a digital signal

STEP 5 — The digital signal processor adjusts and stores the image on a removable media card

STEP 6 — The camera can then be connected to a computer using software provided with the camera, the memory card can be inserted into a memory-card reader for downloading to the computer, or the card can be inserted into another device, such as a digital picture frame

A digital camera captures images and stores them in the camera. You can connect the camera to a computer to transfer the images to it.

From left to right: a digital camera, a digital video camera, and a webcam.

camera's vertical resolution by its horizontal resolution. For example, a 4,096 × 4,096 resolution has a total of 16,777,216 pixels (16.7 megapixels).

Some digital cameras can capture both video and still images, but a **digital video camera** is designed primarily for capturing video. A digital video camera may be portable, like a regular camera, or it may be tethered to a computer (either by a cable or wirelessly). A digital video camera that must be connected to a computer to take pictures is called a **webcam**, because it's commonly used to capture live video that's streamed to the public via a website.

Digital video cameras have other uses too. In some newer cars, the rear-view camera is a digital video camera. Likewise, the active parking assistance feature in some cars relies on input from a video camera to gauge how close the car is to surrounding vehicles.

Many businesses use digital camera and video technology to help make their practices more efficient. For example, many businesses use digital video cameras to record the activities of employees and customers; this is done not only for security purposes but also to collect data for efficiency studies and audits. Businesses may also use digital cameras to make sure employees are authorized to enter certain restricted areas or access sensitive data on a computer. Some manufacturers use video technology for quality control. For example, a product moving along an assembly line can be photographed and instantly compared with a stored photograph of the perfect product. If a missing or broken part is detected, the product is rejected before it's packaged for shipment.

Hotspot

Photo Storage in the Cloud

Transferring pictures from a smartphone's digital camera is a snap. Many smartphones can be set up to automatically upload the photos you take to your own private storage area online, in the cloud. (Chapter 7 explores the cloud in greater detail.) For example, you can set up an iPhone to upload your pictures to iCloud, or an Android device to upload to Google Cloud. You can then easily retrieve your pictures from any other computing device by logging in to the cloud storage service with your user account for that service.

Audio Input Devices

The speech, music, and sound effects that are entered into a computer are called **audio input**. Nearly all PCs today have sound support that's either built in or provided by an add-on circuit board called a **sound card**. A sound card works by going back and forth between the analog sounds that humans hear and the digital recordings of sound that computers store and play back—translating one into the other. Sound support on a computer system also typically includes ports for connecting microphones, digital musical instruments, and other sound input devices. There may also be ports for output devices such as speakers and headphones, which are covered later in this chapter.

Voice input technologies enable users to enter data by talking to the computer. Some operating systems, including Windows, provide basic software for voice input as part of their accessibility tools. Some application software, such as word-processing programs, also support voice input.

There are several types of voice input systems. *Speaker-independent systems* recognize words spoken by an individual and translate them into text, regardless of the person's vocal tone, pitch, or accent. This type of system is best to use when it's important to be able to understand diverse accents and pronunciations, and when the set of words and phrases used is limited. For example, a voice-activated automated teller machine (ATM) enables customers to conduct financial transactions by speaking into the machine. When you call Customer Support for some companies, your call is initially answered by a speaker-independent speech recognition system that routes the call appropriately based on what you say you want.

A headset that includes both headphones (for output) and a microphone (for input) enables two-way communication with the computer and with other people.

Cutting Edge

Biometric Security

Authenticating computer users by analyzing their physical characteristics is known as *biometric security*. All biometric security technologies rely on various types of digital input devices, such as fingerprint readers, cameras and video cameras, and microphones. These devices record input from the person attempting to access the computer, such as his or her fingerprint, face, retina, or voice. That input is then compared against scans, pictures, video clips, and audio clips for allowed users, which are stored in a database. The person is permitted access to the computer only if it finds a match.

Until recently, biometric security was very expensive, because the technology for reliably scanning and analyzing physical characteristics was still new. However, recent technological advances have made this type of security system affordable for almost any business or individual. Many smartphones, for example, have a fingerprint identity sensor built in, enabling password-free security. The iPhone X has facial recognition software instead of a fingerprint sensor, for even more security. It uses the phone's front-facing camera to examine your features to verify your identity.

Biometric security is not perfect, however. Hackers have found ways to fool some biometric devices to gain access to computer systems—not just in the movies, but in real life too. Occasionally, a device will not recognize a legitimate user and will lock him or her out, but workarounds such as alternative authentication methods can allow the user access until a system administrator can correct the problem.

Speaker-dependent systems, on the other hand, can be "trained" to understand a specific person's voice by completing a series of practice drills. The speaker reads many paragraphs of text into a microphone, and the software stores his or her voice patterns as examples of what specific words sounds like. This software can also learn new words. *Speaker recognition software* also analyzes voice patterns, but rather than translate them into words, it uses the sounds to authenticate a speaker's identity.

Sensory and Location Input Devices

A computer can accept input not only from a human user but also from the environment. For example, many smartphones and tablets accept input data from several types of environmental sensors. An **accelerometer** reports how fast an object is moving, a **gyroscope** describes an object's orientation in space, and a **compass** reports the direction of an object's location. Sensors like these can provide data to fitness-tracking applications (apps), as well as apps that need to be aware of the user's activities, such as whether he or she is driving, running, or walking. Some advanced pedometers (which count steps taken) also use sensory input. For example, some models of Fitbits can be worn when sleeping and then connected to a computer to analyze how much the wearer tosses and turns during the night. Mobile devices like tablets and smartphones use a built-in gyroscope to determine screen orientation (that is, which direction is up) and rotate the screen display appropriately.

Global positioning system (GPS) devices also provide location information. A GPS works by orienting the current location of the device to a signal from an orbiting satellite. Based on this information, the device can report its position on Earth to within about 5 meters (16 feet). GPS receivers can be built into smartphones, tablets, and other portable devices, and many vehicles have GPS navigation systems available on a dashboard screen. GPS devices can also be attached to children's clothing or pets' collars and used to report their locations.

Tech Ethics

The Chip Debate

Radio frequency identification (RFID) chips have long been used in labels to track products and packages, and in badges to allow employees into secure areas of a facility. Alzheimer's patients may be fitted with ankle or wrist bracelets containing RFID chips to help care providers locate them if they wander off. People generally accept these uses of RFID. But how would you feel about having such a chip implanted inside your body?

Certain types of RFID chips can be embedded under the skin of living beings. Dog and cat owners already take advantage of this technology, "microchipping" pets to help facilitate their return if they are ever lost. The same technology is used as an alternative to branding on livestock such as cattle. Implanting RFID chips in humans, though, has proved to be controversial.

From 2004 to 2010, PositiveID (formerly VeriChip) marketed a chip that could be implanted in humans and that seemed poised for wide adoption. This glass-encapsulated RFID chip, about the size of a grain of rice, could be implanted in a doctor's office under local anesthetic. When read, it generated a 16-digit number that could then be matched with a database entry to identify the person. That ID number, as unique as a person's Social Security number, could potentially match that person up with a variety of databases containing security clearance, medical, and military information.

PositiveID stopped making the chip, amid widespread international debate over its use. One controversial issue was that the chip's number was unencrypted, so anyone with a chip reader could easily identify the person, with or without his or her permission. However, this technology has recently been revived by BEZH, a Danish company that plans to make its BiChip human microchip compatible with Ripple, a virtual currency that people could use to pay for purchases by having their implants scanned.

2.2 Understanding How Computers Process Data

Watch & Learn
Understanding How Computers Process Data

So far in this chapter, you have read about several ways in which data can enter a computer system. Incoming data is temporarily stored in memory until the software asks the central processing unit (CPU) to perform calculations on it. The CPU delivers the calculated data back to the software, and the software uses it to create output (see Figure 2.6). The output waits in memory until the software is ready to deliver the output to a device or to another program.

You input, process, and output things all the time in your daily life. For example, let's say you want to make a cake. You input the ingredients into a mixer, process the ingredients by mixing them, process the mixture by baking it in an oven, and then output the cake on a plate.

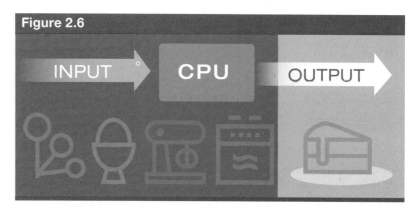

Figure 2.6

With a computer, you input, process, and output data. Suppose you want to multiply two numbers. You input the numbers in a calculator application using the keys on a keyboard, process the numbers by selecting the multiplication key in the calculator application, and then output the result of the calculation in the result area of the calculator application.

At a technical level, data processing is much more complicated than just described, but it follows the same principles. The CPU processes millions of operations per second (MIPS), juggling requests from many different system components.

A computer might have many programs running at once, each sending millions of processing requests to the CPU. All of these requests wait in a queue for their turn to run through the CPU, which performs the actual math operations. When the completed calculations exit the CPU, they are returned to the program from which they were requested.

In addition to the applications you might run, the operating system itself also makes millions of requests of the CPU. These requests are necessary to keep the operating system's user interface up and running.

Bits and Bytes

Computers are electronic devices, which means they operate using electricity. The earliest developers of computer systems used this fact to construct a language for communicating with the CPU. In this language, an electrical charge or pulse is a 1 and a lack of an electrical charge is a 0 (see Figure 2.7). Computers represent all data using binary strings of numbers. (A **binary string** consists of only 1s and 0s.) Each individual 1 or 0 is a **bit**, which is short for *binary digit*.

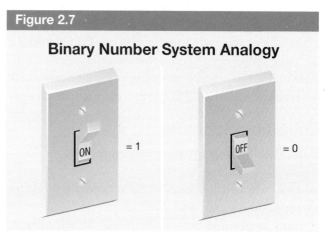

Figure 2.7

Binary Number System Analogy

Computer circuits are toggled on and off by transistors, which operate like light switches. When current flows through the transistor, it's on, and a 1 value is represented. When current does not flow through the transistor, it's off, and a 0 value is represented.

Table 2.1 Measures of Computer Storage Capacity

Term	Abbreviation	Approximate Number of Bytes	Exact Number of Bytes
Byte		1	1
Kilobyte	KB	1 thousand	1,024
Megabyte	MB	1 million	1,048,576
Gigabyte	GB	1 billion	1,073,741,824
Terabyte	TB	1 trillion	1,099,411,627,776
Petabyte	PB	1 quadrillion	1,125,899,906,842,624
Exabyte	EB	1 quintillion	1,151,921,504,606,846,976
Zettabyte	ZB	1 sextillion	1,180,591,620,717,303,424
Yottabyte	YB	1 septillion	1,208,925,819,614,629,174,706,176

A group of 8 bits is a **byte**. A byte is an 8-digit binary number between 00000000 and 11111111. There are 256 possible combinations of 1s and 0s in a string of 8. Most computer programs work with bytes, rather than bits. Computer memory and disk storage capacity is also measured in bytes. A **kilobyte** is approximately 1 thousand bytes. Today's systems work with bytes by the millions and billions; one **megabyte** is 1 million bytes, and one **gigabyte** is 1 billion bytes. See Table 2.1 for other measures of computer memory.

Data transfer rates are usually expressed as a number of bits, rather than bytes. So, for example, you might have a data transfer rate of 8 megabits per second (Mbps). Notice that the abbreviation for the word *bit* is a lowercase *b* and the abbreviation for the word *bytes* is an uppercase *B*: the abbreviation for *megabits* is *Mb* and the abbreviation for *megabytes* is *MB*. To describe bits, you use the same terms as in Table 2.1 except you replace *-byte* with *-bit*: kilobit (Kb), megabit (Mb), gigabit (Gb), terabit (Tb), and so on.

Frequencies are described using the term **hertz (Hz)**. One hertz is one cycle per second. One megahertz (MHz) is one million cycles per second, one gigahertz (GHz) is one billion cycles per second, and so on. The term *hertz* is commonly used to describe the radio frequency (RF) at which wireless networking operates—for example, 2.4 GHz or 5 GHz. It is also used to describe the operational speed of a processor chip—for example, a 3 GHz CPU can process 3 billion data operations per second.

Encoding Schemes

If all of the data in a computer is binary, how does the computer store and display text that humans can understand? It uses an **encoding scheme**, which is a system of binary codes that represent different letters, numbers, and symbols.

Early personal computers employed an encoding scheme called **American Standard Code for Information Interchange (ASCII)**, which assigned a unique 8-bit binary code to each number and letter in the English language plus many of the most common symbols such as punctuation and currency marks. Table 2.2 shows a few examples of ASCII encoding.

Table 2.2 Examples of Characters Encoded with ASCII

Character	ASCII
A	01000001
B	01000010
C	01000011
1	01100001
2	01100010
$	00100100

Although ASCII was widely adopted as a standard for PCs, over time it proved too limited. ASCII's main shortcoming was that it couldn't deal with languages such as Chinese, which has a more complicated alphabet than English. To accommodate a larger selection of characters, a system called **Unicode** was developed. Unicode uses 2 bytes (16 bits) per character and can represent up to 65,536 characters. The first 256 codes are the same in both ASCII and Unicode, so existing ASCII-coded data is compatible with modern Unicode-based operating systems, such as Microsoft Windows and macOS.

2.3 Identifying Components in the System Unit

The main part of the computer is the **system unit**; it houses the major components used in processing data. For a desktop PC, the system unit is a large box. A laptop or tablet contains many of the same components, but the system unit is smaller and packed together more tightly. The following sections discuss the components inside a typical system unit, like the one shown in Figure 2.8.

Power Supply

The **power supply** (sometimes called the *power supply unit*, or *PSU*) converts the 110-volt or 220-volt alternating current (AC) from your wall outlet to the much lower voltage direct current (DC) that computer components require. In a laptop computer, this function is handled by the power converter block that's built into the power cord. In a desktop PC, the power supply is a large, silver box inside the system unit. A desktop power supply usually contains one or more fans, which cool both the power supply and the entire system unit.

Watch & Learn
Identifying Components in the System Unit

Video
External Parts of a PC

Video
Internal PC Components

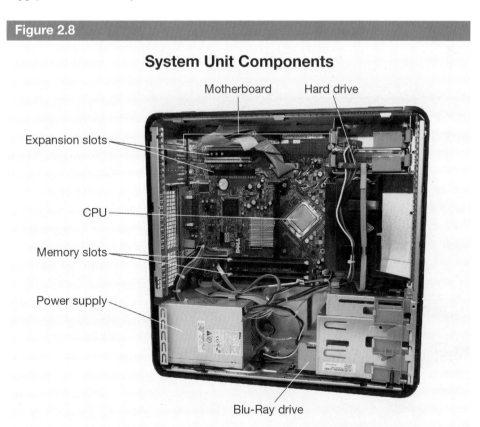

Figure 2.8

System Unit Components

The system unit is the main part of a desktop PC and contains the components necessary for processing data.

Different countries supply different wall-outlet voltages; the Unites States uses 110-volt AC and most of Europe uses 220-volt AC, for example. The computer's PSU must be of the right type for the region in which you are using it, or it must have a switch you can set to control which voltage is in use.

Motherboard

The **motherboard** is the large circuit board that covers almost the entire floor of the system unit (or the side, depending on how the unit is oriented). Like other circuit boards, the motherboard consists of a variety of electronic components (including sockets, chips, and connectors) that are mounted onto a sheet of fiberglass or other material. The electronic components are connected by conductive pathways called **traces**. The motherboard has a controller chip (or set of chips) called a **chipset**, which controls and directs all of the data traffic on the board. The chipset determines the motherboard's capabilities—from the CPUs and types of memory it will support to the speed at which data flows through it.

Video
Motherboard Components

Other components found on a motherboard include the following:
- the CPU, which processes the data
- a coin-style battery (on older systems) for keeping certain chips powered when the system is off
- slots for connecting random access memory (RAM) modules, which provide temporary storage for the data applications and operating system to use while running
- slots for attaching expansion boards, which add various capabilities to the computer (as described in the Expansion Slots and Expansion Boards section)
- ports for connecting input devices (such as a keyboard and mouse) and output devices (such as a speaker, monitor, and printer)
- a power connector for accepting the electricity needed to keep each component powered

Figure 2.9 identifies some of the features of a typical desktop motherboard.

Figure 2.9

Motherboard Components of a Desktop PC

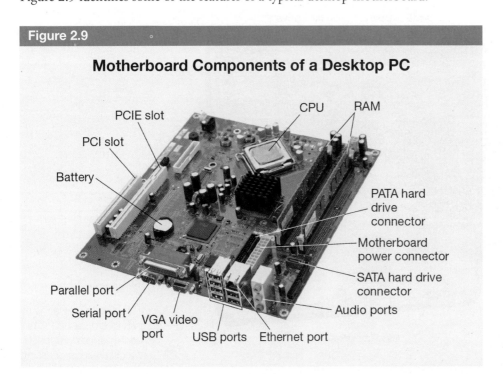

The motherboard holds the major processing and memory components.

Buses

Data moves from one component to another on the motherboard by means of buses. A **bus** is an electronic path along which data is transmitted. Most buses are embedded into the motherboard.

The size of a bus is referred to as **bus width**; it determines the number of bits the bus can transmit or receive at one time. For example, a 32-bit bus can carry 32 bits at a time, whereas a 64-bit bus can carry 64 bits. The larger the bus width, the more data the bus can carry per second.

One way to visualize a bus is to think of it as a highway, with stops along the way (exits) where data is dropped off or picked up. The bus width is like the number of lanes on the highway. The more lanes, the more traffic the highway can handle without delays. Each bus also has a certain speed at which it operates. Think of this as the speed limit on the highway. (On computer buses, however, there are no speeders; all data travels at exactly the speed for which the bus was designed.)

Computers contain two basic bus types: a system bus and an expansion bus. The **system bus** connects the CPU to the main memory, and its speed is governed by the system clock (which is covered later in this chapter). An **expansion bus** provides communication between the CPU and a peripheral device, which may be connected through an expansion board or port (covered in the next two sections of this chapter).

> **Article**
> External Bus Technologies

Expansion Slots and Expansion Boards

An **expansion slot** is a narrow opening in the motherboard that allows the insertion of an expansion board. An **expansion board** (also called an *expansion card* or *adapter*) adds some specific capability to the system, such as sound, networking, or extra ports for connecting external devices. Even if the motherboard already has built-in support for a particular capability, an expansion board is sometimes needed to provide better quality support or support for additional devices. For example, you might add another display adapter to support having more than one monitor.

An expansion board is manufactured to fit in one type of slot, and slot types correspond to different types of buses. Over the years, computers have had several different slot types for expansion boards. When you purchase an expansion board, make sure you get a model that will fit in one of the empty slots in your motherboard.

The most common type of expansion bus in motherboards today is the **Peripheral Component Interconnect Express (PCIe) bus**. PCIe slots come in different sizes.

This photo shows four PCIe slots of different sizes: one ×16 (left) and three ×4 (middle). The fifth slot (right) is a traditional 32-bit PCI slot.

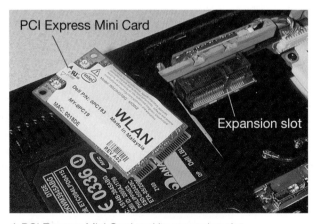

A PCI Express Mini Card and its expansion slot.

The most common interface for display adapters is PCIe ×16, which is a high-speed slot with 16 data channels. The most common interface for other expansion boards is PCIe ×1, which uses only 1 data channel. You may also come across a legacy Peripheral Component Interconnect (PCI) slot, which is an early technology that many motherboards still support. A variant called PCI Extended (PCI-X) that provided enhanced speeds for servers is also mostly obsolete now.

Portable computers, such as laptops and tablets, don't have room for traditional expansion slots and boards. However, some of these devices have a PCI Express Mini Card slot, into which you can insert a miniature version of an expansion board. To install or remove a PCI Express Mini Card, you must shut down the computer and remove a panel on the bottom to access the slot.

Ports

A **port** (sometimes called an *interface*) is an external plug-in socket on a computing device. Ports are used to connect external devices such as printers, modems, and routers to a computer. Some ports are dedicated, or reserved for one specific type of device. Other types of ports are generic and used for a variety of different devices. Table 2.3 shows examples of the types of ports commonly found on personal computers.

Universal Serial Bus (USB) A **universal serial bus (USB) port** is a general-purpose port that's used by many different devices, including input devices (such as keyboards and mice) and external storage devices (such as external hard drives, USB flash drives, and digital cameras). USB technology has many attractive features. For example, you can connect an external hub to a USB port and then connect more than 100 devices to it (in theory, anyway); all of the devices would run off that single USB port in your computer. USB ports are also hot-pluggable, so you can connect and disconnect devices while the computer is running.

Different speeds of USBs have been available over the years. USB 1.1 operates at 1.5 megabits per second (Mbps) and is usually colored white. USB 2.0 (also called Hi-Speed USB) operates at up to 480 Mbps and is usually colored black. USB 3.0 (also called SuperSpeed) transfers data at up to 4 gigabits per second (Gbps) and is usually colored blue.

A USB cable usually has an A-style connector at one end and a B-style connector at the other end. The A-style connector plugs into a USB-A port on the computer, and the B-style connector plugs into a USB-B port on the peripheral device.

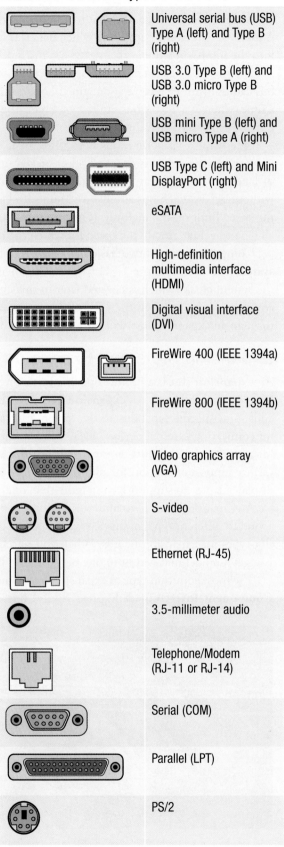

Table 2.3 Common Types of PC Ports

Port	Description
	Universal serial bus (USB) Type A (left) and Type B (right)
	USB 3.0 Type B (left) and USB 3.0 micro Type B (right)
	USB mini Type B (left) and USB micro Type A (right)
	USB Type C (left) and Mini DisplayPort (right)
	eSATA
	High-definition multimedia interface (HDMI)
	Digital visual interface (DVI)
	FireWire 400 (IEEE 1394a)
	FireWire 800 (IEEE 1394b)
	Video graphics array (VGA)
	S-video
	Ethernet (RJ-45)
	3.5-millimeter audio
	Telephone/Modem (RJ-11 or RJ-14)
	Serial (COM)
	Parallel (LPT)
	PS/2

USB connectors also come in mini and micro versions, and these connectors also come in A and B styles. Mini and micro connectors are designed for use with small devices like cameras and smartphones.

A newer USB connector type, USB-C, is symmetrical and therefore can be connected right-side-up or upside-down. The USB-C connector is used for high-speed connections like USB 3.0 and Thunderbolt 3.

Thunderbolt and Lightning A **Thunderbolt port** is a high-speed, multipurpose port found in nearly all Macs and many newer PCs. A Thunderbolt connection can deliver multiple signals in a single cable, including PCI Express (PCIe), DisplayPort, and power signals. There are three types of Thunderbolt ports. Thunderbolt 1 and Thunderbolt 2 use a Mini DisplayPort connector. Thunderbolt 3 uses a USB-C connector. A Lightning port is a small, flat connector used to connect and charge Apple iPhones and iPads.

eSATA eSATA is an external version of the Serial Advanced Technology Attachment (Serial ATA) standard that's used for connecting internal hard drives (as you will read in Chapter 3). (The *e* in eSATA stands for *external*.) Most external hard drives connect to a computer via the USB or FireWire port, but some newer models use the **eSATA port** for faster transfer speeds (up to 6 Gbps).

High-Definition Multimedia Interface (HDMI) High-definition multimedia interface (HDMI) is the interface for high-definition (HD) TV and home theater equipment. It's a digital format and was designed for interfacing with digital devices. Many computers have a **high-definition multimedia interface (HDMI) port** for connecting to HDMI-capable input and output devices.

Digital Visual Interface (DVI) A **digital visual interface (DVI) port** is a very common type of port for display monitors in modern computer systems. A DVI port is used not only to connect computer monitors but also in consumer electronics devices such as home theater systems, game consoles, and TVs. Unlike HDMI, which is the dominant video standard for HDTV, DVI is backward compatible with VGA, the previous standard for connecting computer monitors. An adapter plug is used to change the connector type for older monitors.

Ethernet An **Ethernet port** enables a computer to connect to a router, cable modem, or other networking device using a cable. (Wireless Ethernet doesn't require a port.) An Ethernet port and cable resemble those used for a telephone jack but are a little wider. The official name of the Ethernet connector is **RJ-45 port**. (*RJ* stands for *registered jack*.) Wireless networking is becoming so popular that some new laptops do not include an Ethernet port.

Audio An **audio port** is generally a small, circular hole into which a 3.5-millimeter connector is plugged for analog input devices (such as microphones, analog stereo equipment, and musical instruments) and output devices (such as speakers and headphones). On a desktop PC, these ports are sometimes color-coded, as described in Table 2.4. These audio ports are connected to an audio adapter (also called a *sound*

Table 2.4 Color Coding of Audio Ports

Color	Common Use
Pink or red	Analog microphone input
Light blue	Analog line input (from a cassette tape deck or turntable)
Lime green	Analog output (speakers or headphones)
Black	Analog output (rear speakers in a surround sound system)
Silver	Analog output (side speakers in a surround sound system)
Orange	Analog output (center speaker or subwoofer in a surround sound system)

card or an *audio card*) inside the computer. The audio adapter translates between digital computer data and the analog audio sounds that you record and play back.

Handheld devices like smartphones and mobile media players typically have a single, multipurpose audio port, and the audio adapter is built into the main circuit board of the system itself. Some newer models do not have an audio port at all, relying instead on Bluetooth or other wireless technology to connect to audio output devices like headphones and speakers.

Legacy Ports In computer terms, *legacy* means older technology that enables backward compatibility. A **legacy port** is a computer port that is still in use but has been largely replaced by a port with more functionality. The following kinds of ports are mostly obsolete, but you may still find them on older systems:

- FireWire is a lesser-known competitor to USB. A **FireWire port** is more widely known by its standard number: IEEE 1394. (*IEEE* stands for Institute of Electrical and Electronics Engineers, an international standards organization.) It is a general-purpose port for connecting external drives and video cameras. There are several versions, including FireWire 400 (IEEE 1394a) and FireWire 800 (1394b), which transfer data at 400 Mbps and 800 Mbps, respectively. Faster versions (S1600 and S3200) were also developed, but they were never popular.
- A **video graphics array (VGA) port** is an older kind of monitor connection. It uses a 15-pin female D-sub connector (often abbreviated as DB15F). A D-sub connector is a D-shaped ring with metal pins inside it. D-sub connectors are described based on the number of pins and whether they are male (having pins) or female (having holes).
- A **serial port** (also called a *COM port*, which is short for *communications port*) is a 9-pin male D-sub connector on the PC. Serial ports were used to connect a variety of devices a decade or more ago, but today, almost all of these devices use USB.
- A **parallel port** (also called an *LPT port*, which is short for *line printer*) is a 25-pin female D-sub connector on the PC, which matches up with a 25-pin male connector on the cable. This type of port was used to connect printers before USB became popular.
- A **PS/2 port** is a small, round port that used to be the standard for connecting the keyboard and the mouse to the computer. The keyboard port was light green and the mouse port was purple, but physically, the two ports were identical. The term *PS/2* comes from the name of a very old IBM computer that introduced this connector style.
- A **dial-up modem** is a device for connecting to the internet or a remote network via a landline telephone. It uses a regular telephone cable connector, either an **RJ-11 connector** (for a single-line cable with two wires) or an **RJ-14 connector** (for a dual-line cable with four wires).
- An **S-video port** (also known as an *i480 port*) is a standard for video output for TVs. (The *S* is for *standard*, meaning "not high-definition.") Some computers have an S-video port for connecting with a TV and using it as a monitor.
- A **component video port** (also called a *Red-Green-Blue port*, or *RGB port*) is an older type of display connection consisting of three separate round plugs, color-coded red, green, and blue. Some older televisions and video recording devices use this type of connection, but it is seldom found on computers that are still in use today.

Storage Bays

A **storage bay** (often referred to as a *bay*) is a space in the system unit where a storage device (such as a hard drive or a DVD drive) can be installed. The number of bays in the system unit determines the maximum number of internal storage devices the system can have, so it's an important factor to consider when purchasing a PC. Most portable PCs (that is, laptops and tablets) don't have any empty storage bays, so bays aren't an issue on them. Figure 2.10 shows an empty desktop case with several storage bays.

Storage bays come in two sizes. Small bays are designed to hold hard drives. Large bays are designed to hold CD, DVD, and Blu-ray drives. A case has one or more pop-out panels that you can remove to provide a storage device with outside access. For example, a DVD drive needs outside access so you can insert and eject discs. Storage devices are typically mounted into bays with screws that fit in the holes in the bay housing, but some system units use sliding rails to mount devices.

Figure 2.10

Desktop Computer Storage Bays

- External drive bay for CD, DVD, or Blu-ray drive
- Internal drive bays for hard drives

The case of a typical desktop computer has several storage bays.

2.4 Understanding the CPU

Every computer contains a **central processing unit (CPU)**, which is also sometimes called a *microprocessor* or *processor*. The CPU is responsible for performing all of the calculations for the system—and just about everything that happens in a computer requires performing millions of calculations. These facts make the CPU the most essential (and also the busiest) component in the system.

Watch & Learn
Understanding the CPU

Physically, a CPU is a very small wafer of **semiconductor material**; a complex set of tiny transistors and other electronic components are etched into it. The term *semiconductor* is used because this material is neither a good conductor of electricity (like copper would be) nor a good insulator against electricity (like rubber would be). Semiconductor material therefore doesn't interfere with the flow of electricity in a chip's circuits one way or the other. The most commonly used semiconductor material in CPUs is silicon, which is a type of purified glass. This very complex and expensive wafer is mounted on a small circuit board and protected by a ceramic and metal casing.

Video
PGA and LGA CPUs

From the outside, a CPU may look like a metal or ceramic plate with many short pins (metal spikes) or tiny dots of metal on one side. The type of CPU packaging that has pins is called a *pin grid array* (*PGA*). The pins fit into holes in the CPU socket in the motherboard. The type of CPU packaging that has metal dots is called a *land grid array* (*LGA*). The metal dots line up with metal contacts on the CPU socket, making an electrical connection.

This pin grid array (PGA) chip has a series of pins on its underside. The pins fit into holes in a socket, and each pin transfers a different piece of data into or out of the CPU.

Internal Components

Inside a typical CPU, there are three main components: a control unit, an arithmetic logic unit (ALU), and registers (see Figure 2.11).

The CPU's internal components perform four basic operations that are collectively called a **machine cycle**. The machine cycle includes fetching an instruction, decoding the instruction, executing the instruction, and storing the result.

Control Unit The **control unit** directs and coordinates the overall operation of the CPU. It acts as a "traffic officer," signaling to other parts of the computer system what they should do. The control unit interprets instructions and initiates the action needed to carry them out. First it fetches, which means it retrieves an instruction or data from memory. Then it decodes, which means it interprets or translates the instruction into strings of binary digits that the computer understands.

Arithmetic Logic Unit (ALU) As shown in Figure 2.11, the **arithmetic logic unit (ALU)** is the part of the CPU that executes the instruction during the machine cycle. In other words, the ALU carries out the instructions and performs arithmetic and logical operations on the data. The arithmetic operations the ALU can perform are addition, subtraction, multiplication, and division. The ALU can also perform logical operations, such as comparing two data strings and determining whether they are identical.

Registers A **register** functions as a "workbench" inside the CPU, holding the data that the ALU is processing. Placing data into a register is the storing step of the machine cycle. Registers can be accessed much faster than memory locations outside the CPU.

Various types of registers are available in the CPU, and each serves a specific purpose. Once processing begins, an instruction register holds the instructions currently being executed. A data register holds the data items being acted on by the ALU. A storage register holds the immediate and final results of the processing.

> **Video**
> How Data Flows into a PC
>
> **Video**
> What Is the Machine Cycle?
>
> **Practice**
> Looking Inside a CPU

Figure 2.11

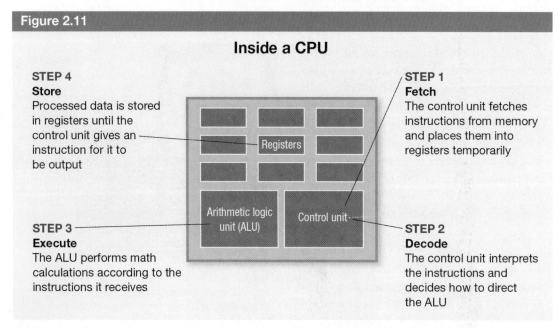

Inside a CPU

STEP 1 Fetch — The control unit fetches instructions from memory and places them into registers temporarily

STEP 2 Decode — The control unit interprets the instructions and decides how to direct the ALU

STEP 3 Execute — The ALU performs math calculations according to the instructions it receives

STEP 4 Store — Processed data is stored in registers until the control unit gives an instruction for it to be output

The CPU contains a control unit, arithmetic logic unit (ALU), and registers. These components interact in a machine cycle to fetch, decode, execute, and store instructions and data.

Chapter 2 Sizing Up Computer and Device Hardware

Caches Because the CPU is so much faster than the computer's other components, it can potentially spend a lot of time idle, waiting for data to be delivered or picked up. The farther the data has to travel to get to the CPU, the longer the delay. To help minimize these delays, modern CPUs have multiple caches. A **cache** (pronounced "cash") is a pool of extremely fast memory that's stored close to the CPU and connected to it by a very fast pathway. Only when the needed data isn't found in any of the caches does the system have to fetch data from storage (for example, the main memory of the PC, or a hard drive).

When the CPU needs some data, it looks first in the L1 (level 1) cache, which is the one closest to the CPU. The L1 cache is an **on die cache**; this means that when the CPU is stamped into the silicon wafer, the L1 cache is stamped into the same piece of silicon at the same time. The L1 cache is quite small, so it can't hold everything the CPU has recently used or may need to use soon.

If the data needed isn't found in the L1 cache, the system looks in the L2 cache. On older systems, the L2 cache was on the motherboard, but on modern systems, it's in the CPU package, next to the silicon chip. If the data isn't in the L2 cache, the system checks the L3 cache, which is larger and slightly farther away from the CPU. On a multicore CPU (discussed on pages 63–64), all of the cores share a common L3 cache.

Speed and Processing Capability

The most obvious quality of a CPU is its speed, which is measured as the number of operations it can perform per second. A CPU's advertised speed is the maximum speed at which it can reliably operate without overheating or malfunctioning. The speed of a CPU is measured in gigahertz. One hertz is one cycle per second, and one gigahertz is 1 billion hertz.

Article
Benchmarking a System

The CPU's speed rating is just a suggestion. The actual speed at which the system components function is determined by a chip on the motherboard called the **system clock** (or *system crystal*). It's a small oscillator crystal that synchronizes the timing of all operations on the motherboard. The CPU is so much faster than the other components in the system that it typically operates at a multiple of the other components' speeds, performing multiple actions per tick of the clock. For example, the

Hotspot

Coming Soon: Transistors on Steroids!

What's in a chip? Millions of microscopic transistors—those tiny pieces of semiconductor material that amplify a signal or open and close a circuit.

What happens when you lace silicon with the chemical element germanium and cool it down to a temperature of absolute zero? You get a supersonic transistor—the building block used to produce supersonic chips. This is what IBM has done. It's created a superfast transistor that will lead to ever-faster computers and wireless networks.

IBM's speedy transistors run 100 times faster than the transistors currently available, reaching speeds of 500 gigahertz (GHz). For comparison, cell phone chips dawdle along at a mere 2 GHz, and desktop computers at around 3.5 GHz.

In initial tests, the IBM transistor attained its highest speed at a temperature near absolute zero (which is -451 degrees Fahrenheit). But even at room temperature, the transistor ran at 300 GHz. Although mass-produced transistors probably wouldn't match the racecar speed of the prototype, they would still be much faster than what is available today.

Superfast transistors might make it possible for a wireless network to download a DVD in five seconds and for buildings to be outfitted with 60 GHz wireless connections. Better yet, cars could come equipped with radar that would automatically adjust speed according to traffic or swerve to avoid oncoming vehicles.

system clock rate might be 300 megahertz, and the CPU might run at 10 times that speed (3 gigahertz).

A CPU's performance depends on more than just its maximum speed. The number of transistors inside the CPU is equally important. Modern CPUs have millions or even billions of transistors. For example, the Intel i7-8700K CPU (introduced in late 2017) has more than 3.5 billion transistors. In contrast, the Intel 80286 (introduced in 1982) had about 134,000 transistors. More modern and more expensive CPUs tend to have more transistors.

Another factor is a CPU's **word size**. In computer terms, a *word* is a group of bits that a computer can manipulate or process as a unit. If a CPU is 32-bit, it can handle 32-bit blocks of data at a time. Most modern desktop systems have 64-bit CPUs, and 32-bit CPUs are found in many mobile devices. The Microsoft Windows operating system is available in versions for both 32-bit and 64-bit systems.

The performance of a CPU also depends on a variety of special data-handling procedures that manufacturers have built into CPU designs over the years. Many of these procedures provide different ways to process multiple instructions at a time. Here are some of the key advances to date in CPU design and manufacturing:

Article
Advances in Chip Architecture

Article
Improvements in Chip Materials and Manufacturing Processes

- **Better manufacturing.** CPU manufacturing has greatly improved, resulting in smaller and closer circuits. Having smaller circuits means that more transistors can be fit on the same size chip and that the CPU will generate less heat as it operates. Having circuits be closer together means the data and instructions don't have to travel as far, and so they reach their destinations faster. CPU manufacturing has also been improved by the use of better materials; for instance, using copper rather than aluminum allows better conductivity.
- **Pipelining.** In older computers, the CPU had to completely execute one instruction before starting a second instruction. Modern computers can perform multiple instructions at once by using a technique called **pipelining**. This technique enables the CPU to begin executing another instruction as soon as the previous one reaches the next phase of the machine cycle. Figure 2.12 illustrates the difference in productivity that pipelining provides.
- **Parallel processing.** A **thread** is a part of a program that can execute independently of other parts. Early computer systems had only one CPU, and it

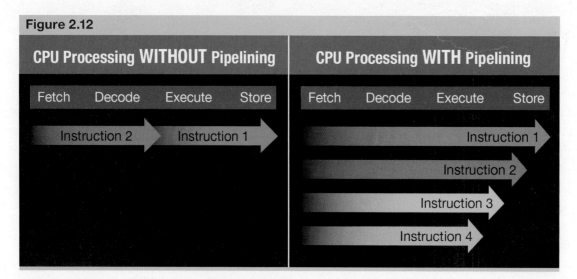

Figure 2.12

Without pipelining, the CPU must complete one instruction before starting a second instruction. With pipelining, the CPU begins executing a new instruction as soon as the previous instruction reaches the next phase of the machine cycle.

could process only one thread at a time. **Parallel processing** allows two or more processors (or cores within a single processor) in the same computer to work on different threads simultaneously.

- **Superscalar architecture. Superscalar architecture** is a type of CPU design that creates better throughput. Specifically, it enables the operating system to send instructions to multiple components inside the CPU during a single clock cycle. For example, a single clock cycle might deliver instructions to an ALU and store data in a register.
- **Multithreading. Multithreading** enables the operating system to address two or more virtual cores in a single-core CPU and share the workload between them. Intel's version of multithreading is called **hyper-threading**.
- **Multicore processors.** The term **core** refers to the essential processor components (that is, the ALU, registers, and control unit). Most modern computers have multicore processors. A **multicore processor** enables a computer to process several instructions at once, as if the system physically contains more than one CPU. Each core has its own L1 and L2 cache, and all of the cores share a single L3 cache.

> Video
> Multicore Processors

CPU and System Cooling

CPUs generate a lot of heat as they operate. Computers need cooling systems to keep their CPU and other components from overheating. When a CPU overheats, the computer shuts down unexpectedly and won't start again until the system has cooled down sufficiently.

The CPU typically has some type of cooling unit mounted on it, with a thermal paste or adhesive used to ensure a tight seal. It might be an active device (such as a fan) or a passive device (such as heat sink, which is a heat-conducting block, usually aluminum, that wicks heat away from the CPU), or both.

In a desktop PC, the power supply's fan pulls heat out of the case, and other case-cooling fans may be installed as well. Laptops may have small fans and air vents strategically placed to dissipate as much heat as possible, especially near the CPU.

Some powerful desktop PCs use liquid cooling systems. A **liquid cooling system** has a water reservoir and a closed system of plastic tubes with a pump that circulates the water between various "hot spots" inside the case, such as the CPU and the larger

A combination of a heat sink and a cooling fan may be mounted to a CPU to provide cooling.

Some desktop computers use a liquid cooling system to remove excess heat from the CPU.

chips on the motherboard and the display adapter, to cool those areas. Liquid cooling does a great job at cooling, but it can be complicated and costly to set up. Some high-end gaming enthusiasts use it to keep their PC from overheating when they **overclock** the CPU (that is, when they run the CPU faster than its speed rating).

2.5 Understanding Memory

At the most basic level, **memory** is an electronic chip that contains a grid of transistors that can be on (1) or off (0). Memory stores data by holding a 1 or 0 value in each transistor and then reporting it when requested to do so.

There are two types of memory: **read-only memory (ROM)** and **random-access memory (RAM)**. ROM chips store the same data permanently, regardless of whether the computer is powered on. ROM can't be rewritten (at least not by the computer in which it's installed). ROM chips are used to store data that never changes. For example, the motherboard may contain a small amount of ROM that provides the startup instructions for the processing component of the system. ROM has the advantage of being very quick to read from, but the fact that it can't be updated easily limits its functionality. In contrast, the values stored on RAM chips can be easily changed as the computer operates, storing first one value and then another as needed. For this reason, RAM is the primary type of memory used in almost all computing devices.

Watch & Learn
Understanding Memory

Video
Memory Installation

RAM Basics

The most common use for RAM is as temporary storage for data and programs when the computer is in use. When people refer to a computer's "main memory," they generally mean the RAM that's installed on the motherboard and available for use by the operating system.

Cutting Edge

Nanotechnology

A *nanometer* (nm) is one-billionth of a meter. How small is that? It's so small that it's atomic. (Individual atoms range in size from 0.25 to 1.75 nm.) *Nanotechnology* is the engineering field of creating useful materials and even machines out of individual atoms. The general idea is that computers can be built at a microscopic level and programmed to perform specific functions. For example, researchers in the medical field hope to create nanotechnology robots (called *nanobots*) that can be injected into a human body. Inside the body, the nanobots would replicate (or make duplicates of themselves) and then hunt down and kill certain viruses or cancers. The same technology could be used to create nanobots that clean up dangerous environmental spills and perform other tasks that aren't well suited to humans.

Nanotechnology has already been used successfully in several fields. For example, scientists have used this technology to create carbon nanotubes that are 100 times stronger and 100 times lighter than steel. Nanotubes can be used to make many types of objects stronger and lighter—from boats and aircraft to golf clubs and bicycle parts.

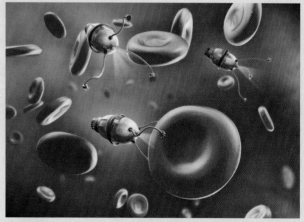

Scientists hope to create nanobots that can be programmed to repair blood cells in a human body.

As your computer starts up, the operating system is loaded into RAM. RAM functions as a work area. Then, when you use the operating system to start an application, the program files are loaded into RAM. When you create a data file using that application, that file is also placed in RAM. The operating system retrieves data from RAM and sends it to the CPU for processing; after that, the operating system accepts the output from the CPU and places it back into RAM.

RAM is a critical part of a computer's functionality. The more RAM a computer has, the more applications and data files you can have open at once. On some computers, you can install additional RAM if needed.

RAM has three main functions:

1. To accept and hold program instructions and data
2. To supply data to the CPU for processing and then temporarily store the results
3. To hold the final processed information until it can be sent to a more permanent storage location (such as disk drive) or sent to an output device (such as a printer)

The CPU must be able to find programs and data once they have been stored in RAM. To enable retrieval, program instructions and data are placed at specific, named locations in RAM. This storage organization method enables the finding of particular data and instructions no matter which part of the RAM they are physically stored in—hence, the term *random access memory*. Each location in RAM has an individual **memory address** that the operating system can work with directly, rather than having to search from beginning to end (like you would do to find information stored on a cassette or VHS tape).

The content of RAM is easily changed. In fact, it changes almost constantly while the computer is running. Data moves freely from storage devices to RAM, from RAM to the CPU, and from the CPU to RAM. Processed data is then discarded from RAM (if it's no longer needed) or saved to a storage device (if it is needed).

Static versus Dynamic RAM

The RAM that comprises the computer's main memory is **dynamic RAM (DRAM)**. This type of RAM requires a constant supply of electricity to keep its contents intact; another term sometimes used for it is *volatile RAM*. If the computer loses power, the content of RAM will be lost because all of the bits will go back to 0. Because of this possibility, you should save your work frequently when working in applications. Doing so will guard against data loss in the event of a power outage or a system error that requires restarting.

In contrast, some RAM is static; it does not lose its data when the power goes off. **Static RAM (SRAM)**, also called **nonvolatile RAM**, is faster than DRAM and much more expensive, so it isn't used as the main memory in PCs. However, many common computing devices use small amounts of SRAM for special purposes. For example, the

> Video
> What Are the Various Types of RAM?

Practical TECH

Upgrading Your RAM

If your computer's performance is sluggish when you run several programs at once, you would likely benefit from having additional RAM installed. If you aren't very tech savvy, you can have this upgrade done at a PC repair shop or the service department of an electronics store. The staff there will be able to determine whether your RAM can be upgraded and if so, what type of RAM you need to buy and how to install it.

caches in a CPU are static RAM, and so are the buffers (temporary data-holding areas) in some hard disks, routers, printers, and display screens.

The DRAM in a typical PC consists of tiny silicon wafers mounted in ceramic chips. These chips are in turn mounted on one or more small, rectangular circuit boards called **dual inline memory modules (DIMMs)**. The word *dual* indicates that both sides of the circuit board contain memory chips. Each DIMM has a particular capacity (the amount of RAM it holds) and speed (the motherboard bus speed it's rated to keep up with).

RAM Speed and Performance

In early computer systems, dynamic RAM operated at a certain speed that was independent of the system bus. In other words, it was asynchronous dynamic RAM. In modern systems, however, most RAM is **synchronous dynamic RAM (SDRAM)**, which means its operations are synchronized with the system clock.

With **single data rate (SDR) SDRAM**, data moves into or out of RAM at the rate of one word per clock cycle. With **double data rate (DDR) SDRAM**, data transfers twice as fast as with SDR SDRAM, because it reads or writes two words of data per clock cycle. DDR2 doubles that rate (reading or writing four words of data per clock cycle), and DDR3 doubles it again (reading or writing eight words). DDR4 doesn't increase the number of words per clock cycle, but it offers some other technical changes that improve performance.

RAM comes mounted on dual inline memory modules (DIMMs), which fit into memory slots on the motherboard.

Another way to describe the speed or efficiency of RAM is as **memory access time**: the time required for the processor to access (read) data and instructions from memory. Access time is usually stated in fractions of a second. For example, a millisecond (*ms*) is one-thousandth of a second. See Table 2.5 for a summary of the terms and abbreviations used to describe memory access times.

Table 2.5 Memory Access Times

Unit of Time	Abbreviation	Speed
Millisecond	ms	One-thousandth of a second
Microsecond	μs	One-millionth of a second
Nanosecond	ns	One-billionth of a second
Picosecond	ps	One-trillionth of a second

RAM Storage Capacity

RAM capacity is measured in bytes, and so is disk storage capacity (covered later in this chapter). Most computers have enough memory to store millions or even billions of bytes. Because of this fact, it's common to refer to storage capacity in terms of megabytes (1 million bytes), gigabytes (1 billion bytes), and terabytes (1 trillion bytes). A typical desktop computer system today might have 8 gigabytes (GB) of RAM, for example.

Table 2.1 (on page 56) lists the names given to various quantities of bytes; these terms are used when describing both RAM and storage capacity. The numbers of bytes given in the table are slightly off, because 1 kilobyte isn't exactly 1,000 bytes but rather 1,024 bytes (2 to the tenth power, or 2^{10}). That means that another 24 bytes should be added to every 1,000. In 1 megabyte, for example, there are 1,048,576 bytes (1,024 × 1,024, or 2^{20}), rather than an even 1 million.

ROM and Flash Memory

A computer's system unit has one or more ROM chips that contain instructions or data permanently placed on them by the manufacturer. A typical PC has ROM chips on which essential startup programs have been stored. These programs are stored on read-only chips because it's critical that they not be erased or corrupted.

One type of startup program on a PC is the basic input/output system (BIOS), which boots (starts) the computer when it's turned on. (On a Mac, there is no BIOS, but there is a system-level utility called *OpenFirmware* that is roughly equivalent.) The BIOS also controls communications with the keyboard, disk drives, and other components. Also stored in ROM and activated with the startup of the computer is the power-on self-test (POST) program. The POST checks the physical components of the system to make sure they are all working properly.

When ROM was originally developed, it was unchangeable—true to its name. Replacing the BIOS in a system required installing a new chip. Because ROM chips were expensive, engineers developed a technology called **erasable programmable ROM (EPROM)** that allowed chips to be erased with a strong flash of ultraviolet light and then reprogrammed. EPROM enabled the reuse of a system chip but required removing the chip from the computer and placing it in a special machine, which wasn't very convenient. The use of ultraviolet light to erase ROM chips was referred to with the phrase "flashing the BIOS." This phrase continues to be tech lingo for updating a computer's BIOS, even though EPROM technology is no longer used.

As computer technology evolved, users wanted to be able to rewrite certain ROM chips on their own. A technology called **electrically erasable programmable ROM (EEPROM)** was developed that allows reprogramming a ROM chip electronically using only the hardware that comes with the computer. EEPROM allows upgrading the computer's BIOS without removing the BIOS chip from the motherboard and placing it in a special machine. A utility program activates the erase function for the ROM chip and then writes new data to it. If you want to

USB flash drives hold flash ROM, which is technically a form of read-only memory but can be written and rewritten.

Cutting Edge

Goodbye to Flash Memory?

For many years, flash memory has been the dominant technology for solid-state storage in everything from cell phones to digital cameras—but that may soon change. A technology called *phase-change RAM (PRAM)* is faster and more efficient than flash and may replace it altogether.

The idea behind PRAM has been around for several decades. The chip contains a chemical compound called *chalcogenide*; heating this compound to a very high temperature changes its physical state. The two states serve as the 1s and 0s of the binary code used for data storage. The benefit of PRAM technology is that the chip doesn't need to erase a block of cells before writing new data; the bits can be changed individually. Because of this, PRAM chips can read and write data 10 times faster than their flash counterparts and use less power in the process. In a cell phone, for example, PRAM memory could increase the battery life by up to 20 percent.

update the BIOS of a modern PC, you can do so in just a few minutes. All you need to do is download and install a BIOS update utility and the latest BIOS software from the PC's manufacturer.

Flash memory is a type of EEPROM. It may seem like RAM, because it's so easily written and erased. But the technology behind it is much more similar to EEPROM than to static RAM. Flash memory is read and written in blocks (or pages), rather than individual bytes (as RAM is). Also, like EEPROM, flash memory can be reprogrammed only a limited number of times: 1 million write operations. Because of this limitation, flash memory is used primarily for storage—not as a substitute for RAM in the main memory of a computer.

Flash memory is used for storage in a variety of devices, including USB flash drives, solid state hard drives, digital cameras, digital audio players, cell phones, and video game systems. A variety of flash memory cards are also available and plug into slots on devices such as printers, cameras, scanners, and copiers to enable easy data exchange. Types of flash cards include CompactFlash (CF), SmartMedia, Secure Digital (SD), MultiMediaCard (MMC), xD-Picture, and Memory Stick.

2.6 Exploring Visual and Audio Output

Watch & Learn
Exploring Visual and Audio Output

Output is one of the four steps of the information processing cycle. Specifically, output is processed data that exits from the computer. Output can be temporary or permanent. For example, the output that appears on your display screen is temporary, as is the output that you hear through speakers or headphones. Output that still exists after the computer has been turned off is permanent—for example, the printout from a printer (sometimes called a *hard copy*).

An **output device** is any type of hardware that makes information available to a user. Popular output devices for visual and audio output include displays (or monitors), projectors, and speakers and headphones.

Displays

Nearly every computer has some sort of display on which users can view output, such as a stand-alone monitor or built-in screen. Displays are available in a wide variety of shapes, size, costs, and capabilities. The term **screen** refers to the viewable area of any display device. For example, laptops, tablets, and cell phones all have built-in screens. When the display device is separate from the computer and has its own power supply and plastic housing, it's called a **monitor**.

Display Types The original type of display for a desktop computer was the **cathode ray tube (CRT) monitor**. A CRT monitor used the same technology as an old TV set and was large, heavy, and bulky.

The modern flat-screen monitors in use today are thin and lightweight. Most flat-screen monitors show information onscreen using a **liquid crystal display (LCD)**.

CRT monitors are no longer popular, but you may occasionally see one in use with an old computer system.

A flat-screen monitor, such as this LCD model, is lightweight and thin.

Chapter 2 Sizing Up Computer and Device Hardware

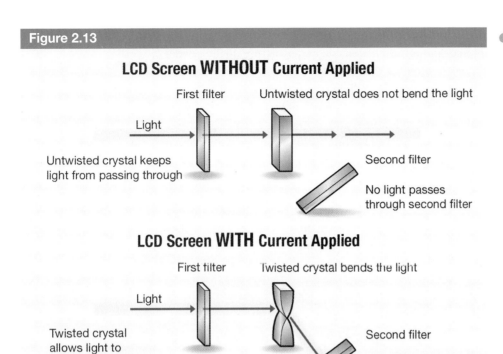

Figure 2.13

Practice
How Liquid Crystals Work

The liquid crystals in an LCD screen twist when an electrical current is applied to them. This changes the angle of the light passing through them, allowing it to reach the second filter and display an image on the screen.

Figure 2.13 illustrates how an LCD crystal produces an image on the display. An LCD screen has two polarized filters, and between them are liquid crystals. For light to appear on the display, it must pass through the first filter, then through the crystals, and finally through the second filter. The second filter is at an angle to the first, so by default, light doesn't reach it after passing through the crystals and that area of the display remains black. However, when current is applied to the crystals, they twist; this makes the light passing through the crystals change angle and point in the direction of the second filter. When the light passes through the second filter, it lights up an area of the display. Color LCD screens use a filter that splits the light into separate cells for red, green, and blue.

LCD and LED Flat-screen monitors may be advertised as LCD or LED, with the best and most modern ones usually being LED. However, LCD and LED are not alternative technologies; they are complementary. All LCD monitors require backlighting to make the image bright enough to see. First-generation LCD monitors used a technology called *cold cathode fluorescent lamps (CCFL)* for backlighting and are advertised as LCD monitors. Most modern LCD monitors use *light-emitting diodes (LEDs)* for backlighting and are advertised as LED monitors.

Some higher-end display monitors and TVs for sale today use organic light-emitting diode (OLED) technology. OLED is very similar to standard LED

Figure 2.14

and LCD except the semiconductor layer is an organic (carbon-based), light-emitting compound, with slightly different properties that enable better picture quality, wider viewing angles, deeper blacks, faster response time, thinner and lighter hardware, and lower power consumption. Those are some compelling benefits!

Disadvantages of OLED as it exists today include shorter lifespan (with brightness degrading significantly after about 10,000 hours of use), degrading color balance over time, and susceptibility to water damage. OLED monitors also don't perform well in bright sunlight, and the power consumption can spike when displaying a screen that is mostly white.

Display Performance and Quality Factors Displays have a **maximum resolution** (sometimes called *native resolution*), which is the highest display mode they can support. Images are displayed on the monitor screen using pixels (see Figure 2.14). Maximum resolution is expressed as the maximum number of pixels that will display horizontally multiplied by the maximum number of pixels that will display vertically. High-definition (HD) video has a resolution of 1,920 × 1,080, and ultra-high-definition (UHD) video has a resolution of 3,840 × 2,160. Monitor resolutions continue to increase over time; 4K monitors support up to 4,000 pixels horizontally, and 8K monitors support up to 8,000 pixels horizontally.

Images look best at the native resolution, because there is a one-to-one relationship between the monitor's physical pixels and the number of pixels that the operating system's display mode tells the monitor to use. Images are sharp and clear at this resolution but may appear fuzzy at lower resolutions. So, if you prefer a certain resolution, you should use a monitor that has that particular resolution as its maximum.

New technologies have improved monitor performance and appearance in the last few years. High-dynamic-range (HDR) technology improves color range and quality, especially in the brightest and darkest areas of an image.

Another important feature of a computer monitor is the **aspect ratio**, which is the ratio of width to height. In the past, the most common aspect ratio was 4:3, the same as for a standard-definition TV. Many newer monitors are widescreen models and have an aspect ratio of 16:9, the same as for a widescreen high-definition TV. When

Video
Display Resolution

Hands On
Are Your Onscreen Items Tiny?

Cutting Edge

New Applications for OLED

OLED is used in innovative ways that move beyond TVs and computer monitors. For example, OLED's ability to be used on curved surfaces has been harnessed to create immersive 3-D tunnels that users can walk through in zoos and museums.

Even more OLED innovation is coming. In the near future, expect to see thin displays using OLED technology that can be rolled up into a scroll for storage. These roll-up monitors aren't just for ultra-portable computers; there are many potential uses for a thin flexible display surface. Look for flexible displays on wristbands and T-shirts, Halloween masks and other children's amusement items, and window decals that display active content.

selecting a monitor resolution in your operating system, keep in mind the physical dimensions of the screen you are using. For example, the resolution 1,024 × 768 has an aspect ratio of 4:3, and the resolution 1,024 × 600 has an aspect ratio of 16:9. If the resolution you choose has a different aspect ratio than your monitor, images may appear squashed or elongated.

Monitors have a **refresh rate**, which is the number of times per second that each pixel is refreshed with new data. Refresh rate is measured in hertz (Hz).

On CRT monitors, refresh rate was important because each pixel decayed in brightness quickly and refreshing pixels frequently (at the rate of at least 80 Hz, or 80 times per second) prevented display flicker. On modern monitors, the pixels stay at a consistent brightness all the time without being re-energized. Refreshing involves checking for updates to the pixels' colors. LCD monitors typically have a refresh rate of about 60 Hz (or 60 times per second), and that's sufficient for most uses. Higher-end monitors may refresh faster, at a rate of up to 120 Hz or even 240 Hz.

> **Hands On**
> Changing Your Display's Resolution, Refresh Rate, and Color Depth

Display Adapters A display doesn't have much (or any) processing power of its own; it relies on the computer to feed it the information it shows. Providing that information is the purpose of the **display adapter** (sometimes called *video card*, *video adapter*, or *graphics card*). A display adapter is either built into the motherboard (as is the case for most laptops and other portable devices) or added as an expansion board. It converts the computer's digital output to instructions and sends them to the display. If the display adapter is a separate board, it has its own RAM, chipset, and data bus (to carry data from place to place on its own circuit board). If it's built into the motherboard, it uses the bus, chipset, and RAM of the motherboard.

This display adapter is an expansion board and thus separate from the motherboard. The model shown here has DVI, S-Video, and VGA ports. The large, circular object is a cooling fan for the adapter's processor.

A display adapter might connect to a monitor using a variety of connectors, including HDMI, DVI, VGA, DisplayPort, or Thunderbolt. A display adapter has its own processor, called a **graphics processing unit (GPU)**; the GPU works in cooperation with the computer's CPU.

The display adapter requires memory because it must hold all of the data needed to display the contents of the screen. For instance, the display adapter keeps track of what color each individual pixel should show at any given moment and continuously feeds that information to the display. Some older displays show output in only one color; this type of display is called a **monochrome display**. You might occasionally use a monochrome display, such as the LCD display on a microwave or other appliance. Most computer displays, however, support thousands or even millions of color choices for each pixel.

Each color has a numeric value. The number of binary digits needed to describe a particular color uniquely depends on the color depth of the operating system's display mode. The **color depth** (also called *bit depth*) is the number of bits used in a display adapter to store information about each pixel. The greater the color depth, the more data the operating system must send to the display adapter and the display adapter must send to the monitor every second. For example, in a four-digit binary number, there are 16 possible values. It follows that 4-bit color allows for 16 different colors, 8-bit color allows for 256 colors, and so on. Most displays on modern computers operate at 16-bit (high color) or 32-bit (true color) color depth.

The higher the resolution and the greater the color depth, the more RAM the display adapter requires. However, modern display adapters use RAM for much more than storing the values of individual pixels. Display adapters also use RAM to support the many special features they have for improving display performance. For instance, display adapters have processors that help render the complex 3-D objects that appear in the latest games and drawing software. The exact capabilities of the display adapter are determined by its chipset.

Projectors

A **video projector** (or *screen projector*) is a device that captures the text and images displayed on a computer screen and projects them onto a large screen so an audience can see them clearly. Projectors are often used in classrooms, conference and meetings rooms, and convention halls. A meeting held using remote video is called a *video teleconference (VTC)*.

A projector can be connected to a PC through the same port as a monitor (HDMI, VGA, or DVI, for example). Some projectors can also be connected via USB or another port, and many PCs allow you to use a regular monitor and a projector at the same time, as long as there are sufficient ports available.

A projector's brightness is measured in **lumens (lm)**. The brighter the projector, the better it will work when projecting a large image in a big, well-lit room. For example, in a small, dark room, 1,500 to 2,500 lm is an adequate brightness, but in a large, well-lit room, more than 4,000 lm is optimal. Most projectors have a replaceable **lamp** (like a lightbulb) inside that generates the brightness. Some projectors use a light-emitting diode (LED) lamp, which never needs to be replaced.

Using a projector allows multiple people to view the output from a computer on a large screen at a meeting or conference.

Virtual Reality and Augmented Reality Hardware

Viewing output on a monitor or projection screen is a two-dimensional experience. The picture might be clear, but nobody would mistake it for part of the real world. But what if you could view computer output in three dimensions, like you view the rest of the world, so that you feel immersed in that environment? That's the idea behind *virtual reality (VR)*.

VR headsets place the user's entire field of view within an enclosed visual environment—the headset screen—creating a visually immersive experience. Some examples of the VR experiences available today include viewing tourist attractions and museum collections, analyzing sports plays from different perspectives, and, of course, playing games.

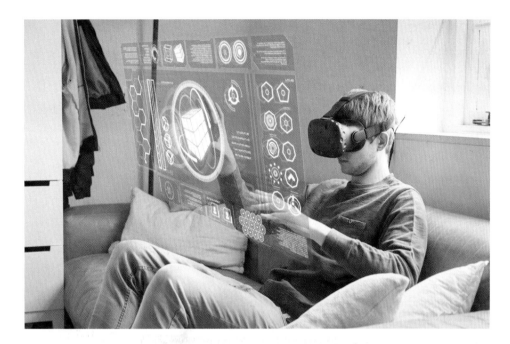

Augmented reality headsets can project instructions or other graphics on top of your real-world vision.

Augmented reality can superimpose text labels or other content over your view of a machine or product.

Some VR headsets include hand controls that allow the user to interact with the virtual environment. Some advanced models even include haptic feedback—that is, physical sensation generated by the VR device. For example, you might wear a glove that stimulates your fingers so they feel like they are touching something.

Augmented reality (AR) is a variant of VR that enables you to see digital content on top of your real-world vision. You have already experienced AR if you have played *Pokémon GO* on a mobile phone, or if you've enhanced pictures with Snapchat. AR is responsible for the virtual first-down lines and other status lines you see on the field when watching football on TV. AR is used in industrial settings to help workers with their jobs. For example, a worker wearing an AR headset might see usage or safety instructions for a machine, projected onto the machine's control panel.

Speakers and Headphones

Most computers provide a means of producing audio output, which can include music playback, system sound effects and game sounds, the narration of a presentation, and the audio part of video chatting. Having audio output can also enable people with visual impairments to control and interact with the computer.

As noted earlier in this chapter, the sound card (or audio adapter) is both an input device and an output device. It translates the digital instructions from the computer into analog output, such as sound waves that a person can hear through a speaker. It also does the reverse, translating analog input (such as the sound waves produced by a person speaking into a microphone) into digital input for an application. Depending on the computer system, the sound card may be built into the motherboard or it may be an actual card (that is, an expansion board).

A sound card typically has a series of color-coded RCA jacks for plugging in different input and output devices. These jacks are called *RCA jacks* because they were invented by the Radio Corporation of America (RCA). The term *jack* is often used to refer to audio connectors because of their origin in home stereo systems, whereas *port* is used for most other computer connectors. Either term can be used for audio connectors on a computer. Refer back to Table 2.4 on page 61 to review the colors and meanings of these different types of connectors.

Sound exits the computer via speakers or headphones. Most computers have separate ports for these two output devices, so you don't have to disconnect one to use the other. When a set of headphones is attached, the system may automatically direct the sound output to that port, muting the sound to the speakers.

Simple speakers usually come in pairs (left and right, for stereo output). High-end speaker systems have more speakers; for example, the Dolby Digital 5.1 (Surround Sound) system uses six speakers.

Some speaker systems are wireless, which means they can connect to the computer via a wireless technology (such as Bluetooth). Some speakers can also function as docking stations for mobile devices, such as iPods. A **docking station** is like a home base for a portable device and gives it extra capabilities, such as better speakers, more ports, and power recharging.

Practical TECH

Connecting Speakers to Your Computer

Most speaker systems have multiple speakers, but most computers have only one speaker port. Fortunately, this isn't a problem with the speaker systems designed for use with computers. In a system with two speakers, one speaker connects to the computer and the other speaker connects to the first speaker. If a subwoofer unit is part of the speaker system (as it is for Dolby Digital 2.1 or 5.1), the subwoofer often connects to the computer and all the speakers in the system connect to the subwoofer. (A subwoofer is a speaker that produces very low sounds; the term *subwoofer* is also popularly used to refer to the central connecting speaker in a multispeaker system, regardless of the tones it produces.)

Some speaker systems have a subwoofer unit to which individual speakers connect.

2.7 Differentiating between Types of Printers

A **printer** is the most common type of device used to produce hard-copy output on paper or another physical medium, such as transparency film. Printers are categorized into two broad groups based on how they affix or attach the print to the medium: nonimpact and impact.

A **nonimpact printer** forms the characters and images without actually striking the output medium; it creates print using electricity, heat, or photographic techniques. Most printers sold today are nonimpact, including inkjet printers, laser printers, thermal printers, and plotters. All of these printers use ink or toner to create print, whether black or color. The ink or toner cartridge must be replaced or refilled to keep this type of printer operating.

An **impact printer** is a lot like an old-fashioned typewriter; it forms characters and images by physically striking an inked ribbon against the output medium. Impact printers aren't commonly used anymore, but they are still found in some business and factory settings that require the same image to be printed on each page of a multipart form. Impact printers do a good job of printing text on multipart forms, but they don't print graphics very well and they usually don't print in color. Impact printers are economical to use, however; the ribbons they use don't cost much compared with the ink and toner used by other printers.

Most printers sold today connect to the computer using a USB port, but some older printers may use a legacy parallel port. Some printers can also connect to computers via wireless networking, such as Bluetooth or Wi-Fi (both covered in Chapter 6).

Inkjet Printers

An **inkjet printer** is a nonimpact printer. It forms characters and images by spraying thousands of tiny droplets of ink through a set of tiny nozzles and onto a sheet of paper as the sheet passes through it.

Two technologies are used to force the ink through the nozzle: thermal and piezoelectric. A **thermal inkjet printer** heats the ink to about 400 degrees Fahrenheit; doing so creates a vapor bubble that forces the ink out of its cartridge and through a nozzle. A vacuum is created inside the cartridge, which draws more ink into the nozzle. Figure 2.15 illustrates the operation of a thermal inkjet printer.

Watch & Learn
Differentiating between Types of Printers

Hands On
What Printers Are Installed on Your Computer?

Practice
Thermal Inkjet Printer Operation

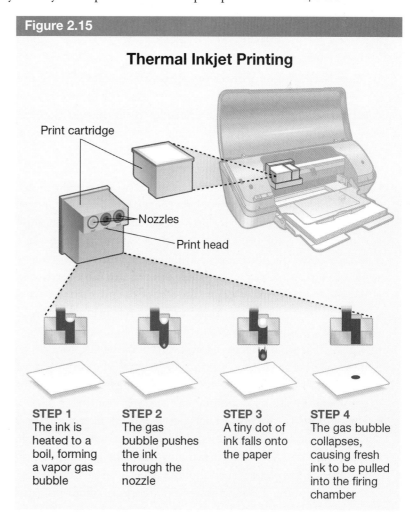

Figure 2.15

Thermal Inkjet Printing

STEP 1 The ink is heated to a boil, forming a vapor gas bubble

STEP 2 The gas bubble pushes the ink through the nozzle

STEP 3 A tiny dot of ink falls onto the paper

STEP 4 The gas bubble collapses, causing fresh ink to be pulled into the firing chamber

A thermal inkjet printer produces output by heating the ink until a gas bubble forces it out of the cartridge through the nozzle. Doing this creates a spray of tiny ink droplets that form text and images on a sheet of paper or other medium.

Inkjet printers provide good-quality color output. They are inexpensive to buy, but the per-page cost of operation can be high because of the price of ink cartridges.

Most inkjet ink cartridges are more than just ink reservoirs; they also contain the print heads. The dots on the strip at the top of each cartridge function as electrical contacts; they allow the printer to instruct the print head how and when to discharge ink.

A **piezoelectric inkjet printer** (or *piezo printer*) moves the ink with electricity instead of heat. Each nozzle contains piezoelectric crystals, which change their shape when electricity is applied to them and then force out the ink. The electrical process is easier on the printer, because it doesn't require heating the ink. It also produces better output, because the ink used is less prone to smearing.

Inkjet printers vary in the following ways:

- **Ink cartridges.** Most inkjet printers use four cartridges with different colors of ink: black, cyan (similar to blue), magenta (similar to red), and yellow. However, some high-end printers have more than four cartridges. In most cases, the cartridges contain the print heads, which makes them expensive. On a high-end inkjet printer, the print heads may be separate cartridges, which can be replaced less often than ink cartridges. These printers can use a better-quality print head but still be economical because the cost of the ink cartridges is lower.
- **Speed.** Print speed is measured in **pages per minute (ppm)**. The advertised speed of a printer is the fastest speed at which it can print once the print cycle begins. This means that the actual print speed may be slower, because additional time may be needed for a print job to reach the printer and for the printer to warm up so it can print the first page. High-quality printing and complex graphics may also make printing take longer (resulting in a lower ppm rate).
- **Resolution.** The number of dots per inch (dpi) determines the printer's resolution. A typical resolution for an inkjet printer is between 600 and 2,400 dpi.
- **Photorealism.** Some inkjet printers can print photo-quality images right to the edge of the paper if special photo paper is used. To produce the best-quality photo printouts, some printers require special photo-ink cartridges. Specialty photo printers are available that are designed to print photos on specially sized and treated paper.
- **Interface.** A typical inkjet printer uses a USB connection to the computer; some inkjet printers also have Ethernet (wired network) or Wi-Fi (wireless network) capability.

Figure 2.16

Photo Printing from a Memory Card

STEP 1 Insert the memory card into the digital camera and take a photograph with the camera

STEP 2 Remove the memory card from the digital camera and insert it into the card slot on the printer

STEP 3 Select desired image to print, number of copies, and size of print by pushing buttons on the printer

STEP 4 Remove the photo from the printer

A printer that has a memory card slot can print photographs without a computer. It processes the data directly from the memory card removed from a digital camera.

- **Memory card slots.** Some printers accept flash memory cards, so you can print images from a digital camera or other source without using a computer. Figure 2.16 shows how a printer with a memory card slot can accept mages and produce photo prints from a digital camera.

Inkjet printers are popular for use in homes and small businesses, because they are inexpensive. As mentioned earlier, though, the cost of the replacement ink cartridges is high, so the cost per page with inkjet printers is higher than with laser printers. Another disadvantage is that liquid ink tends to dry out over time, especially if the printer isn't used frequently, and the print heads can get clogged with dried ink. This problem can be often solved by running a cleaning cycle, which forces more ink through the heads to break up the clogs. Unfortunately, running a cleaning cycle also wastes some ink, creating further expense. For these reasons, larger businesses often use laser printers.

Laser Printers

A **laser printer** works much like a photocopier. The main difference is that a photocopier scans a document and then prints the scanned copy, whereas a laser printer receives digitized data from a computer and then prints the data. Typically, a laser printer looks like a big, square box with a paper tray inside. Laser printers are used extensively in business and industry because of their durability, speed, and low cost per page.

A laser printer is typically taller than an inkjet printer, because it has to accommodate the drum and one or more toner cartridges inside.

A laser printer contains at least one large cylinder (known as a **drum**) that carries a high negative electrical charge. One drum is needed for each color of ink, which means a color laser printer has four drums—one each for black, cyan, magenta, and yellow. The printer directs a laser beam to partially neutralize the electrical charge in certain areas of the drum. A rotating mirror helps move the laser across the drum horizontally. The drum rotates past a reservoir containing powdered **toner**, which is a combination of iron and colored plastic particles. The toner clings to the areas of lesser charge, and the image is formed on the drum. Next, the drum rotates past the positively charged paper, and the toner jumps onto the paper. Finally, the paper passes through a **fuser**, which melts the plastic particles in the toner and causes the image to stick to the paper. Figure 2.17 illustrates the laser printing process.

Toner comes in cartridges that fit into the laser printer. A color laser printer has four toner cartridges: black, cyan, magenta, and yellow.

Video
Laser Printer

Practice
The Laser Printing Process

Laser printers cost more than inkjet printers in terms of purchase price, but their operational cost per page is lower. A toner cartridge might cost more than an ink cartridge, but it will print many times more pages. In addition, laser printers aren't subject to the problem of dried-up ink that might occur if the printer isn't used frequently.

Figure 2.17

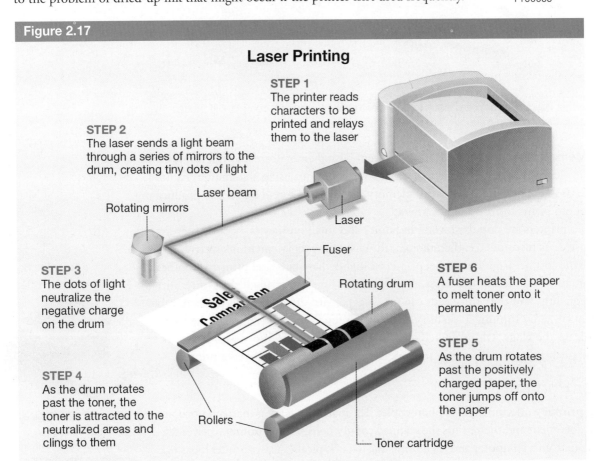

A laser printer produces output by neutralizing the negative electrical charge on certain areas of a rotating drum; the drum then picks up toner in these neutralized areas.

Multifunction Devices and Fax Machines

A **multifunction device (MFD)** is a printer that also functions as a copier and a scanner (and sometimes a fax machine). Because MFDs serve all these purposes, they are also called *all-in-one devices*. They are popular with small businesses, because they provide the key functions needed in a single machine that takes up less space than would be needed for separate machines. Almost all MFDs use inkjet or laser technology.

A multifunction device (MFD) typically includes a printer, a copier, and a scanner in one unit. Some models also have fax capabilities.

A **fax machine** can send and receive copies of documents through a telephone line, similar to the way a dial-up modem sends network data. (The word *fax* is short for *facsimile*, which means "exact copy.") A fax machine can be a stand-alone device or part of an MFD.

Microsoft Windows includes a Fax and Scan utility that enables the computer's modem to send and receive faxes. If you have this utility and a dial-up modem that connects your computer to a telephone line, you don't need a fax machine or MFD to send and receive faxes.

Thermal Printers

A **thermal printer** uses heat to transfer an impression onto the paper. Thermal printers are not as common as inkjet and laser printers, but they are ideal for certain uses. Three main types of thermal printers are available:

- A **direct thermal printer** prints an image by burning dots into a sheet of coated paper when it passes over a line of heating elements. Early fax machines used direct thermal printing. The quality produced by this type of printer isn't very good, and it can't produce shades of gray or colors—only black and white. Direct thermal printers are used in a variety of places where print quality is less important than inexpensive, high-volume output, such as on cash register receipts, bar code labels, rental car paperwork, and theater tickets.
- A **thermal wax transfer printer** adheres a wax-based ink onto the paper. During the printing process, a thermal print head melts this ink from a ribbon onto the paper. When the wax has cooled down, it is permanently affixed. Images are printed as dots, which means they must be dithered to produce shades of colors. (*Dithering* is a technique that creates the illusions of new colors and shades by using varying patterns of dots. As a result, the quality of images produced by a thermal wax transfer printer can't compete with that of images produced by a modern inkjet or laser printer.)
- A **dye sublimation printer** (also called a *thermal dye transfer printer*) produces an image by heating ribbons containing dye and then dispersing the dyes onto a specially coated paper or transparency. This type of printer is the most expensive and slowest type of thermal printer; it also requires special paper, which is quite expensive. Even so, dye sublimation printers produce exceptional high-quality, continuous-tone images that are similar to actual photographs. A new type of dye sublimation printer called a *snapshot printer* produces small, photographic snapshots and is much less expensive than a full-size thermal printer.

Plotters

A **plotter** is a type of printer that produces large-size, high-quality precision documents, such as architectural drawings, charts, maps, and diagrams. Plotters are also used to create engineering drawings for machine parts and equipment.

Plotters use a variety of technologies. For example, an electrostatic plotter produces a high-quality image using a series of tiny dots that are tightly packed together. The printing mechanism consists of a row of fine, electrically charged wires. When the wires contact the specially coated paper, an electrostatic pattern is produced that causes the toner to be fused onto the paper. An inkjet plotter (also called a *wide-format inkjet printer*) may be used when large color images need to be produced.

Engineers often use plotters to produce high-quality, large-size prints of plans and designs for buildings, processes, and machines.

Special-Purpose Printers

Some printers are designed for specific tasks, rather than for general use. For example, there are special-purpose printers that print labels, badges, photo stickers, postage, and bar codes. Other printers produce full-color advertisements for mass mailing, and some even print on cardboard boxes or etch images into glass or metal.

A **label printer** is a small device that holds a roll of labels and feeds them continuously past a print head. After the labels have been printed, they are peeled from their backing and pressed on letters, packages, or other items. Label printers can produce color or black-and-white labels, and can use inkjet or thermal technology.

A **postage printer** is similar to a label printer but may include a scale for weighing letters and packages. The postage amount that's printed depends on the weight of the item and its destination, which is typically input using software or a numeric keypad on the printer. To use a postage printer, you sign up with an authorized service (such as a branch of the US Post Office). Each time you print postage, the amount is subtracted from your account balance.

A label printer is designed specifically to print adhesive labels.

A **portable printer** is a lightweight, battery-powered printer that can be easily transported. This type of printer can connect to a laptop or tablet computer, as well as a smartphone. Repair and service technicians that travel to remote sites use portable printers to generate invoices and receipts for customers.

3-D printing is an exciting technology that is changing the way we think about printing. A **3-D printer** uses a special kind of plastic, metal, or other material to create a three-dimensional model of just about any object you can design in a 3-D modeling program on a computer. The possible uses for 3-D printing are many. For example, a company designing a new object can create a working prototype of it on a 3-D printer much more inexpensively than having the object milled or fabricated in a machine shop. 3-D printing can also be used to reproduce parts for old machines for which replacement parts are no longer being manufactured. And, instead of buying a new toy for a child, parents can purchase the design for the toy online and print it on their own 3-D printer. Even houses and human organs can be printed. Chapter 13 provides more information about 3-D printing technology.

Chapter Summary

2.1 Working with Input Technology

An **input device** is any hardware component that enables users to input information. **Keyboards** are used with almost all desktop and laptop computers; for tablet PCs and smartphones, software-generated **virtual keyboards** substitute for physical keyboards. **Pointing devices** move the onscreen pointer in a graphical user interface (GUI); common pointing devices include **mice**, **trackballs**, **touchpads**, and **touchscreens**. Input hardware designed for gaming includes **joysticks**, **steering wheels**, and **game controller pads**. Some other input devices include **graphics tablets**, **scanners**, **digital cameras**, microphones, and **global positioning systems (GPSs)**. Scanners store material as **bitmap**, which is a matrix of rows and columns of pixels. A **pixel** is the smallest picture element a monitor can display (an individual dot on the screen) and is represented by one or more bits of binary data.

2.2 Understanding How Computers Process Data

Binary strings of data consist of individual **bits** that are either on (1) or off (0), like light switches. A group of 8 bits is a **byte**. One thousand twenty-four bytes (which is 2^{10} bytes) is a **kilobyte**, and 1,024 kilobytes (2^{20}, or approximately 1 million bytes) is a **megabyte**. Larger measurements include a **gigabyte** (2^{30}, or approximately 1 billion bytes) and a terabyte (2^{40}, or approximately 1 trillion bytes). Frequencies are described using the term **hertz (Hz)**. Letters, numbers, and symbols are represented using encoding schemes such as **ASCII** and **Unicode**.

2.3 Identifying Components in the System Unit

The **system unit** is the part of the computer that houses the major components used in processing data. The main circuit board is the **motherboard**; all other components connect to it. The motherboard has a controller chip (or set of chips) called a **chipset**, which controls and directs all of the data traffic on the board. Data moves around on the motherboard by way of pathways called **buses**. The **power supply** provides electricity in the proper voltages to each component. Add-on capabilities are provided by **expansion boards**, which are inserted into **expansion slots** in the motherboard. The most common type of expansion bus today is the **Peripheral Component Interconnect Express (PCIe) bus**. The computer case may contain one or more **storage bays** for holding disk drives.

The **USB port** is the most popular general-purpose port, and it comes in several speeds. A **Thunderbolt port** is a high-speed, multipurpose port found in nearly all Macs and many newer PCs. **HDMI** and **DVI** are digital interfaces for monitors; **VGA** is an older, analog interface for monitors. Network cables plug into **RJ-45 ports** (also called **Ethernet ports**). **Audio ports** use 3.5-millimeter round plugs. A **legacy port** is a computer port that is still in use but has been largely replaced by a port with more functionality; examples include the **parallel port** and the **PS/2 port**. Dial-up modems use **RJ-11 connectors** or **RJ14 connectors** (telephone connectors).

2.4 Understanding the CPU

A **central processing unit (CPU)** processes data in four steps: fetching an instruction, decoding the instruction, executing the instruction, and storing the result. The **control unit** directs the overall process; it fetches and decodes. The **arithmetic logic unit (ALU)** executes the instruction. **Registers** store the data temporarily until it exits the CPU. To minimize the CPU's idle time waiting for data to be delivered, **caches** are used; they store data temporarily just outside the CPU, where it can be accessed quickly.

A CPU is rated for its maximum speed in GHz, but its actual speed (number of operations per second) is determined by the **system clock** on the motherboard. The amount of data processed per clock cycle depends on the **word size** of the CPU (32-bit or 64-bit). Advanced CPUs have many special features that improve processing efficiency, such as **pipelining**, **parallel processing**, **superscalar architecture**, **multithreading**, and **multicore processors**. Some powerful desktop PCs use **liquid cooling systems**, which have a water reservoir and a closed system that circulates the water between hot spots (such as the CPU) inside the case.

2.5 Understanding Memory

There are two types of **memory: read-only memory (ROM)**, which is unchangeable, and **random access memory (RAM)**, which is changeable. **Static RAM (SRAM)** doesn't require electrical refreshing to hold its contents, but dynamic RAM does. **Dynamic RAM (DRAM)** can be synchronous (operating in time with the system clock) or asynchronous (not dependent on the system clock). **Synchronous dynamic RAM (SDRAM)** can complete one operation (**single data rate**, or **SDR**), two operations (**double data rate**, or **DDR**), or three operations (DDR3 and DDR4) per clock tick. RAM storage capacity is measured in bytes (kilobytes, megabytes, gigabytes, and so on). **Flash memory** is a type of **electrically erasable programmable ROM (EEPROM)**, which maintains its data without power but can be easily rewritten.

2.6 Exploring Visual and Audio Output

An **output device** is any type of hardware that makes information available to a user. Some devices, such as laptops and tablets, have built-in display **screens**. Display screens that are separate from computers are called **monitors**. A modern, flat-screen monitor is a **liquid crystal display (LCD)**; it uses transistors to define the color of each pixel. Light-emitting diode (LED) and organic LED (OLED) are modern types of LCD displays with improved display quality and reduced power usage. A display has a **maximum resolution**, which is the number of pixels it can display; the resolution is expressed as horizontal × vertical pixels (for example, 1,600 × 900). Displays also have an **aspect ratio**, which is the ratio of width to height. The number of times per second the display is re-energized is its **refresh rate**. The **color depth** is the number of bits needed to describe the color of each pixel in a particular display mode, such as 16-bit or 32-bit. The monitor connects to the computer via a **display adapter** using an interface such as HDMI, DVI, or DisplayPort.

Video projectors can serve as monitors too; projector brightness is measured in **lumens**. Virtual reality (VR) is an experience in which you can view computer output in three dimensions so you feel immersed in the environment. Augmented reality (AR) is a variant of VR that enables you to see digital content on top of your real-world vision.

Audio output goes through the audio adapter to speakers or headphones. Some speakers function as **docking stations** for mobile devices, giving them extra capabilities such as more ports and power recharging.

2.7 Differentiating between Types of Printers

A **printer** is the most common type of device used to produce hard-copy output. Printers are categorized into two broad groups: nonimpact and impact. Both of the printer types commonly used today are **nonimpact printers**. An **inkjet printer** sprays small dots of liquid ink onto a page. A **laser printer** transfers powdered **toner** to a **drum** and then from the drum to paper, where it's heated and transferred onto the page via a **fuser**. Inkjet printers are inexpensive to purchase, but the ink cartridges are expensive in terms of cost per page. Laser printers are more expensive to purchase, but the toner cartridges last longer than ink cartridges, so the cost per page is lower.

Other types of printers include **thermal printers** and **plotters**. A **multifunction device (MFD)** combines printing functionality with copying, scanning, and sometimes faxing. **3-D printers** use plastic, metal, or other materials to create solid objects based on 3-D digital models by layering material on a platform.

Key Terms

Numbers indicate the pages where terms are first cited with their full definition in the chapter. An alphabetized list of key terms with definitions is included in the end-of-book glossary.

Chapter Glossary

Flash Cards

2.1 Working with Input Technology

input device, 42
keyboard, 42
alphanumeric keyboard, 42
QWERTY layout, 42
virtual keyboard, 44
touchscreen, 45
pointing device, 46
mouse, 46
trackball, 46
touchpad, 46
joystick, 47
steering wheel, 47
game controller pad, 47
graphics tablet, 48
stylus, 48
scanner, 48
digitized information, 48
charge-coupled device (CCD), 48
bitmap, 48
pixel, 48

resolution, 49
dots per inch (dpi), 49
optical character recognition (OCR), 49
flatbed scanner, 49
3-D scanner, 50
bar code, 50
universal product code (UPC), 50
bar code reader, 50
digital camera, 51
megapixel, 51
digital video camera, 52
webcam, 52
audio input, 53
sound card, 53
voice input, 53
accelerometer, 54
gyroscope, 54
compass, 54
global positioning system (GPS), 54

2.2 Understanding How Computers Process Data

binary string, 55
bit, 55
byte, 56
kilobyte, 56
megabyte, 56
gigabyte, 56

hertz (Hz), 56
encoding scheme, 56
American Standard Code for
    Information Interchange (ASCII), 56
Unicode, 57

2.3 Identifying Components in the System Unit

system unit, 57
power supply, 57
motherboard, 58

trace, 58
chipset, 58
bus, 59

bus width, 59
system bus, 59
expansion bus, 59
expansion slot, 59
expansion board, 59
Peripheral Component Interconnect
    Express (PCIe) bus, 59
port, 60
universal serial bus (USB) port, 60
Thunderbolt port, 61
eSATA port, 61
high-definition multimedia interface
    (HDMI) port, 61
digital visual interface (DVI) port, 61
Ethernet port, 61

RJ-45 port, 61
audio port, 61
legacy port, 62
FireWire port, 62
video graphics array (VGA) port, 62
serial port, 62
parallel port, 62
PS/2 port, 62
dial-up modem, 62
RJ-11 connector, 62
RJ-14 connector, 62
S-video port, 62
component video port, 62
storage bay, 63

2.4 Understanding the CPU

central processing unit (CPU), 631
semiconductor material, 63
machine cycle, 64
control unit, 64
arithmetic logic unit (ALU), 642
register, 64
cache, 65
on die cache, 65
system clock, 65
word size, 66

pipelining, 66
thread, 66
parallel processing, 67
superscalar architecture, 67
multithreading, 67
hyper-threading, 67
core, 67
multicore processor, 67
liquid cooling system, 67
overclock, 68

2.5 Understanding Memory

memory, 68
read-only memory (ROM), 68
random-access memory (RAM), 68
memory address, 68
dynamic RAM (DRAM), 68
static RAM (SRAM), 68
nonvolatile RAM, 68
dual inline memory module (DIMM), 70
synchronous dynamic RAM
    (SDRAM), 70

single data rate (SDR) SDRAM, 70
double data rate (DDR) SDRAM, 70
memory access time, 70
erasable programmable ROM
    (EPROM), 71
electrically erasable programmable
    ROM (EEPROM), 71
flash memory, 72

2.6 Exploring Visual and Audio Output

output device, 72
screen, 72
monitor, 72
cathode ray tube (CRT) monitor, 72
liquid crystal display (LCD), 72
maximum resolution, 74
aspect ratio, 74
refresh rate, 75
display adapter, 75

graphics processing unit (GPU), 75
monochrome display, 75
color depth, 75
video projector, 76
lumen (lm), 76
lamp, 76
docking station, 78

2.7 Differentiating between Types of Printers

printer, 79
nonimpact printer, 79
impact printer, 79
inkjet printer, 79
thermal inkjet printer, 79
piezoelectric inkjet printer, 80
pages per minute (ppm), 80
laser printer, 81
drum, 82
toner, 82
fuser, 82

multifunction device (MFD), 83
fax machine, 83
thermal printer, 83
direct thermal printer, 83
thermal wax transfer printer, 83
dye sublimation printer, 83
plotter, 84
label printer, 84
postage printer, 84
portable printer, 84
3-D printer, 84

Chapter Exercises

Complete the following exercises to assess your understanding of the material covered in this chapter.

Tech to Come: Brainstorming New Uses

In groups or individually, contemplate the following questions and develop as many answers as you can.

1. As touchscreens have become more affordable, more devices have started including them. Touchscreens are even provided with some devices you might not normally think of as computers, such as kitchen appliances and car radios. What devices besides traditional computer systems have you seen—either in person or on the internet, or in print media—that have touchscreens?

2. Bluetooth is a short-range, wireless networking technology that was mentioned several times in this chapter. One of the most common uses for Bluetooth is to connect a cell phone to a wireless headset. What other uses can you think of for Bluetooth technology in your home, school, and workplace? (Keep in mind that Bluetooth has a range limit of about 300 feet.)

Tech Literacy: Internet Research and Writing

Conduct internet research to find the information described, and then develop appropriate written responses based on your research. Be sure to document your sources using the MLA format. (See Chapter 1, Tech Literacy: Internet Research and Writing, page 37, to review MLA style guidelines.)

1. Suppose you have a friend who has a movement disability that makes it difficult to use a regular mouse; specifically, she can't position the pointer precisely enough. Research the availability of alternative input devices that would provide greater functionality, and write your friend a letter telling her about your findings.

2. What is Moore's law, and how does it affect the PCs that you use and purchase? Write a paragraph that explains the concept and that provides an estimate of CPU speed on a computer purchased three years from now.

3. Conduct internet research about impact printers. Where are they still available for sale and at what prices? Supply web addresses for at least two online stores that carry them. Who buys them, and what are they used for? After finding out the answers to these questions, write a fictional scenario of a company in which impact printers are important to day-to-day business.

4. Go to the Intel website (**https://CUT7.ParadigmEducation.com/IntelProducts**) and locate information about the latest Intel Core CPUs for desktop PCs. Choose one of the CPUs and find the following facts about it:
   - Number of cores
   - Number of threads
   - Clock speed
   - Word size (may be labeled as "Instruction Set")

   Use the System Information utility in the Microsoft Windows operating system (or an equivalent utility, if using a different operating system) to determine the CPU used in your own computer. Compare its performance stats to those of the Intel CPU you researched.

5. Suppose you need a motherboard that meets the following specifications:
   - Will support an Intel Core i7 processor (LGA1155 processor socket)
   - Will support at least 32 GB of DDR3 RAM
   - Has at least 2 ×16 PCIe slots
   - Has at least 3 ×1 PCIe slots
   - Has at least 1 PCI slot
   - Has built-in sound support
   - Has at least 2 USB 2.0 ports and at least 2 USB 3.0 ports

   Find at least two motherboards for sale online that meet or exceed these specs. Use a spreadsheet application to compare the two boards.

Tech Issues: Team Problem-Solving

In groups, develop possible solutions to the issues presented.

1. Schools and businesses lose millions of dollars a year because of computer thefts. What are some ways that schools and businesses can prevent desktop computers from being taken from their buildings without authorization?

2. Suppose you have been tasked with choosing new printers for your school's computer lab. Would you buy several of the same kind, or a variety? What type(s) of printers would you get, and why? Consider factors such as initial cost, cost per page, reliability, speed, and compatibility with your computer lab's network. Create a spreadsheet that summarizes your findings.

3. Computers have become so inexpensive that it may cost you more to have your old PC repaired than it does to buy a new one with better features. However, discarding your old computer every few years can be seen as wasteful and environmentally damaging. What could you do to lessen the negative impact on the environment? Suppose you had an old but still repairable computer to get rid of. Use the web to find charities or businesses in your area that will take your old computer and dispose of it responsibly or repair it and find it a new home.

Tech Timeline: Predicting Next Steps

Early computers were huge mainframes in special rooms that couldn't be moved without a forklift. The invention of desktop PCs enabled each user to have a computer on his or her desk, but the user was still tethered to the desk. Laptops, notebooks, tablets, and smartphones have taken personal computing to progressively smaller and more portable levels. What do you think is next? The following timeline outlines some of the major events in the development of computer portability. Research this topic to predict at least three important milestones that could occur between now and the year 2040, and then add your predictions to the timeline.

Mainframes

1952 IBM introduces the IBM 701 Electronic Data Processing Machine, one of the earliest mainframe computers, which uses vacuum tubes for processing and memory.

1959 IBM introduces the IBM 7090, a mainframe that uses transistors for processing and magnetic core memory.

Desktop Computers

1981 IBM introduces the original IBM PC, one of the first commercially successful desktop personal computers.

1984 Apple introduces the Macintosh desktop computer.

Laptop and Notebook Computers

1982 The Kaypro Corporation releases the Kaypro II, one of the first transportable computers; it weighed 26 pounds and was built into an aluminum case.

1984 IBM releases the IBM Portable Personal Computer 5155; it was essentially a PC in a carrying case and weighed 30 pounds.

1986 IBM releases the IBM Convertible Personal Computer 5140; it weighed just 12 pounds and was the first PC to run on batteries.

1988 NEC releases the UltraLite, a true notebook computer; it's just 1.4 inches thick and weighs 4.4 pounds.

2018 Dell releases the XPS 13 notebook, which is 0.3 inches thick at its thinnest point and weighs 2.7 pounds.

PDAs and Tablets

1993 Apple releases the Newton MessagePad, one of the first hand-held personal digital assistants (PDAs).

1996 Palm Computing introduces the Palm Pilot (later called a *Palm*), one of the most successful PDAs ever.

2000 Microsoft launches the Pocket PC, which uses a touch-capable version of Windows called *Windows CE 3.0*.

2010 Apple releases the iPad Version 1 with up to 64 GB of storage, 256 MB of RAM, and a 1 GHz CPU.

2017 Apple releases the 10.5-inch iPad Pro, with 512 GB of storage, 4 GB of RAM, and a 2.38 GHz CPU.

Smartphones

1994 IBM and BellSouth release the Simon Personal Communicator, the first smartphone; it could make calls and send and receive faxes and emails.

1999 Research in Motion (RIM) releases the first BlackBerry smartphone, integrating multiple communication types (phone, text, and email) into a single inbox.

2007 Apple releases the iPhone, featuring a phone, a camera, audio and video capabilities, and personal organizer utilities.

2008 T-Mobile releases the first Android-powered smartphone, the G1.

2010 Microsoft releases Windows Phone 7, with a tile-based user interface that resembles the Start screen in Windows 8.

2017 Microsoft announces that the Windows Phone will be discontinued.

Wearable Computers

1972 Hamilton Watch Company introduces the Pulsar, the first digital programmable watch.

1983 Casio releases a line of computer watches that store data, play simple games, and have built-in calculators, alarms, and timers.

2013 Samsung launches the Galaxy Gear Smartwatch, an Android-based watch with gesture-based navigation.

2014 Google makes available for consumer purchase Google Glass, a head-mounted, display-based computer that uses natural language commands to communicate with the internet. Google suspended the product in 2015, promising to continue to work on the technology.

2015 Apple introduces Apple Watch, a wrist-wearable computer that runs many of the same apps as the iPhone.

2017 Fitbit releases the Ionic smartwatch, which automatically tracks user activity and includes a GPS.

Ethical Dilemmas: Group Discussion and Debate

As a class or within an assigned group, discuss the following ethical dilemma.

As computer technology becomes increasingly portable, having a computer with you 24 hours a day is feasible. Many people carry a computer or smartphone with them wherever they go.

Do you think that employers have a right to limit employees' use of personal smartphones or tablet PCs during work hours? If you were an employer, what would your policy be and why? If you were an employee of this employer, how would you feel about the policy? What sort of compromise policy would be fair to both the employee and the employer?

What about in school? Should school administrators be able to ban or limit students' use of smartphones during classes? What about during free times, such as between classes, during lunch, and during before-school and after-school organized activities and clubs?

Chapter 3

Working with System Software and File Storage

Chapter Goal

To understand how operating systems and data storage devices contribute to computing productivity

Learning Objectives

- **3.1** Define *software* and identify four types of system software.
- **3.2** Explain the purpose of the system BIOS.
- **3.3** Explain the main functions of an operating system.
- **3.4** Differentiate between types of operating systems.
- **3.5** Describe the major types of utility programs and their purposes.
- **3.6** Explain how software is used to create computer programs.
- **3.7** Explain how file storage works.
- **3.8** Differentiate between types of storage devices and media.
- **3.9** Describe the uses and functions of large-scale storage.

Online Resources
The online course includes additional training and assessment resources.

Tracking Down Tech
Operating Systems and You

3.1 Defining System Software

Watch & Learn
Defining System Software

Chapter 1 described how a computer system consists of both hardware (the physical parts) and software (the programming instructions). Software gives orders, and hardware carries them out.

Application software helps users perform both useful and fun tasks, such as word processing, accounting, and game playing. You buy application software because you want to do the activities it enables you to do. (Application software is covered in detail in Chapter 4.)

System software, on the other hand, exists as a support platform for running application software. System software starts up the computer, keeps it running smoothly, and translates human-language instructions into computer-language instructions. There are four categories of system software:

- **BIOS programs.** A **basic input/output system (BIOS)** is software on a chip on the motherboard that helps the computer start up. Chip-stored software is sometimes called **firmware** because it is hybrid of hardware and software.
- **Operating systems.** An **operating system (OS)** is software that provides the user interface, manages files, runs applications, and communicates with hardware.
- **Utilities.** A **utility program** is software that performs troubleshooting or maintenance tasks that keep the computer running well and protect the computer from security and privacy violations.
- **Translators.** A **translator** is software that translates programming code into instructions (software) the computer will understand.

Operating systems are found not only in personal computers, but also in automobile dashboard computers and other devices.

Chapter 3 Working with System Software and File Storage

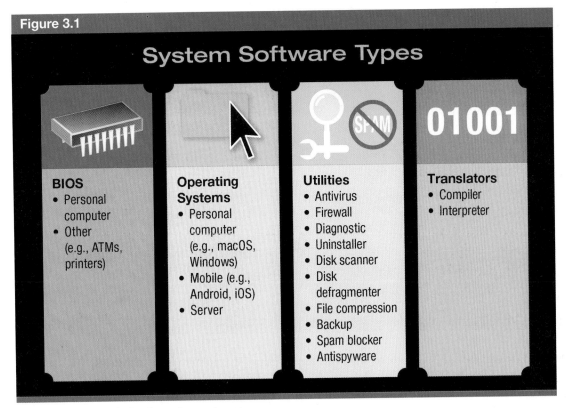

System software includes these four categories.

See Figure 3.1 for examples of each type of system software. You will read about each type in the next several sections of this chapter.

3.2 Understanding the System BIOS

Watch & Learn
Understanding the System BIOS

The low-level programs that help hardware start up and communicate with other pieces of hardware are stored on read-only memory (ROM) chips on the circuit boards. That way, they can never be accidentally erased or changed. This ROM-stored software is known as the BIOS, or sometimes ROM-BIOS. The BIOS for the motherboard is the **system BIOS**. Other devices have BIOSs too. For example, a printer has software that starts it and runs a self-test when the power comes on.

The motherboard's BIOS is able to automatically detect some aspects of the system configuration, such as the amount of random-access memory (RAM) installed, the type of central processing unit (CPU), and the capacity of the hard drives. However, since the motherboard manufacturer can't anticipate all the customization that a given user will perform, the user needs to be able to modify the BIOS settings. This modification is performed through the **BIOS Setup utility**. (This is also sometimes called a CMOS Setup utility because in older systems, the settings were stored on complementary metal-oxide semiconductor chips.) Many newer computers have a Unified Extensible Firmware Interface (UEFI) instead of a BIOS Setup utility. UEFI and BIOS Setup are similar, but UEFI is 64-bit (whereas BIOS is 32-bit) and offers some more advanced disk management and security features.

The exact procedure for entering the BIOS Setup utility varies depending on the PC. As the PC starts, the screen briefly displays a message telling you what key to press to enter Setup. For example, you might see a message like this: *Press F2 for Setup*. The

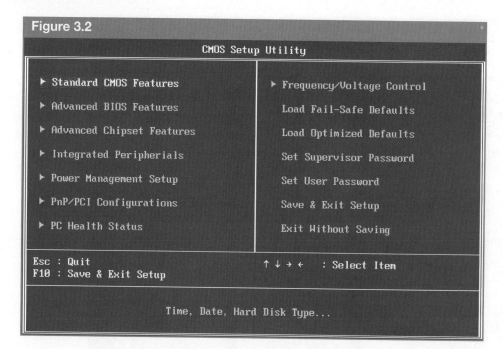

Figure 3.2

This generic BIOS Setup utility screen is typical of many such screens. It displays instructions that provide shortcut keys and navigation guidance.

Hands On
Checking Out Your BIOS Setup

Tech Career Explorer
Working with BIOS Setup

key that's specified could be F1, F2, Delete, or any other key, depending on the manufacturer and version of the BIOS. If you don't press the specified key, the message will go away after several seconds and the PC will proceed to start up normally. If you press the named key while the message is still on the screen, however, the Setup program that's stored on the BIOS chip will appear. On some newer systems (such as those that use UEFI and have Safe Boot enabled), you might not be able to enter the setup program by pressing a key.

Figure 3.2 shows a generic version of a BIOS Setup program. Notice that the program calls itself CMOS Setup Utility, which is the same thing. The program you see on your computer may look very different from this, as the look and interface vary greatly across BIOS manufacturers and models.

Why would you want to enter the BIOS Setup utility and make modifications? Here are some of the most common reasons:

- **To check the system specifications.** You may want to check the system specs to make sure the BIOS is correctly seeing the hardware you have. If you install some new RAM, for example, BIOS Setup will tell you how much it detects, allowing you to see whether the RAM is working. If you are having a problem with a certain disk drive, BIOS Setup will tell you whether the motherboard sees the drive.
- **To change the boot order.** The boot order is the order in which the computer tries the drives to find one with a valid operating system. If you want to boot from a DVD (for example, to run a diagnostic utility) rather than your hard drive, setting the DVD drive as the first boot device will allow the system to do so.
- **To disable certain built-in components.** You may want to disable certain components on the motherboard that you don't use. Disabling these components will prevent memory and other system resources from being wasted by being assigned to them. For example, if your motherboard has a built-in display adapter but you prefer to use a different display adapter installed in one of the motherboard's expansion slots instead, you can turn off the built-in one.

3.3 Understanding What an Operating System Does

> **Watch & Learn**
> Understanding What an Operating System Does

The operating system (OS) is the most important piece of software on a PC. When the computer starts up, the BIOS completes its initial testing and then searches the **boot drive** (a drive that contains a valid operating system). Next, the BIOS passes control to the operating system, which loads itself into memory and completes the startup process. Once loaded, the OS manages the computer and performs a variety of interdependent functions related to input, processing, output, and storage, such as these (see also Figure 3.3):

- booting (starting up) the computer
- providing a user interface
- running programs
- configuring and controlling devices
- managing essential file operations, such as saving and opening files

These five functions are described in greater detail later in this chapter.

Operating systems differ from one another in several important ways. For example, they can be graphics based or text based. They can also support a single user or multiple users at a time, and they can run on different sizes and types of computers.

One defining characteristic of an OS is the hardware it will run on—in other words, its **platform**. The most popular platforms for desktop and laptop PCs are IBM compatible and Macintosh. An **IBM-compatible platform** is based on the same standard as the original IBM PC back in the 1980s, which ran an operating system called MS-DOS. The IBM-compatible platform is also called the **Intel platform**. This platform comes in two versions: 32-bit and 64-bit. The 32-bit version is called **x86** (a reference to the numbering system of Intel CPUs that was used in early computers: 286, 386, and 486). The 64-bit version is called **x64**.

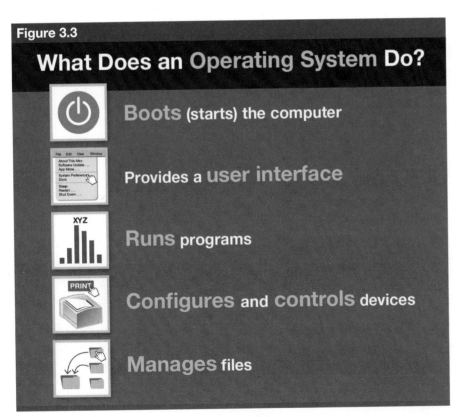

Figure 3.3 What Does an Operating System Do?

Despite the differences across operating systems, they all perform the same five basic functions.

Windows is the most common OS installed on the IBM-compatible platform. Other operating systems, such as Linux and UNIX, also work on the IBM-compatible platform. A platform can also be web based; for example, applications can be written to run via a web interface through a cloud service. See Section 3.4, "Looking at Operating Systems by Computer Type," on page 104, for more details about operating systems and platforms.

Starting Up the Computer

Booting (or starting) a computer after the power has been turned off is referred to as a **cold boot**. Restarting a computer without powering off is a **warm boot**. When a PC starts up, the BIOS software automatically executes its programming. It completes a **power-on self-test (POST)** that checks the essential hardware devices to make sure they are operational. It also looks for a storage device such as a hard drive that contains a usable operating system. When the BIOS software finds one, it transfers control to that operating system. The operating system loads itself and finishes starting up the computer.

The OS loads into memory the system configuration and other necessary operating system files, such as the kernel. The **kernel** manages computer components, peripheral devices, and memory. It also maintains the system clock and loads other operating system and application programs as they are required. The kernel is a **memory resident** part of the operating system, which means that it remains in memory while the computer is in operation. Other parts of the operating system are nonresident; a **nonresident** part of the operating system remains on the hard disk until it is needed. The loaded (memory resident) portion of the operating system contains the most essential instructions for operating the computer, controlling the monitor display, and managing RAM efficiently to increase the computer's overall performance. The process of starting up a computer is illustrated in Figure 3.4.

Practice How a PC Starts Up

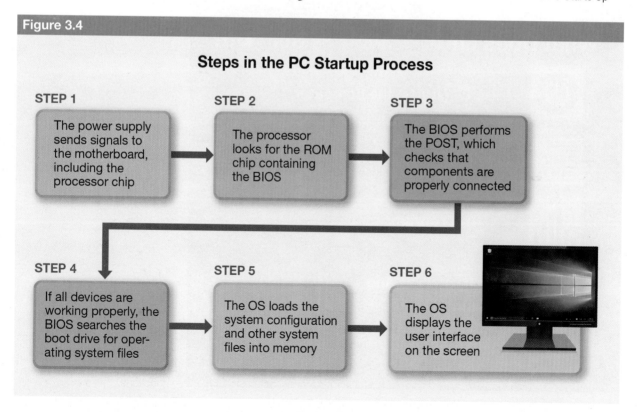

Figure 3.4

Providing a User Interface

Any type of software, including an operating system, contains a **user interface** that allows communication between the software and the user. Another name for a user interface is a **shell**. The user interface controls how data and commands are entered, as well as how information and processing options are presented as output on the screen. There are two main types of user interfaces: text-based and graphical user interface.

Early personal computers, such as the original IBM PC, used an OS called MS-DOS. MS-DOS has a text-based interface, which is also called a **command-line interface**. A command-line interface presents the user with a group of characters called a *command prompt* (for example, *C:\>*). The command prompt indicates the computer is ready to receive a command. In response to the command prompt, the user types a line of code that tells the computer what to do. For example, the command *copy C:\ income.doc D:* instructs the computer to copy a file named income.doc located on drive C to drive D. The command-line interface isn't common anymore, but neither is it obsolete. It's still the native interface for the UNIX and Linux operating systems, for example, but users can choose a graphical shell if they prefer. Windows users see a graphical interface by default. However, they have the option of opening a command prompt window to interact with the operating system through a command-line interface, as shown in Figure 3.5.

Tech Career Explorer
Working with a Command-Line Interface

Figure 3.5

```
C:\>dir /w
 Volume in drive C is OS
 Volume Serial Number is 8043-EB9D

 Directory of C:\

2108FP.TXT            accelmagsetup.log    autoexec.bat
[Books]               config.sys           [dell]
[Drivers]             [Ebook files]        [EMCP]
[Intel]               [IUware Online]      mini-agent.log
mini-agent.txt        [Nancy]              [PerfLogs]
[Program Files]       [Snagit]             [Users]
vcredist_x86.log      [Windows]            [Workgroup Templates]
               7 File(s)      1,593,164 bytes
              14 Dir(s) 273,125,781,504 bytes free

C:\>_
```

A command-line interface for the Windows operating system.

Practical TECH

Command Line: What Is It Good For?

Graphical user interfaces are attractive and easy to figure out, but many computer professionals prefer to use command-line interfaces for some tasks. That's why server operating systems such as Linux make the GUI optional. If you know what you want the computer to do, expressing that command in text form can sometimes be the easiest approach. For example, suppose you have 100 files that all start with *BX (BX001.doc, BX002.doc,* and so on), and you want to rename them so they all begin with *AG* instead (*AG001.doc, AG002.doc,* and so on). In a GUI, you will have to rename each file individually by clicking the name and editing it. But at a command prompt, you can type *REN BX*.doc AG*.doc* and the files will all be renamed at once, saving you at least 30 minutes of work.

This is just one example. Some network-related commands are also available only at the command prompt. For instance, the command PING checks whether a particular network location is reachable, and TRACERT traces the path between your PC and a network location.

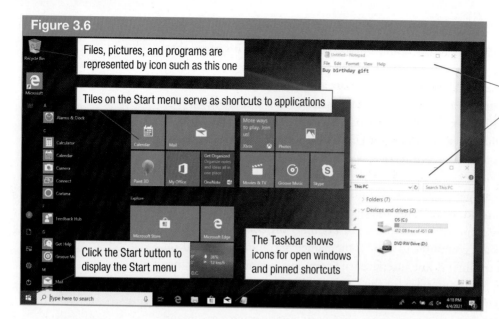

The graphical user interface (GUI) desktop for Windows 10.

A **graphical user interface (GUI)** enables the user to select commands by pointing and clicking with a mouse or other pointing device or by tapping with a finger. With this type of interface, the keyboard is used primarily for data entry. Microsoft Windows and macOS are both GUIs. Figure 3.6 shows the Microsoft Windows 10 desktop. The **desktop** is the background for the GUI—the environment in which all the windows and applications open and close. Files, folders, and programs are represented by small pictures called **icons**. File listings and applications open in **windows**, which are well-defined rectangular areas on the computer screen. Most windows are movable, so you can arrange multiple windows side by side to view them simultaneously. On the Start menu, large rectangular **tiles** serve as shortcuts to applications you can run.

When graphical user interfaces were introduced, almost all of them were based on the desktop metaphor. The user operated a mouse to select and move objects on the desktop, to run programs, and to manage files. However, many portable devices now have touchscreens, and most operating systems are designed to take advantage of that capability.

In user interfaces designed primarily for use via a touchscreen, the graphics have an entirely different look and feel. They are typically based on pages or menus of icons. The user taps an icon to select it. The user can also drag across the screen with a finger to perform various actions, such as scrolling to different pages of the display and shutting down an application. (You read about these actions in Chapter 2.) Figure 3.7 shows the Apple iOS operating system, which is designed for easy use on a touchscreen device such as an iPad tablet or an iPhone.

Hands On
Comparing GUIs and Command-Line Interfaces

Apple iOS operating system on an iPhone.

Running Applications

The main job of any computer is to run applications—that is, programs that perform tasks the user wants to accomplish, whether for work or for entertainment. The user can access applications by browsing the file system for them or by using shortcuts set up on the desktop, on a toolbar, or in the case of Windows, on the Start menu. For example, many of the tiles on the Windows 10 Start menu in Figure 3.6 are shortcuts to applications you can run, such as Calendar, Mail, and Microsoft Edge.

Most operating systems enable you to run multiple applications at a time and switch freely between them. In Windows 10, shown in Figure 3.6 on page 102, you can click an application's button on the Taskbar to switch to that application. In iOS, shown in Figure 3.7, you can browse open apps by swiping up from the bottom of the screen (iPhone X and later) or by pressing and releasing the Home button two times quickly (iPhone 8 and earlier).

Hands On
Opening and Closing Applications

Configuring and Controlling Devices

Configuring and controlling computer components and attached devices is a major function of the OS. Each device speaks its own language, so the OS uses a driver to communicate with each one. A **driver** is a small program that translates commands between the OS and the device. Windows comes with a large collection of drivers for popular device types, such as keyboards, mice, network adapters, and display adapters. Most devices come with their own driver disks as well.

Hands On
Viewing Devices and Drivers

Video
Device Drivers

Managing Files

An OS provides an interface that enables users to work with files and folders (see Figure 3.8). File management includes moving, copying, renaming, and deleting files and folders. In addition, it includes formatting and copying disks and viewing file listings in various ways (for example, sorted by a certain property or filtered to show only files of a certain type).

The name of the file management utility is different not only in different operating systems but even in different versions of the same operating system. In Windows 10, the

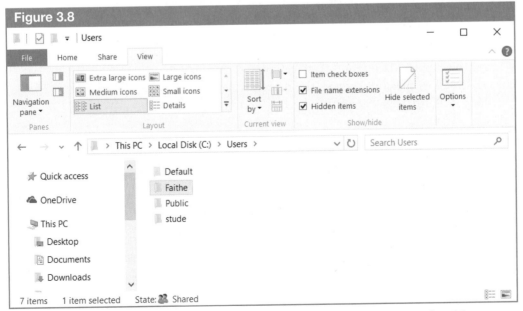

Figure 3.8

The Windows 10 File Explorer displays files and folders in a window. Here you see a list of the operating system's Users folders.

file management utility is called File Explorer; in some earlier versions of Windows it was called Windows Explorer. In macOS, the file management feature is called Finder.

Command-line interfaces (such as UNIX) don't have a separate file management application. The user types file management commands directly into the command prompt.

3.4 Looking at Operating Systems by Computer Type

The previous section described how operating systems can be differentiated by the types of functions they perform. Another way to differentiate OSs is by the types of computers on which they are designed to run. Operating systems are designed for three main types of computers: personal computers, mobile devices, and servers. Each type of computer has its own special needs based on what tasks it performs and who uses it. The following sections look at the most popular OSs for each type of computer.

> **Watch & Learn**
> Looking at Operating Systems by Computer Type

Personal Computer Operating Systems

A personal computer (PC) is one that's designed to be used by one person at a time, such as a desktop or laptop PC. Smaller mobile devices, such as tablets and phones, also qualify as PCs according to this definition.

Microsoft Windows Microsoft Windows is the most popular PC operating system in the world. In 2018, a NetMarketShare survey reported that more than 88.5 percent of all desktop and laptop PCs used some version of Windows. Windows 10 is the most recent version. Earlier versions include Windows 8.1, Windows 8, Windows 7, Windows Vista, and Windows XP.

> **Video**
> The History of Windows

The operating system version can be important for running specific applications, particularly on Windows systems. Some new applications may run only on Windows 10, for example, and some very old applications might have problems running on newer versions of Windows. Windows offers Compatibility mode for troubleshooting version-related application issues.

> **Hands On**
> Exploring Compatibility Mode Settings

As you saw in Figure 3.6 on page 102, the Windows 10 desktop provides a user-friendly environment in which to work. The Taskbar provides access to open applications and windows. The Start button opens the Start menu, which provides access to all installed applications.

macOS macOS is used on Apple Macintosh (Mac) computers. Apple assigns names to differentiate versions of its operating system. The current version of macOS is Mojave. Earlier versions included High Sierra, El Capitan, and Yosemite. Like

Practical TECH

Resetting Your Windows PC

If your Windows desktop or laptop PC is running poorly, reinstalling the OS could make it run much faster. By wiping out everything on the hard drive and starting fresh, you get rid of any unwanted software that you may have inadvertently downloaded or imperfectly removed over the years. If you have Windows 10, you can refresh or reinstall Windows via the Settings app. Click *Update & security*, click *Recovery* in the Navigation pane, and then click the Get Started button in the *Reset this PC* section. Make sure you back up any files you want to keep beforehand.

Chapter 3 Working with System Software and File Storage

Windows, macOS uses a desktop metaphor for its user interface. Figure 3.9 shows a Macintosh desktop. Notice the similarities between this desktop and the Windows desktop in Figure 3.6 on page 102. The functionality of the macOS is also similar to that of the Windows operating system, although the features have different names, as noted in Figure 3.10.

Macintosh computers and the macOS are popular among graphics professionals. In fact, some of the most powerful software for page layout and graphics editing (such as Adobe Photoshop) was originally designed for the Mac. Today, however, much of the same software is available for both Windows and macOS.

Linux Linux (pronounced "LIN-uks") is a UNIX-based operating system that runs on a number of platforms, including Intel-based PCs, servers, and handheld devices. The Linux kernel (that is, the central module) was developed mainly by computer programmer Linus Torvalds, and the name Linux is a combination of Linus and UNIX. UNIX is a command-line operating system for servers that has been popular for many years; this chapter will cover Linux in more detail later.

Video
Linux Timeline

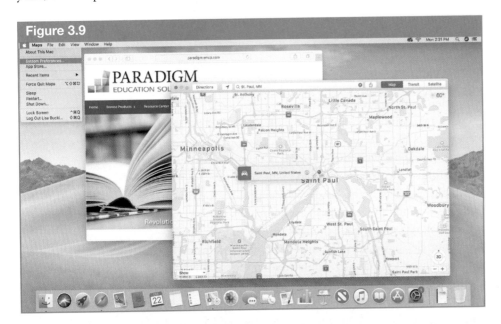

Figure 3.9

The macOS desktop looks similar to the Windows desktop.

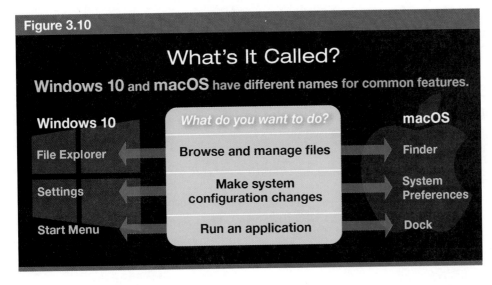

Figure 3.11

The popular Linux distribution includes an attractive GUI.

Article
Open Source Software

Torvalds designed Linux as **open source software**. This means that the developer retains ownership of the original programming code but makes the code available free to the general public. People are encouraged to experiment with the software, make improvements, and share the improvements with the entire user community. Open source software is quite different from **proprietary software**, which is software that an individual or company holds the exclusive rights to develop and sell. Proprietary software uses code and file formats that are designed exclusively for that software; its developers don't standardize it or make its source code available to the general public.

Fans of Linux praise its stability, flexibility, security, and generally low cost. The Linux kernel is free, but vendors usually package it with various tools, utilities, and shells and then charge users for the package. This package is called a **distribution (distro)**. A number of commercially available Linux distros are available for personal and business computers, including Red Hat Linux, Debian, SUSE Linux Enterprise, and Ubuntu Linux. One important feature of most distros is a GUI shell that enables the user to interact with the operating system using a graphical desktop. Figure 3.11 shows the shell for Ubuntu Linux, one of the most popular distributions available today.

Hotspot

OS Simulators

Are you interested in trying out some old operating systems without having to download and install them? Here are links to simulators that you can use with a web browser:

- Windows 1.01 (1985)—https://CUT7.ParadigmEducation.com/Windows1.01
- Mac OS System 7 (1991)—https://CUT7.ParadigmEducation.com/MacOS7
- Windows 3.1 (1992)—https://CUT7.ParadigmEducation.com/Windows3.1

For slightly newer versions of Windows and macOS, check out https://CUT7.ParadigmEducation.com/OSsimulators, where you will find simulators for a dozen or so Windows versions (starting with Windows 95) and macOS versions (starting with Mac OS 8.6).

The popularity of Linux on PCs is growing, especially as the number of software programs available on the Linux platform increases. Word-processing, spreadsheet, and presentation programs are available in an open source format called OpenOffice. Software companies are also developing programs that will allow Windows-based applications to run on Linux-based computers.

Chrome OS Chrome is the name of both an operating system and a web browser, and both are produced by Google. Chrome OS is based on Linux. It's a commercial distribution of an open source project called Chromium OS.

Chrome OS is a simple operating system that was designed primarily for thin clients. A **thin client** is a computer that has only minimal processing capabilities of its own, and is designed primarily to interface with other computers (such as on the internet). A netbook is a small laptop computer that's equipped as a thin client. The most common way to acquire Chrome is to buy a netbook with Chrome preinstalled as its operating system. This configuration is sometimes called a Chromebook.

The Chrome OS was intentionally designed to have few features. It includes the Google Chrome web browser, of course, and a simple file manager and media player. Having a smaller, leaner operating system on a computer that performs only simple tasks can be an advantage. It can start up, shut down, and respond to commands more quickly than a more full-featured operating system.

Mobile Operating Systems

As computing devices have gotten smaller, what they need in an operating system has changed. As you read in the previous section about the Chrome OS, some thin clients designed mostly for web use don't need the full-featured power of an operating system such as Microsoft Windows or macOS. In fact, they run better and faster with a smaller, more agile OS.

A mobile device such as a tablet or smartphone typically doesn't include a mechanical hard drive. Instead, it uses solid-state memory (similar to the memory in a USB flash drive) to store both the operating system and any data files the user might save.

Cutting Edge

Fuchsia OS

Someday, a single operating system may run every device in your home, from your refrigerator to your phone, and your television to your desktop PC. Which OS will it be? It's hard to imagine any of the OSs available today working well across such a wide variety of devices. Windows is too bulky and mouse-reliant, and Apple won't license macOS or iOS to run on third-party hardware. Android and Chrome are possibilities, but they weren't designed to be all-purpose and would need to be redesigned at a base level.

Google's Fuchsia OS, a new open-source OS project under development, may be a strong candidate for the job. Fuchsia's main appeal is that it's designed specifically to be a *universal OS*, running on a wide variety of devices, including embedded systems, tablets, smartphones, laptops, and desktop PCs. It is a *lightweight OS*, meaning it requires very little in terms of CPU and memory, so even an inexpensive device with not much computing power can use it.

In addition, Fuchsia is a *capability-based OS*, meaning it is able to communicate securely between mutually untrusting devices. That's ideal for an Internet of Things (IoT) environment, where devices need to communicate with one another well enough to accomplish their tasks without being able to harm one another by stealing data or spreading malware. When an OS is designed from the ground up to handle security concerns, it does so more efficiently than when security is an afterthought built on top of the basic kernel.

Figure 3.12 The Apple iPad and iPhone use the iOS operating system.

Solid-state memory and storage is expensive, so having a smaller operating system means more memory will be available for user data.

Most mobile devices have a touch-sensitive screen (rather than a keyboard and mouse) as the primary user interface. As a result, the operating system's user interface relies heavily on tapping onscreen icons or tiles and performing various swipes, drags, stretches, and pinches with the fingers.

iOS iOS is the operating system used in Apple mobile devices, such as iPhones and iPads. Apple doesn't license this operating system for any third-party devices, so you will find it only on Apple hardware. Figure 3.12 shows an iPad running iOS Version 9. To open an application in this interface, you tap its icon.

One benefit of iOS is that hundreds of thousands of applications (apps) are available for it—some for free and others for small fees. You can extend the basic default capabilities of your Apple device by downloading apps from the Apple Store. Use the App Store icon to access the App Store app, and from there, browse the available apps by category or keyword.

Android Android is an open source OS created by Google. Android is commonly used on smartphones and tablets because it's simple and easy to use, and a large number of apps are available for it. Because the Android OS is free, devices that run it are often less expensive than iOS-based devices. Figure 3.13 shows an Android-based smartphone screen.

Figure 3.13 The Android OS on a smartphone.

Other Smartphone Operating Systems

Blackberry 10 is the latest OS for Blackberry smartphones. Blackberry phones differ from most modern smartphones in that they may have a hardware keyboard built in, as shown in Figure 3.14. Blackberry devices were very popular in the early 2000s, before touchscreen phones became the norm, but are less common now.

Windows Phone OS was the operating system for smartphones based on Windows 8 (and later, Windows 10), used on many Nokia smartphones from 2009 through 2016. Windows Phone OS never gained widespread adoption, partly because of the scarcity of apps available, and Microsoft discontinued the Windows Phone OS in 2017.

Figure 3.14 BlackBerry is one of the few smartphones available with a hardware keyboard.

© 2018 BlackBerry Limited. All Rights Reserved. Used with permission.

Server Operating Systems

Some operating systems are designed specifically for use with network servers. These OSs allow multiple users to connect to the server and share network resources, such as files and printers. A server version of an operating system contains many special features that aren't included in a PC operating system—for example, software for managing connections, authenticating users, prioritizing requests from multiple users, and providing multiuser access to databases. Some servers are specialized and provide a certain type of service on the network, such as database server, file server, mail server, print server, or internet server.

Article
Server Operating Systems

The three most popular server OSs are UNIX, Linux, and Windows Server. Each of these operating systems actually comes in many variants:
- UNIX servers include BSD, HP-UX, Aix, and Solaris
- popular Linux distros designed for server use include Debian, CenOS, and RHEL
- Windows Server comes in many different versions

Some server OSs don't have a graphical user interface. Many IT professionals prefer to use a command prompt when managing a server. UNIX and Linux, for example, don't include a GUI as part of the main kernel, although a GUI can be added to Linux if desired. Windows Server, on the other hand, is mainly administered through a GUI, although a command prompt is available when needed.

Of the three main server operating systems, **UNIX** (pronounced "YOO-niks") has been in existence the longest and is the only one that was developed from the ground up for servers. UNIX was created in the 1970s by programmers at Bell Laboratories to run servers and large computer systems. It uses a complex command-line interface and offers some superb capabilities, including allowing simultaneous access by many users to a single powerful computer. From its inception, UNIX has been a **multiuser operating system**: one that allows many people to use one CPU from remote stations. UNIX is also a **cross-platform operating system**: one that runs on computers of all kinds, from PCs to supercomputers. Because UNIX was used on most servers at universities and laboratories, most early internet activities were UNIX-based, and many internet service providers (ISPs) continue to use UNIX for their servers.

Linux is a variant of UNIX, although it has developed so much on its own that it's now considered a separate operating system. Because Linux is open source, many variants of this OS are available for server use. One of the most popular server configurations is a **solution stack**, which is a set of complementary applications. It's called LAMP, which is an acronym of the features it includes: the Linux kernel; a web server such as Apache; a database such as MariaDB, MySQL, or MongoDB; and CGI scripting tools such as Perl, PHP, and/or Python. Most Linux distributions include all these tools and many more like them, so any user with the right knowledge and skills can set up a Linux server without buying any commercial software.

Windows Server is a variant of Microsoft Windows. Microsoft has released many versions of Windows Server since the original Windows NT 3.1 Advanced Server in 1993. Each version has provided a GUI that resembles the equivalent desktop Windows version; the Windows Server 2019 GUI interface closely resembles Windows 10, for example. Windows Server versions are available in several editions. Each targets the needs of a specific size of company, such as Essentials (for small companies), Standard (for medium-sized companies), and Datacenter (for large companies that maintain a lot of data). Windows Server can be configured to run in Server Core mode (command-line interface) or in Server with a GUI mode (GUI interface).

Servers like this one use operating systems such as UNIX, Linux, and Windows Server.

3.5 Exploring Types of Utility Programs

A utility program performs a maintenance or repair task, such as checking for viruses, uninstalling programs, or deleting data that's no longer needed. An operating system typically includes several utility programs that are preinstalled at the factory. Several companies, including Symantec and McAfee, produce software suites that contain a variety of utility programs to monitor and improve system health and functionality.

Utility programs are useful for avoiding and correcting many of the problems that computer users are likely to encounter. These programs are usually stored on the hard drive, along with the basic operating system, and activated when needed by the user. Table 3.1 lists popular kinds of utility programs, some of which are described in detail in the following pages.

Watch & Learn
Exploring Types of Utility Programs

Table 3.1 Utility Programs and Their Functions

Utility Program	Function
Antivirus	Protects the computer system from a virus attack
Firewall	Protects a personal computer or network from access by unauthorized users, such as hackers
Diagnostic	Examines the computer system and corrects problems that are identified
Uninstaller	Removes programs, along with related system files
Disk scanner	Identifies and fixes errors in the file storage system
Disk optimizer (defragmenter)	Identifies disk problems (such as separated files) and rearranges files so they run faster
Disk toolkit	Recovers lost files and repairs files that may be damaged
File compression	Reduces the sizes of files so they take up less disk space
Backup	Makes a backup copy of files on a separate disk
Spam blocker	Filters incoming spam messages

continues...

Table 3.1 Utility Programs and Their Functions ...*continued*

Utility Program	Function
Antispyware	Protects the computer system from software that tracks the activities of internet users
Device driver	Allows hardware devices (such as disk drives and printers) to work with the computer system
Extender	Adds new programs and fonts to the computer system
File viewer	Quickly displays the contents of a file
Screen capture	Captures as a file the content shown on the monitor

Antivirus Software

Antivirus software (also called a *virus checker*) is one of the most important types of utility programs. Examples include Norton AntiVirus and McAfee AntiVirus Plus; Windows 7 and higher also includes basic antivirus protection in the Windows Defender utility.

A virus is harm-causing code that's buried within a computer program; it's transferred to a computer system without the user's knowledge when he or she runs the program to which the virus has attached itself. A worm is a variation that doesn't require running an **executable file** (a file that runs a program) to infect; it's spread through networks and email systems or by visiting an infected website.

Virus contamination of a computer system can have consequences that range in severity from being mildly annoying to disastrous. Antivirus utilities search for viruses, worms, and other **malware** (malicious software) on a system and delete or quarantine them when they are found. Antivirus utilities can also monitor system operations for suspicious activities (such as the rewriting of system resource files) and alert users when they are occurring.

Video
Antivirus Software

Firewalls

A **firewall** is a security system that acts as a boundary to protect a computer or network from unauthorized access (see Figure 3.15). It's designed to work like the firewalls

Practice
How a Firewall Works

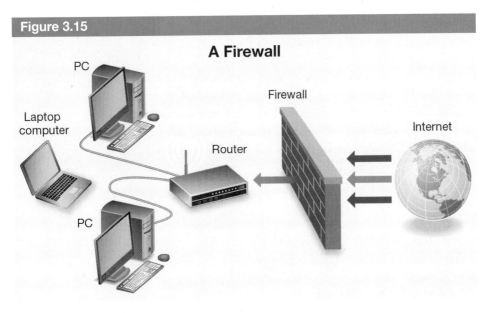
Figure 3.15
A Firewall

A firewall is designed to prevent unauthorized users from accessing a PC or a network connected to the internet.

between individual units in an apartment building, whose purpose is to prevent fire from spreading from one apartment to another. A computer firewall may consist of hardware, software, or a combination of the two.

> Hands On
> Examining Firewall Settings

The data traffic between your computer and the internet may carry information for multiple programs, including web browsers and email programs. Your PC keeps this data organized by assigning different port numbers to different kinds of traffic. For example, port 110 is usually used for incoming mail messages. There are many ports, and most of them are unused most of the time. If an intruder knows that a particular port number is available, he or she might be able to use it to gain access to your computer and steal passwords and personal data. A firewall prevents such intrusions by allowing port access only to the specific users and applications you set up.

A **personal firewall** is a software-based system designed to protect a PC from unauthorized users attempting to access other computers through an internet connection. Symantec and McAfee both make personal firewall software, both as stand-alone products and as part of security suites. Windows also includes a personal firewall called Windows Defender Firewall that's enabled by default and adequate for most PCs.

A **network firewall** typically consists of a combination of hardware and software. In addition to installing firewall software, a company may add a hardware device (such as a dedicated firewall device or **proxy server**) that screens all communications entering and leaving networked computers to prevent unauthorized access. For example, the device or server may check an incoming message to determine whether it's from an authorized user. If the message is not from an authorized user, it is blocked from entering the network.

A firewall provides a first line of defense against unauthorized access and intrusion. Although most firewall systems are effective, users should practice other security measures as well, such as safeguarding passwords.

Diagnostic Utilities

A **diagnostic utility** analyzes a computer's components and system software programs and creates a report that identifies the problems found. The utility also provides suggestions for correcting these problems, and in some situations, it can repair problems automatically. The Windows operating system contains many different troubleshooter utilities; you can access them from the Help system. More advanced diagnostic utility software can be purchased separately from software vendors.

Uninstallers

An **uninstaller** is a utility program for removing (deleting) software programs and associated entries in the system files. When an application program is installed, the operating system stores additional files related to the program. Those files may remain on the hard disk and waste valuable space if you remove a program without using an uninstaller utility. An uninstaller utility locates these additional files and removes them along with the program, freeing up valuable disk space.

Disk Scanners

A **disk scanner** examines the hard disk and its contents to identify potential problems, such as physically bad spots and errors in the table of contents. For example, a disk scanner checks to make sure that each sector of data is claimed by only one file. Check Disk is the disk scanner utility included with Microsoft Windows. Disk scanners are useful only for mechanical hard drives, not for solid-state drives.

> Hands On
> Checking a Disk for Errors

Disk Defragmenters

A **disk defragmenter** utility scans the hard disk and reorganizes files and unused space; doing this allows the operating system to locate and access files and data more quickly. The operating system stores a file in the first available sector on a disk, but sometimes there isn't enough space in one sector to store the entire file. If the sector already contains data, then the remaining portions of the file will be stored in other available sectors. Splitting up and storing the file like this may result in portions being stored in **noncontiguous sectors** (that is, sectors that are not connected or adjacent). A fragmented file takes longer to load because the operating system must locate and retrieve all the various pieces of the file.

This problem can be solved by defragmenting the disk so files are stored in contiguous sectors. Microsoft Windows includes a disk defragmenting utility called Optimize Drives. If your operating system doesn't come equipped with a disk defragmenter, you can purchase a utilities package that contains one. Defragmenting should be done only on mechanical hard disk drives; solid-state drives don't benefit from it.

File Compression Utilities

A **file compression utility** compresses (or shrinks) the size of a file so it occupies less disk space; some types of this utility also combine multiple files into a single compressed file for easier transfer. Files are compressed by reducing redundancies (that is, instances of repeated bits). For example, suppose that the text in a file has 28 zero (0) bits in a row. Rather than write this sequence as 28 separate digits, a file compression program might rewrite it as 28 × 0—a space savings of about 85 percent. When the file is decompressed, the long string of 0 bits is restored.

One of the most popular extensions for compressed files is .zip; thus, the term *zip file* has come to mean a compressed archive file. Windows supports the zip file format for compressed archives and can open zip files as if they are folders. Third-party compression utilities are also available with more features, such as WinZip and StuffIt.

Backup Utilities

A **backup utility** allows the user to make copies of the contents of storage media. The utility can be directed to back up the entire contents or only selected files. Some backup utilities compress files so they take up less space than the original files. Because compressed files are unusable until they have been uncompressed, many backup utilities include a restore program for uncompressing files.

You don't have to use a backup utility to back up files. You can manually copy them to a backup location using File Explorer or whatever file management utility is provided by your operating system. For individual PCs that don't store critical data, this informal type of backup may be all you need. Businesses, on the other hand, may find backup utilities useful in creating automated backups at regular intervals.

Another way to back up your files is to save them to an online cloud location, such as Microsoft OneDrive, iCloud, or Google Drive. Cloud storage is discussed later in this chapter.

Spam Blockers

Unwanted commercial email is known as **spam**, or *junk mail*. According to Symantec, a company that blocks spam for some of the United States' top internet service providers, spam now accounts for roughly 55 percent of all internet email traffic. A utility program called a **spam blocker** is often used to filter incoming spam messages. Most email programs include a basic spam blocking utility, and third-party programs are also available with more features and more intelligent methods of differentiating spam from legitimate email.

Antispyware

Spyware is a form of malware that tracks the activities of an internet user for the benefit of a third party. Spyware is secretly downloaded to the user's computer to collect keystrokes or trace website activity for malicious purposes, such as password interception, fraudulent credit card use, and identity theft. One particular type of spyware, called **adware**, is more annoying than harmful. Adware tracks the websites that a user visits to collect information for marketing and advertising companies. Some adware presents users with pop-up advertisements that contain contests, games, or links to unrelated websites.

Not all antivirus software protects internet users from spyware, so users should consider getting separate antispyware and adware protection software. The utility Windows Defender, which comes with Microsoft Windows, contains both antimalware and antivirus protection. Third-party products are also available, such as adaware, ad block, and Webroot SecureAnywhere.

3.6 Programming Translation Software

Watch & Learn
Programming Translation Software

The final type of system software we will look at falls under the broad heading of *translators*. A translator converts programming code to machine language. Translators are needed because computers can't understand programming code written in a human language, such as English or Spanish. They understand only binary code written in zeroes (0s) and ones (1s), which is sometimes called **machine language**. (The concept of binary code was explained in Chapter 2.) Machine language is considered a **low-level language**.

Machine language is difficult to learn, and programmers find that writing machine language programs is boring and time consuming. To get around these problems, application programs are usually written using an English-like programming language called a **high-level language**. Examples of high-level languages are COBOL, Java, and BASIC (which has several versions). Figure 3.16 shows a sample of programming code in DOS BASIC.

A high-level language must be translated into machine language format before the CPU can execute it. The two major types of language-translating software are interpreters and compilers. Each programming language generates code that needs to be compiled, or interpreted for execution.

A **compiler** translates an entire program into machine language once and then saves it in a file that can be reused over and over. Each programming language has its own compiler. After reading and translating the program, the compiler displays a list of errors it encountered in the compiling process. The programmer corrects the errors and recompiles the program until there are no more errors. Compiling it creates an executable file, which usually has a .com or .exe extension. Most programming languages are compiled, including C++ and Visual Basic.

Some programming languages are interpreted, not compiled. An **interpreter** reads, translates, and executes one line of instruction at a time, and identifies errors as they are encountered. In this sense, interpreters are somewhat more user friendly than compilers. An interpreted program must be re-interpreted each time it's run; it doesn't become an executable file. JavaScript is an example of an interpreted programming

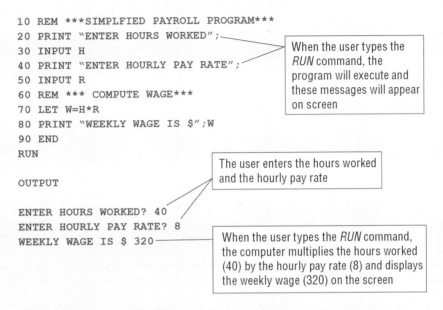

Figure 3.16

A DOS BASIC Payroll Program

This simple payroll program is written in DOS BASIC, a high-level language.

language. JavaScript code exists in the web page file in human-readable form; you can open a web page in a text-editing program and see it. The web browser serves as the interpreter, running the JavaScript code when it loads the page.

Compiled programs run faster than interpreted programs, because they don't have to be reprocessed each time they are run. For this reason, most applications are written in compiled languages, not interpreted ones. You will read much more about programming languages, both compiled and interpreted, in Chapter 11.

3.7 Working with File Storage Systems

Watch & Learn
Working with File Storage Systems

As you read in Chapter 2, a computer system consists of input, processing, output, and storage devices. Now that you know something about how computers use system software to perform the first three functions, let's take a look at the last function: storage.

How File Storage Works

Information is stored on your computer in files and folders that are saved on a drive—either a disk drive or a solid-state drive. A **disk drive** is a mechanical device that reads data from and writes data to a disk, such as a hard disk or DVD. A **solid-state drive (SSD)** is a device that stores data in nonvolatile memory, which is memory that persists even if the power is turned off. A familiar example of a solid-state drive is a USB flash drive. (Storage devices and media are discussed in more detail in section 3.8, "Understanding File Storage Devices and Media," on page 119.)

Each physical drive has one or more logical volumes. A **volume** is a storage unit with a letter assigned to it, such as *C* or *D*. Some drives are set up to divide the available space into multiple volumes; other drives allow the entire storage area to be addressed as a single volume letter.

Within each volume, the top level of the storage hierarchy is the **root directory**. For example, if you display the contents of drive C in your file management utility, the files and folders that appear are located in the root directory of the C drive. The root directory is like the lobby of a building: it provides an entryway to the disk, but you don't want to store much in it because it will get crowded in there (metaphorically, anyway).

Most files are stored in folders. A **folder** is an organizing unit, like a box or a physical file folder. A volume can have hundreds or even thousands of folders. For example, a volume on which Windows is installed may have folders named *Windows*, *Program Files*, and *Users*, among others. A folder that is located within another folder is called a **child folder** (or *subfolder*), and a folder that contains a child folder is called a **parent folder**.

One advantage of a graphical user interface over a command-line interface is that it allows you to see the folder structure more easily. To see this for yourself, compare Figures 3.17 and 3.18. With a graphical user interface, you can browse different folders by clicking or double-clicking their icons onscreen. With a command-line interface, however, you must type a command to display the contents of a particular folder.

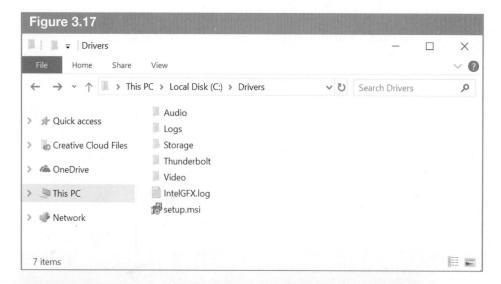

This screen capture shows the contents of a folder called Drivers as it appears in a GUI Windows 10. Notice that the Drivers parent folder contains five child folders and two files.

This screen capture shows the same contents displayed in Figure 3.17 but as it appears in a command-line interface.

Chapter 3 Working with System Software and File Storage

The **path** to a file provides the information you need to locate the file in the storage medium. It includes the letter of the volume in which the file is located followed by a colon, the names of the folders to go through to get to the file, and the name of the file. Each location name is followed by a backslash (\). For example, the Notes.txt file shown in Figures 3.17 and 3.18 has a path of C:\Drivers\Notes.txt. Figure 3.19 identifies the individual pieces of the path.

A **folder tree** is often helpful in understanding the folder structure of a volume. In Windows 10, you can browse a folder tree in the Navigation pane on the left side of the File Explorer window. Figure 3.20 shows the path C:\Drivers\audio\R283235\W7-32 in a folder tree. Notice that you don't see any files in the folder tree—only folders.

Figure 3.19

The file path is the same whether you view the file through a GUI or a command-line interface.

File Types and Extensions

There are hundreds of file types, and operating systems generally handle them differently. How does the OS know what type a file is? It depends on the OS. Some OSs examine a file's content to determine its type, but most rely on the file's extension. A **file extension** is a short code (usually, three or four characters) that appears at the end of the file name, separated from the name by a period. In the file name *Notes.txt*, for example, *Notes* is the file name and *.txt* is the file extension.

Microsoft Windows uses extensions to determine if a file is a system file, an executable file that runs an application, a helper file for an application, or a data file (and if so, what type). Windows maintains an internal database containing the **file association** for each file extension. File association defines a default application and/or action. You can customize the entries in this database, so that a particular type of file opens in a different application than the one assigned by default. Doing this is useful if you have multiple applications that will open a certain file type and you prefer one over the others. For example, suppose both Windows Media Player and Apple iTunes are installed on your computer. Windows Media Player is the default application for MP3 files, but you can change the file association to Apple iTunes. Table 3.2 lists some of the most common file extensions and applications that might be set up to open those files.

Figure 3.20

Here you see the tree structure in the Navigation pane of the File Explorer in Windows 10.

Table 3.2 Common File Extensions and Associated Applications

File Extension	File Type	Associated Applications
.app .bat .com .exe	Executable file	
.doc .docm .docx	Word-processing document	Microsoft Word WordPad
.pdf	Portable Document Format	Adobe Acrobat Adobe Reader
.txt	Plain text file	Notepad

continues…

Table 3.2 Common File Extensions and Associated Applications ...*continued*

File Extension	File Type	Associated Applications
.rtf	Rich text format (a generic word-processing format)	Microsoft Word WordPad
.bmp .gif .jpg .pcx .png .tif	Graphics files	Adobe Photoshop Paint Photos (Windows app) Windows Photo Viewer
.ppt .pptm .pptx	Presentation	Microsoft PowerPoint
.xls .xlsm .xlsx	Spreadsheet	Microsoft Excel
.au .mp3 .wma	Audio clips	Apple iTunes Groove Music Windows Media Player
.m4v .mod .mov .mp4 .qt .wmv	Video clips	Apple QuickTime Movies & TV (Windows app) VLC Media Player Windows Media Player
.htm .html .mht	Web pages	Firefox Google Chrome Microsoft Edge Microsoft Internet Explorer
.accdb .mdb	Access database	Microsoft Access

Files and Shortcuts

A file can't be in two places at once, but a shortcut makes it seem that way. A **shortcut** is an icon that points to a file and that you can place in a handy location, such as on your computer desktop or on the Windows taskbar. Any time you want to use the file, you can double-click the shortcut to access it, rather than having to navigate to the file's actual location and double-click the file itself. The shortcut looks like the icon for the file to which it points except that most shortcuts have a small arrow in the lower left corner, as shown in Figure 3.21.

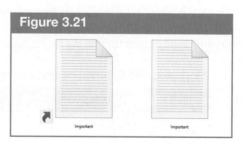

Figure 3.21

The shortcut (on the left) includes an arrow, indicating it points to a file. The icon for a file (on the right) does not include an arrow.

Basic File Management Skills

File management refers to manipulating files using the operating system interface. These manipulations can include editing, saving, moving, copying, renaming, and deleting files. Every operating system has its own commands and methods for file management, but the procedures are similar among most GUI operating systems.

Table 3.3 describes methods for accomplishing basic file management tasks in Windows. Note that some of the methods include keyboard shortcuts, such as *Ctrl + X*. Keyboard shortcuts are also known as *hot keys*. All desktop operating systems have dozens of hot keys; check the Help system for your OS to get a list of hot keys available to you. Most applications also display commands for accomplishing tasks as buttons and options in ribbons and menus. You can use those commands instead of hot keys.

Table 3.3 Managing Files in Windows 10

Task	Method
Open a file	Double-click the file.
Edit a file (data files only)	Open the file and then make changes to the data.
Save a file	Press Ctrl + S.
Cut a file to the Clipboard	Select the file and then press Ctrl + X.
Copy a file to the Clipboard	Select the file and then press Ctrl + C.
Paste a file from the Clipboard	Display the destination location and then press Ctrl + V.
Move a file	Select the file, press Ctrl + X, display the destination location, and then press Ctrl + V.
Delete a file	Select the file and then press Delete.
Rename a file	Select the file, press F2, type the new name, and then press Enter.
Search for a file	Click in the search box in the upper right corner of the File Explorer window, type the search text, and then press Enter.
Sort a file listing	Right-click the file listing, click the *Sort By* option at the shortcut menu, and then select the sort type desired.
See a file's size	Right-click the file, click the *Properties* option at the shortcut menu, and then note the file size on the General tab.
Change a file's permissions (simple)	Right-click the file, click the *Properties* option at the shortcut menu, click the *Read-only* check box at the Properties dialog box to insert or remove a check mark, and then click the OK button.
Change a file's permissions (advanced)	Right-click the file, click the *Properties* option at the shortcut menu, click the Security tab, and then set up permissions for different users and groups.

3.8 Understanding File Storage Devices and Media

Watch & Learn
Understanding File Storage Devices and Media

The computer's RAM is **primary storage**, because it's where the data is placed immediately after it's input or processed. Primary storage is by nature temporary. **Secondary storage**, on the other hand, is permanent. Secondary storage devices include all the devices represented by volume letters in your computer's file storage system: hard drives, removable flash drives, CD and DVD drives, and so on.

There are two types of secondary storage: fixed and removable. **Fixed storage** is mounted inside the computer; to get it out, you would have to open up the computer's case. An internal hard disk drive is an example of fixed storage. **Removable storage** can easily be separated from the computer. CDs, DVDs, external hard drives, and USB flash drives are all examples of removable storage. External storage devices can use a variety of different interfaces. USB is the most common, followed by FireWire (IEEE 1394a) and eSATA (external SATA). (Refer to the "Ports" section in Chapter 2 for more information about these different types of ports.)

Some storage is disk based—that is, the data is stored on disk platters that rotate past a read/write head. Hard disk drives (HDDs), CD drives, and DVD drives are all disk based. Other storage is solid state—that is, the data is stored on nonvolatile memory chips. USB flash drives and solid-state drives (SSDs) are examples.

Finally, some storage media can be separated from its drive and some cannot. For example, in an HDD or SSD, the storage media and the means of reading it are permanently sealed together in a metal cartridge. In contrast, a CD or DVD can be popped out of the drive that reads it.

The following sections examine the technologies behind several storage device types and weigh their pros and cons.

Hard Disk Drives

A **hard disk drive (HDD)** is the traditional type of mechanical hard drive that has dominated the computer market for several decades. Most desktop and laptop computers have at least one HDD because the hard disk drive is the most economical form of high-capacity storage for personal computers.

An HDD consists of one or more rigid metal platters (disks) mounted on a spindle in a metal box with a set of read/write heads—one for each side of each platter. The box is sealed to prevent contamination from dust, moisture, and other airborne particles. Storage capacity ranges from a few hundred gigabytes to more than 10 terabytes.

The drive mechanism is integrated with the platters; you can't physically separate the drive from the platters without destroying the drive. For this reason, the terms *hard disk* and *hard drive* are mostly synonymous; there is little reason to refer to one separately from the other, since they are never apart. But technically, a hard disk is a platter and a hard drive is the device consisting of the mechanical components that spin a set of platters and read from and write to the platters. Figure 3.22 shows the inner workings of a hard drive.

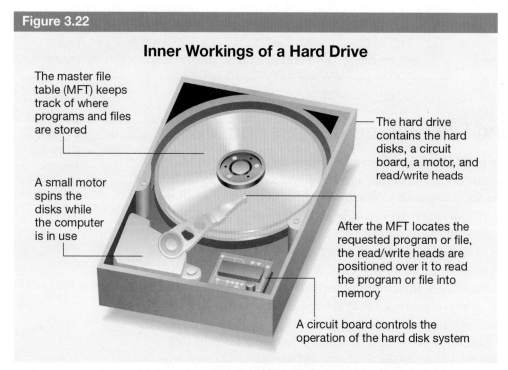

Figure 3.22

Inner Workings of a Hard Drive

- The master file table (MFT) keeps track of where programs and files are stored
- A small motor spins the disks while the computer is in use
- The hard drive contains the hard disks, a circuit board, a motor, and read/write heads
- After the MFT locates the requested program or file, the read/write heads are positioned over it to read the program or file into memory
- A circuit board controls the operation of the hard disk system

A hard drive contains one or more hard disks on which data is stored. When activated, read/write heads move in and out between the disks to record and/or read data.

Data is stored along the tracks and sectors of hard disks. A **track** is a numbered, concentric ring or circle on a hard disk. A **sector** is a numbered section of a track. One sector holds 512 bytes of data. Because most files are larger than 512 bytes, a disk is organized logically into groups of sectors called **clusters**. The OS addresses clusters, rather than individual sectors, to keep the disk's list of content at a manageable size.

The disk's table of contents is called the **master file table (MFT)** or **file allocation table (FAT)**, depending on the file system in use. The file system used in modern versions of Windows—called the **New Technology File System (NTFS)**—uses a master file table.

As the platters spin, they rotate past the read/write heads. The heads are attached to a metal arm that moves in and out to position the heads so that specific tracks pass underneath them. When the user wants to access a particular file, the computer determines its location on the disk and tells the hard drive to retrieve it.

Data is stored magnetically on the surfaces of the platters. **Magnetic storage** creates transitions by magnetizing areas of the disk with a positive or negative polarity. The transitions between positive and negative areas are read as *1*s, and the areas that lack transitions are read as *0*s. Although the read/write head is a single unit, it actually performs two different functions: it magnetizes areas of the disk, and it reads changes in polarity and relays them to the drive controller. The drive controller specifies whether the head should read or write at any given moment.

The disk surface is coated with a thin layer of iron oxide particles. The read/write head has a wire coil around it, and electricity passes through the coil. This generates a magnetic field, which polarizes the surface of the disk as positive or negative. The flow of electricity through the wire then reverses, changing the polarization and creating a transition point on the disk, as shown in Figure 3.23. This is known as a **flux transition**.

An HDD spins continuously while the computer is on, whereas a CD or DVD in its drive spins only when data is being stored or accessed. The continuous spinning of a hard disk provides faster access to data, because there's no wait while the drive gets up to speed.

One measurement of HDD performance is **average access time**—that is, the average amount of time between the operating system requesting a file and the HDD delivering it. A key factor involved in access time is rotational speed, which is measured in revolutions per minute (rpm). An HDD running at 7,200 rpm will likely have a faster access time than one running at 5,400 rpm, for example. The faster the platters rotate, the quicker a certain sector can be placed under the read/write head. Another key factor in access time is the speed and accuracy at which the actuator arms can move the read/

Figure 3.23

Data Storage on a Magnetic Disk

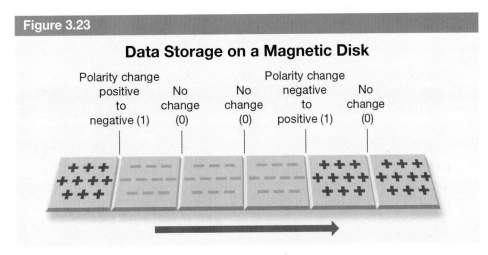

The transition on a magnetic disk between areas of positive polarity and negative polarity.

write heads. This type of drive is known as Integrated Drive Electronics (IDE) because the media on which data is written and the controller are physically inseparable.

Another measurement of HDD performance is **data transfer rate**, which is the speed at which data can be moved from the HDD to the motherboard and then on to the CPU. Data transfer rate is dependent on the interface used to connect the HDD to the motherboard. The traditional interface for many years was **parallel ATA (PATA)**; it uses a 40-pin ribbon cable between the HDD and the motherboard. That interface had a maximum transfer rate of 133 megabytes per second (MB/s). The modern interface is **serial ATA (SATA)**; it transfers data at up to 600 MB/s. (HDDs are limited to about 200 MB/s, but solid-state drives that use SATA as the interface can achieve higher speeds.)

Solid-State Drives and USB Flash Drives

A solid-state drive (SSD) stores data in a type of electrically erasable programmable ROM (EEPROM). As you may recall from Chapter 2, EEPROM can be erased and written with electricity. EEPROM is a nonvolatile type of memory, which means it retains whatever you put in it—even when the power is off. Thus, in a solid-state drive, electricity controls the on/off state of the transistors within the semiconductor material inside the memory chips. A solid-state drive has no moving parts; the term *drive* is used metaphorically.

USB flash drives use solid-state storage, and so do solid-state hard drives. Solid-state hard drives are solid-state equivalents of HDDs in terms of storage capacity and physical size of the box.

Network and Cloud Drives

Not all storage is local—that is, directly connected to your computer. If you are connected to a network or the internet, you may have other storage options in addition to your own volumes.

A **network share** is a drive or folder that's been made available to users on computers other than the one on which the content actually resides. You can

Practical TECH

How Fast Is *Your* SSD?

How much performance improvement can you expect when moving from HDD to SSD? Quite a bit. For example, on a system that boots up Windows in 40 seconds from an HDD, a boot time of about 22 seconds can be expected from an SSD. A large Excel workbook that takes 14 seconds to open from an HDD can open in about 4 seconds from an SSD. People who work with large files will notice a big difference when moving from HDD to SSD, and so will people who play graphics-intensive games. People who mostly use computers to check email and surf the internet won't notice as big a difference, since these activities are not very storage intensive.

Cutting Edge

Hybrid Drives

SSDs are fast and quiet, whereas traditional HDDs are inexpensive and high capacity. So, why not have the best of both! A dual-drive hybrid system combines two separate drives in the same computer: one SSD and one HDD. You can store your most frequently used files (such as operating system files) on the SSD and use the HDD for less frequently accessed files (such as data files). In addition, some operating systems can combine the two drives into a hybrid volume to automatically optimize the computer's usage of each.

Some hard drive manufacturers also produce a *solid-state hybrid drive (SSHD)*. It incorporates solid-state memory into a modified version of a traditional HDD, and the result is a single drive that merges the capabilities of the two types of drives. The most frequently accessed data is stored in the solid-state portion, and the less frequently accessed data is placed on the mechanical hard disk platters. A hybrid drive boosts disk performance without the expense of a full-sized solid state hard drive.

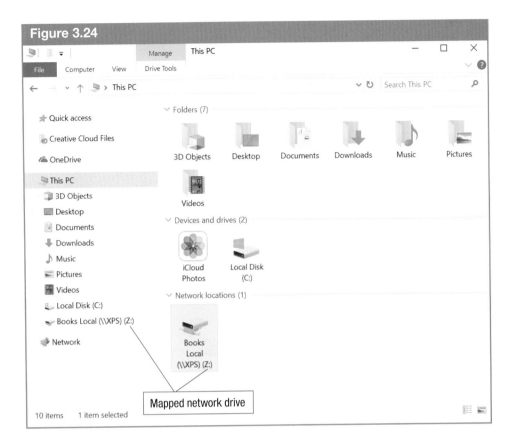

Figure 3.24

A mapped network drive shows up along with the local drive letters in the This PC window in Windows 10.

browse the available network locations through the file management utility in your operating system (File Explorer in Windows, for example). You might see a network share as a volume letter on your computer, even if it's an individual folder and not really a whole volume at its source. Assigning a volume letter to a network folder is called *mapping a network drive*; Figure 3.24 shows a mapped network drive in Windows 10. Some network shares don't allow users access until they have gone through authentication (positive identification), such as by providing a password or a preapproved user ID.

A **cloud drive** isn't technically a drive but rather a secure storage location on an internet-accessible remote server. For example, Microsoft's OneDrive service provides a free cloud drive to anyone who signs up for a Microsoft account. Similarly, Apple offers iCloud storage to users of iPads, iPhones, and other Apple devices, and Google Drive offers free storage to many types of users and devices. Storing data files on a cloud drive keeps them safe because their existence doesn't depend on the health of your computer's own hardware. Cloud drives can also be accessed from multiple devices. You can upload a photo you took with your smartphone to your cloud drive and then print it from your desktop PC.

Hands On
Mapping a Network Drive

Optical Storage Devices and Media

CDs and DVDs are optical discs. An **optical disc** stores data in patterns of greater and lesser reflectivity on its surface. They are read and written in **optical drives**, which contain a laser that shines light on the surface and a sensor that measures the amount of light that bounces back. On a writeable drive, the laser also can change the surface of the disc, altering its reflectivity in certain areas. CD and DVD technologies were originally developed for music and video storage, but today CDs and DVDs can store music, videos, and computer data. Most optical drives in computers today can

Article
DVD Technology

accommodate both CDs and DVDs, and some can even read and write to Blu-ray discs (BD) as well.

Optical drives are usually backward compatible. For example, a DVD drive can read and write CDs and DVDs, and a Blu-ray drive can read and write Blu-ray discs, DVDs, and CDs.

How Optical Storage Works An optical disc is divided into areas of greater reflectivity (called **lands**) and areas of lesser reflectivity (called **pits**). The drive shines a laser beam onto the disc and then a sensor measures the amount of light that bounces back. From this measurement, the sensor determines whether the area is land or pit. When the sensor detects a change in reflectivity—either from more to less or less to more—it sends a pulse indicating a 1 value. When it doesn't detect a change, it sends a 0 value. See Figure 3.25.

Unlike HDDs, which store data in concentric circles (tracks), optical discs typically store data along a single track that spirals outward from the center of the disc to the outer edge. The data is stored in sectors, similar to the sectors on an HDD (see Figure 3.26).

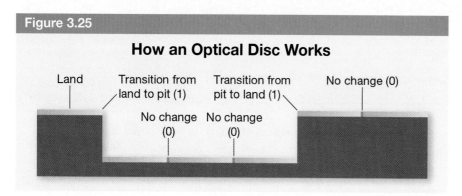

Figure 3.25

An optical disc stores data by differentiating between areas of greater and lesser reflectivity.

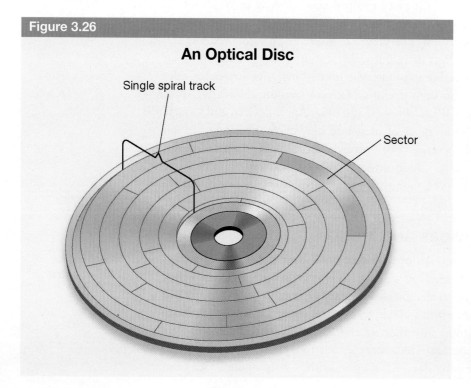

Figure 3.26

On an optical disc, information is stored in a series of sectors along a single spiral track. The actual tracks and sectors are much smaller than shown here—so small you would not be able to see them if drawn to scale in this figure.

Types of Optical Discs Most optical discs have the same diameter and thickness, but they differ widely in capacity and features, depending on their specifications (see Table 3.4). These are the three most common types of optical discs:

- A **compact disc (CD)** can store up to 900 MB of data. CDs are commonly used for distributing music and small applications and for inexpensively storing and transferring data.
- A **digital versatile disc (DVD)**, also called a *digital video disc*, can store up to 17 gigabytes (GB) of data, although the most common type (single-sided, single layer) stores up to 4.7 GB of data. DVDs are most often used to distribute large applications and standard-definition movies. DVDs can be single layer or dual layer (DL).
- A **Blu-ray disc (BD)** can store up to 128 GB of data in up to four layers. BDs are used to distribute high-definition movies and to store and transfer large amounts of data.

Table 3.4 Types, Storage Capacities, and Features of Optical Discs

Type	Storage Capacity	Features
Compact disc (CD)	**650–900 MB**	
CD-ROM		The data is stamped into the CD at the factory and can't be changed. Used for distributing digital data, such as computer software and copyrighted music.
CD-R		Can be written to only once. Used for creating one-of-a-kind CDs, such as music compilations and presentations for business distribution.
CD-RW		Allows rewriting up to 1,000 times. Used for backing up important files that change over time.
Digital versatile disc (DVD)	**4.7–17.0 GB (depending on number of sides and layers)**	
DVD-ROM	4.7 GB per side (single layer)	The data is stamped into the DVD at the factory and can't be changed. Used for distributing copyrighted movies and computer software.
DVD-R DVD+R		Can be written to only once. Used to create one-of-a-kind DVDs that won't be changed later.
DVD-RW DVD+RW		Recordable and rewriteable up to 1,000 times. Used for backing up important files that change over time.
DVD-RAM	4.7 GB–9.4 GB (depending on the number of sides)	Has a longer life than other types, up to 30 years. Can be rewritten up to 100,000 times. Is available in two sizes and several types, some of them removable or nonremovable cartridges.
Blu-ray disc (BD)	**25–128 GB (depending on number of layers)**	
BD-ROM		The data is stamped into the disc at the factory and can't be changed. Used for distributing high-definition movies.
BD-R		Can be written to only once.
BD-RE		Can be erased and rewritten up to 10,000 times.

All three types of discs can be read-only (ROM), recordable one time (R), or rewriteable multiple times (RW for CDs and DVDs; RE for Blu-ray). Read-only discs are manufactured with their content built in and can't be modified. Recordable discs can be written once but not changed after that. Rewriteable discs can be rewritten multiple times. There are two competing standards for writeable and rewriteable DVDs, and they are indicated by plus and minus signs like this: DVD+R and DVD-R.

DVDs can be single sided or double sided. On a double-sided DVD, both sides contain data. Each side of the DVD can be single layer or dual layer in terms of its storage system. A dual-layer model can store twice the data of a single-layer model by using a semitransparent top layer to store data that's read using a laser at a different angle. A Blu-ray disc can contain up to four layers.

Caring for Optical Discs Table 3.5 lists guidelines for the handling and care of optical discs. Dirt and other foreign substances on optical discs can cause read errors and/or prevent the discs from working at all. Even so, optical discs don't require routine cleaning and should be cleaned only when necessary. Using a commercially available cleaning kit is recommended, but other methods can also be used.

Hands On
Cleaning an Optical Disc

Table 3.5 Care Instructions for Optical Discs

Do
Store each disc in a jewel case or disc sleeve when not in use.
Use a felt-tip, permanent marker to write on the nonshiny side of the disc, if applicable.
Hold the disc only by its edges.
Use the recommended disc cleaning method to remove dirt and other substances.

Don't
Allow anything to touch the shiny (data) side or sides of the disc.
Stack disks that aren't stored in jewel cases or disc sleeves.
Place objects on the disc.
Expose the disc to direct sunlight or excessive heat.
Place food or beverages near a disc.

Tech Ethics

The Stored Communications Act

Do you know that different legal standards apply to the privacy of your information stored online versus on your local hard disk drive? Law enforcement authorities require a warrant and probable cause to search your home and your local hard drive. However, they may need only a subpoena or a court order and prior notice to look at your online storage.

The Stored Communications Act (SCA), which was part of the Electronic Communications Privacy Act of 1986, enacted some privacy protections for electronic data held by a third party (such as your online storage provider), but the act goes only so far. The SCA protections for electronic data are similar to the Fourth Amendment protections against unreasonable search and seizure. Legal precedents establishing the limits of the SCA protections are still evolving, and it has even been suggested that the SCA is unconstitutional because it permits a lower standard of privacy than the Fourth Amendment would normally provide.

For a summary of recent legal decisions about this issue, see https://CUT7.ParadigmEducation.com/SCA. As more people move toward cloud storage and cloud-based applications, protecting the privacy of information held in those services will become increasingly relevant.

3.9 Understanding Large-Scale Storage

> **Watch & Learn**
> Understanding Large-Scale Storage

Large businesses must store huge amounts of data—both for internal use (such as personnel records) and customer use (such as product listings). Imagine all the data required for an airline reservation system, for example, or an online merchant such as Amazon. And imagine the impact on the company's business if some of that data was lost because of a defective storage device.

Large computer systems, such as mainframe and server systems, typically use the same basic storage technologies as smaller computer systems: HDDs, SSDs, and optical discs. However, because of the large volume of the data and the need to store and back it up safely, extra technologies are employed to work with these basic storage components.

There are two main concerns in large-scale data storage: how quickly data can be stored and retrieved, and how easily and reliably lost data can be restored when hardware failures or other problems occur.

Online retailers and other large businesses maintain entire rooms and even buildings full of servers and related equipment.

Local versus Network Storage

With a personal computer, data is usually stored locally—that is, on the computer itself or on an external device directly plugged into it. Local data storage is known as **direct-attached storage (DAS)**.

In a large company, it isn't practical to have multiple users store their data locally. Data should be stored in a central location, where everyone who needs it can access it, regardless of which PCs are in use. **Network-attached storage (NAS)** is storage that's made available over a network, such as on a centrally accessible file server. The file server holds the files that multiple users need to access, and the users retrieve the data via the company's network or via the internet.

An ordinary server (running UNIX or Windows Server, for example) can be used as NAS. It's more common, however, to use a **NAS appliance**, which is a specialized computer built specifically for network file storage and retrieval. A NAS appliance connects directly to the network and can be configured and administered through another computer on the network; a NAS appliance doesn't need to have its own keyboard, mouse, and monitor.

NAS appliances provide centrally accessible storage space for many computers at once through a network connection.

When a company has a lot of data, it's typically stored on many different drives and computers, but the users should be shielded from that fact. For example, when you shop at an online retailer like Amazon, you shouldn't have to know which of Amazon's file servers stores a particular category of product information. A **storage area network (SAN)** enables users to interact with a large pool of storage (including multiple devices and media) as if it were a single local volume. For large companies, SANs offer many of the advantages of DAS and NAS, including less network traffic, easier disaster recovery, and simpler administration.

Article
Storage Area Networks

RAID

In business storage, both speed and reliability are important. Unfortunately, traditional HDDs have neither of these qualities. When data is stored on a single HDD, the data retrieval speed is limited by the device's access time and interface speed. And if that drive fails, all the data on it will be lost.

A technology called **redundant array of independent disks (RAID)** attempts to correct both of these problems by combining multiple physical HDDs. RAID is a group of related disk management methods, numbered 0 through 6. Each method employs a different combination of data-handling features.

RAID0: Striping RAID0 addresses the issue of performance speed by **striping** disks (see Figure 3.27). In striping, each write operation is spread over all the physical drives, so the file can be written and read more quickly. For example, suppose you have four drives in your RAID, and each drive can store and retrieve data at about 100 MB/s. If you are writing a 400 MB file, it will take 4 seconds on a single drive, but if you write one-fourth of the file to each of four drives, the write operation will be accomplished in only 1 second. The file is distributed among the drives at an individual bit level: one bit gets written to disk 1, the next bit to disk 2, and so on. RAID0 doesn't have any means of protecting data from loss; it's purely a performance-enhancing tool.

RAID1: Mirroring RAID1 addresses the issue of data protection by **mirroring** the drive (see Figure 3.28). In RAID1, each HDD has an identical backup (a mirror), and the same data is written to both drives simultaneously. That way, if the main drive fails, the backup drive can supply the data. RAID1 has double the hardware cost, because you pay for twice as many drives as you actually use. Also, it does nothing to enhance performance.

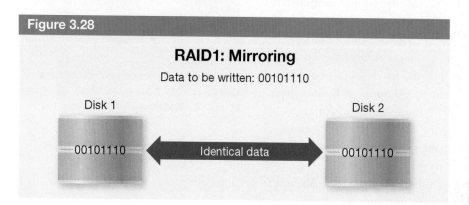

Figure 3.27

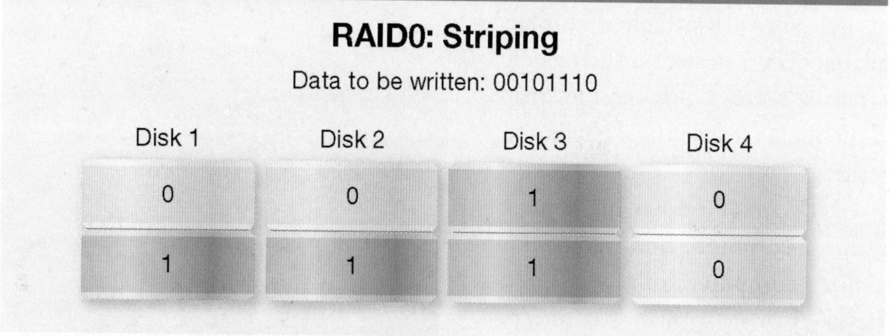

RAID0: Striping

RAID0 spreads out data evenly across all the available disks, one bit at a time.

Figure 3.28

RAID1: Mirroring

RAID1 mirrors the data on a second drive for data protection in case one drive goes bad.

Figure 3.29

RAID5 stripes the data, like RAID0 does, but it also includes a parity bit for data recovery.

RAID2–RAID6: Striping with Distributed Parity RAID2, RAID3, and RAID4 were early attempts at combining the benefits offered by both RAID0 (performance) and RAID1 (data protection) in a single method. The attempt that ultimately became popular is RAID5.

RAID5 combines striping (for performance) with distributed parity (for data protection). Here's a very basic explanation of how distributed parity works: Let's say you have five disks in your **disk array** (set of disks). When you write some data to the disks, four of them each receive a bit (either 0 or 1). The fifth drive receives a **parity bit**. The parity bit is determined by the values of the other bits, which are added up. If the result is an odd number, the parity bit is 1; if the result is even, the parity bit is 0. The parity bit is used to reconstruct data if one of the disks goes bad.

Figure 3.29 shows a simple example of distributed parity using a five-disk array. The data to be written is 00101110. The first four numbers are written to the first four disks, like this:

> Disk 1: 0
> Disk 2: 0
> Disk 3: 1
> Disk 4: 0

These numbers add up to 1, which is an odd number, so the parity bit is 1. It's placed on disk 5.

The disks alternate as to which one contains the parity bit. The next four numbers are written to disks 2 through 5, reserving disk 1 for the parity bit:

> Disk 2: 1
> Disk 3: 1
> Disk 4: 1
> Disk 5: 0

These numbers add up to 3, which is an odd number, so the parity bit is 1. It's placed on disk 1.

Now, let's say that disk 2 fails, but all the others are still functioning. Software can determine what disk 2 held by looking at the parity bit (1). In the first set, because 0 + 1 + 0 = 1, the software knows that the missing number from disk 2 has to be a 0 to make the overall sum odd (parity bit of 1). In the second set, because 1 + 1 + 0 = 2, the software knows that the missing number from disk 2 has to be a 1. Using the parity bits, the entire contents of the failed drive can be reconstructed, and the RAID can continue to function until a replacement drive is installed and populated with the correct bits.

RAID6 is just like RAID5 except it has double-distributed parity; it can tolerate up to two failed drives and still function.

Chapter Summary

3.1 Defining System Software

System software is designed to keep the computer up and running so it can run useful applications. System software includes the **basic input/output system (BIOS)** (which is considered **firmware**), the **operating system (OS)**, **utility programs**, and **translators** (compilers and interpreters).

3.2 Understanding the System BIOS

The BIOS stores low-level programs on a non-volatile memory chip. The **system BIOS** is stored on a chip on the motherboard, and helps start up the computer. You can modify the startup settings for the motherboard by making changes in the **BIOS Setup utility** or, in newer computers, the Unified Extensible Firmware Interface (UEFI).

3.3 Understanding What an Operating System Does

An OS boots the computer, provides a user interface, runs programs, configures and controls devices using **drivers**, and manages essential file operations. The OSs a computer will accept depend on its **platform**. The **IBM-compatible platform** (also called the **Intel platform**) comes in two versions: the 32-bit version (**x86**) or the 64-bit version (**x64**).

Starting from a turned-off condition is a **cold boot**; restarting an already running computer is a **warm boot**. When starting the computer, after the BIOS completes its **power-on self-test (POST)**, the operating system loads the **kernel** into memory and the user interface appears. The kernel is **memory resident** and remains in memory while the computer is in operation; however, some parts of the OS are **nonresident** and loaded only when needed.

The **user interface** (or **shell**) allows communication between the software and the user. The user interface can be a **command-line interface** or a **graphical user interface (GUI)**. Both Microsoft Windows and the macOS use a **desktop** metaphor, with **windows** and **icons**. The Windows Start menu and all tablet operating systems use a touch-driven user interface, with **tiles** or icons you can tap.

3.4 Looking at Operating Systems by Computer Type

A personal computer (PC) is a desktop or laptop computer designed for use by one person at a time. Personal computer operating systems include **Microsoft Windows**, **macOS**, **Linux**, and Chrome OS. Linux and Chromium OS are **open source software**; Windows, macOS, and Chrome OS are **proprietary software**. The Linux kernel is a command-line interface, but there are many **distributions (distros)** that contain attractive graphical shells for the user interface. Chrome OS is a **thin-client** OS designed for netbooks.

Mobile operating systems run on tablets and smartphones. Popular examples include **iOS** (for iPad and iPhone) and **Android** (an open source OS developed by Google).

Server operating systems support multiple simultaneous users and manage many tasks at once. Server operating systems include **Windows Server**, **UNIX**, and Linux (with an appropriate **solution stack**).

3.5 Exploring Types of Utility Programs

A utility program performs a single maintenance or repair task. **Antivirus software** is a common type of utility program; this type of program detects and eliminates **malware** such as viruses and worms. A **firewall** is a security system that blocks unused ports from being used to gain access to a computer or network. A **diagnostic utility** analyzes the computer's problems and creates a report about them. An **uninstaller** removes software and updates the system files. A **disk scanner** checks for disk storage errors. A **disk defragmenter** reorganizes files to optimize how they are stored. A **file compression utility** shrinks file sizes and packages multiple files in a single archive. A **backup utility** allows users to make copies of files and folders. A **spam blocker** is a junk email filter. Antispyware software detects and removes **spyware** and **adware**, which are both forms of malware.

3.6 Programming Translation Software

A computer programmer uses **high-level language** to write a program, and then he or she uses a **compiler** to turn the program into **machine language**. Compiled code results in an executable file. Some programming languages, such as JavaScript, are interpreted rather than compiled. Interpreted code is re-interpreted line by line each time the program is run. It doesn't result in an executable file. Web browser software is an example of an **interpreter** utility.

3.7 Working with File Storage Systems

A drive letter is a **volume**, and each volume has a **root directory**. Within the root directory are **folders**. A folder within a folder is a **child folder**; the folder that holds the child folder is its **parent folder**. The **path** to a file is the volume and folders that contain the file. Using a **folder tree** is one way of understanding and browsing a folder structure on a volume.

The **file extension** at the end of a file's name indicates its type, as in myfile.txt. The operating system knows what program to open a file with based on its extension. For example, a file with the .txt extension opens in Notepad in Windows. A file with an .exe, .com, or .bat extension is an executable file that runs a program. A **shortcut** is an icon that points to a file and that you can place in a handy location, such as on your computer desktop.

3.8 Understanding File Storage Devices and Media

Primary storage is RAM; **secondary storage** is storage that holds data more permanently. **Fixed storage** is mounted inside a computer, such as a hard drive; **removable storage** can easily be removed, such as a USB flash drive.

A **hard disk drive (HDD)** is a traditional, mechanical hard drive. Data stored on a hard disk is organized into concentric rings called **tracks**, each track is broken up into sections called **sectors**, and each sector holds 512 bytes. Multiple sectors are grouped into **clusters**. The disk's table of contents is the **master file table (MFT)** or **file allocation table (FAT)**, depending on the file system. The most common file system used in Windows is **New Technology File System (NTFS)**.

Magnetic storage, such as an HDD, stores data in transitions of positive and negative magnetic polarity. A transition between polarities is a **flux transition**. Measurements of HDD performance include **average access time** and **data transfer rate**. The two common interfaces for hard disk drives are **parallel ATA (PATA)** and **serial ATA (SATA)**.

A solid-state drive (SSD) stores data in a type of EEPROM. A solid-state drive is the solid-state equivalent of an HDD. A **USB flash drive** is a smaller-capacity, more portable SSD.

A **network share** is a drive or folder that's been made available to users on other computers. Some networks require authentication before granting users access to a network share. A **cloud drive** is not really a drive, but a secure storage location on an internet-accessible remote server.

An **optical disc** stores data in patterns of reflectivity called **pits** and **lands**. They are read and written by **optical drives** with lasers in them. Types of optical discs include **Blu-ray disc (BD)**, **digital versatile disc (DVD)**, and **compact disc (CD)**.

3.9 Understanding Large-Scale Storage

Direct-attached storage (DAS) is local storage. **Network attached storage (NAS)** is a folder or volume that is shared on a network. NAS can be a file server or a **NAS appliance**. A **storage area network (SAN)** enables users to interact with a large pool of storage as if it were a single local volume.

A **redundant array of independent disks (RAID)** helps to solve difficulties with data access speed and/or data safety and reliability. **RAID0** increases performance by **striping**. **RAID1** increases data safety by **mirroring**. **RAID5** combines the two by striping multiple disks in a **disk array** and including a **parity bit**.

Key Terms

Numbers indicate the pages where terms are first cited with their full definition in the chapter. An alphabetized list of key terms with definitions is included in the end-of-book glossary.

● Chapter Glossary

● Flash Cards

3.1 Defining System Software
system software, 96
basic input/output system (BIOS), 96
firmware, 96
operating system (OS), 96
utility program, 96
translator, 96

3.2 Understanding the System BIOS
system BIOS, 97
BIOS Setup utility, 97

3.3 Understanding What an Operating System Does
boot drive, 99
platform, 99
IBM-compatible platform, 99
Intel platform, 99
x86, 99
x64, 99
cold boot, 100
warm boot, 100
power-on self-test (POST), 100
kernel, 100
memory resident, 100
nonresident, 100
user interface, 101
shell, 101
command-line interface, 101
graphical user interface (GUI), 102
desktop, 102
icon, 102
window, 102
tile, 102
driver, 103

3.4 Looking at Operating Systems by Computer Type
Microsoft Windows, 104
macOS, 104
Linux, 105
open source software, 106
proprietary software, 106
distribution (distro), 106
thin client, 107
iOS, 108
Android, 108
UNIX, 109
multiuser operating system, 109
cross-platform operating system, 109

Chapter 3 Working with System Software and File Storage

solution stack, 110
Windows Server, 110

3.5 Exploring Types of Utility Programs

antivirus software, 111
executable file, 111
malware, 111
firewall, 111
personal firewall, 112
network firewall, 112
proxy server, 112
diagnostic utility, 112
uninstaller, 112
disk scanner, 112
disk defragmenter, 113
noncontiguous sectors, 113
file compression utility, 113
backup utility, 113
spam, 113
spam blocker, 113
spyware, 114
adware, 114

3.6 Programming Translation Software

machine language, 114
low-level language, 114
high-level language, 114
compiler, 114
interpreter, 114

3.7 Working with File Storage Systems

disk drive, 115
solid-state drive (SSD), 115
volume, 116
root directory, 116
folder, 116
child folder, 116
parent folder, 116
path, 116
folder tree, 117
file extension, 117
file association, 117
shortcut, 118

3.8 Understanding File Storage Devices and Media

primary storage, 119
secondary storage, 119
fixed storage, 119
removable storage, 119
hard disk drive (HDD), 120
track, 121
sector, 121
cluster, 121
master file table (MFT), 121
file allocation table (FAT), 121
New Technology File System (NTFS), 121
magnetic storage, 121
flux transition, 121
average access time, 121
data transfer rate, 122
parallel ATA (PATA), 122
serial ATA (SATA), 122
network share, 122
cloud drive, 123
optical disc, 123
optical drive, 123
land, 124
pit, 124
compact disc (CD), 125
digital versatile disc (DVD), 125
Blu-ray disc (BD), 125

3.9 Understanding Large-Scale Storage

direct-attached storage (DAS), 127
network-attached storage (NAS), 127
NAS appliance, 127
storage area network (SAN), 127
redundant array of independent disks (RAID), 128
RAID0, 128
striping, 128
RAID1, 128
mirroring, 128
RAID5, 129
disk array, 129
parity bit, 129

Chapter Exercises

Complete the following exercises to assess your understanding of the material covered in this chapter.

- Terms Check
- Knowledge Check
- Key Principles
- Tech Illustrated Operating Systems
- Tech Illustrated File Path
- Chapter Exercises
- Chapter Exam

Tech to Come: Brainstorming New Uses

In groups or individually, contemplate the following questions and develop as many answers as you can.

1. The 2013 movie *Her* was a fantasy about an operating system with a human-like personality that the user could interact with as he or she would with a real person. What if having that type of user interface were an option for a personal computer? Would you like such an operating system? Why or why not?

2. Linux, Chromium OS, and other operating systems are free, whereas Microsoft Windows and macOS (the most popular operating systems) are not. What would happen if all operating systems were free? How might the software industry change?

3. In an interpreted program language, the source code is available to anyone who wants it. For example, anyone can look at the source code for JavaScript in a web page. Source code for some compiled software is also freely available, such as the Linux kernel. What would happen if all software were open source, so that anyone could go in and change anything in any program (provided he or she had the programming skills to do so)? How might the software industry change?

Tech Literacy: Internet Research and Writing

Conduct internet research to find the information described, and then develop appropriate written responses based on your research. Be sure to document your sources using the MLA format. (See Chapter 1, Tech Literacy: Internet Research and Writing, page 37, to review MLA style guidelines.)

1. You have been tasked to copy the Birthdays.txt file from a USB flash drive to the Backup folder on the C: volume on a computer that runs UNIX, but you don't know the UNIX command for copying files. How can you find out what command to type? Research UNIX commands online, and write a step-by-step procedure with the exact commands you would use. Assume that the volume letter for the flash drive is F.

2. Suppose you want to try out Linux by installing it on an older computer and using it to browse the web, send and receive email, and run Apache OpenOffice (an open source suite of applications that's similar to Microsoft Office). Which distribution will you get and why? Research the available distributions, determine the best one for your needs, and then write an explanation of your choice. As part of your research, find out whether OpenOffice includes an email program. If it doesn't, identify a free email program that will work with the Linux distribution you have chosen.

3. Suppose you want to put your family's music collection on a NAS device so it's available to every computer in your house all the time. Assume that you have 500 music CDs, each containing about 700 MB of files. Calculate how much storage you will need, and then select a NAS appliance with at least 2.5 times

that much storage. Compare prices for the same model across three different online shopping sites, and write a report detailing your findings.

Tech Issues: Team Problem-Solving

In groups, develop possible solutions to the issues presented.

1. Suppose a friend gives you a computer, but when you try to start it, a message appears asking for a BIOS password. You can't contact the friend to find out the password. How should you proceed? Start by using the internet to determine what other people have done in the same situation.

2. A friend's Android phone won't start up properly; when she turns it on, nothing happens. Tech support has recommended a factory reset of the operating system. Write a note to your friend explaining what a factory reset is, how to perform one, and what the results will be. Use the internet to get the information you need, and document your sources.

3. Suppose you own a small accounting company (20 employees), and so far, you have allowed each employee to choose his or her own operating system. Your technical support staff has run into problems supporting so many different OSs and versions (some of which they aren't very familiar with). You plan to standardize all of the company computers to a single operating system and version. Which one will it be and why? Write a short report that explains the operating systems you considered, announces your decision, and explains the factors that went into making it.

Tech Timeline: Predicting Next Steps

Many improvements have been made to the Windows operating system since Microsoft first introduced it. The timeline below shows a number of versions of Microsoft Windows for the PC and the year each was introduced. Microsoft has stated that Windows 10 will be the final version of Windows, but that doesn't mean the company won't continue adding features to it over the course of many years, much as Apple has done with macOS.

Visit Microsoft's website at https://CUT7.ParadigmEducation.com/Microsoft along with other sites to read more about Windows and its features. Prepare a list of at least three features that you believe Microsoft will (or should) add to Windows in the future.

1983	Windows 1.0 (released to the public in 1985)
1992	Windows 3.1
1995	Windows 95
1998	Windows 98
1999	Windows 2000
2000	Windows Millennium Edition (ME)
2001	Windows XP
2006	Windows Vista
2009	Windows 7
2012	Windows 8
2013	Windows 8.1
2015	Windows 10

Ethical Dilemmas: Group Discussion and Debate

As a class or within an assigned group, discuss the following ethical dilemma.

When you buy a software product like Microsoft Windows, you aren't buying unlimited use of it; rather, you are buying a license to use it on one computer (or a specific number of computers). Yet each year, thousands of illegal copies of Microsoft Windows are made and distributed. Windows includes a Product Activation feature that's intended to cut down on this activity, but thieves still find ways of circumventing the system and making illegal copies available.

Software publishers lose money when their software is copied, because the individuals who are getting the illegal copies would otherwise have had to pay for legitimate copies. Software publishers pass on the cost of such losses to consumers, and higher software prices are the result.

What's your position on software theft? Can copying copyrighted software be justified in any situation? Why or why not? What are your ethical obligations, if any, concerning this matter? Would you report your employer for copyright violation? Would you accept or use illegally produced copies of software on your own computer?

Chapter 4

Using Applications to Tackle Tasks

Chapter Goal

To learn about the types of applications available and be able to select an appropriate application for the task you need to complete

Learning Objectives

4.1 Explain the ways in which application software is classified, sold, and licensed.

4.2 Differentiate between types of business productivity software and select the right tool for a task.

4.3 Explain how personal productivity and lifestyle software can assist individuals and families.

4.4 Describe and differentiate between the types of graphics and multimedia software available.

4.5 Explain the types of game software available and how to determine a game's age appropriateness.

4.6 Explain the ways that communication software enables people to work together.

Online Resources
The online course includes additional training and assessment resources.

Tracking Down Tech
Everyday Applications

4.1 Distinguishing between Types of Application Software

Watch & Learn
Distinguishing between Types of Application Software

Application software enables users to do useful and fun things with a computer. Applications are available for almost every computing need you can imagine, and you will read about many of them in this chapter. You can start to gain a deeper understanding of software by seeing how it's classified according to some basic characteristics, such as the number of intended users, the type of device on which it will function, and how it's sold and licensed.

Individual, Group, or Enterprise Use

One way to categorize applications is according to the numbers of people they serve at once. **Individual application software**—such as a word processor, accounting package, or game—generally serves only one person at a time. Most of the software on a typical home computer falls into this category.

Groupware (also called *collaboration software*) enables people at separate computers to work together on the same document or project. For example, a team of developers might use groupware on their local area network to collaborate on documents, communicate (via messaging, email, and video chat), and share calendars and databases.

Enterprise application software is a complete package of programs that work together to perform core business functions for a large organization. A multi-function application suite might serve many departments. For example, a hospital might have an integrated information system that manages data about patients and their medical care, employees, equipment, and facilities. A scientific research lab might have a centralized computer system that automatically records experiment results and suggests new testing parameters.

Desktop Applications versus Mobile Apps

Another way to distinguish applications is to look at what types of devices they are designed to be used on. A **desktop application**, as the name suggests, is designed to be run on full-featured desktop and laptop computers. Desktop applications can take up a lot of storage space—sometimes 1 gigabyte (GB) or more—and may include additional software tools that support the main program. In contrast, an application meant primarily for tablets and smartphones—popularly known as an **app**—is designed to be compact and to run quickly and efficiently on minimal hardware. Whereas desktop applications may be installed from a DVD, apps are usually downloaded from an online store (such as the Apple App Store or Microsoft Store) through a store app that comes with the device.

Some desktop applications were written for an earlier version of Windows, and may not run well on Windows 10. You can use Compatibility Mode to tell Windows to emulate an earlier version when running such a program; doing that will solve most compatibility problems.

Software Sales and Licensing

Application software can also be categorized by the way it's developed, sold, and distributed. The three main types of software distribution are commercial software, shareware, and freeware (see Figure 4.1).

Chapter 4 Using Applications to Tackle Tasks

Commercial Software **Commercial software** is created by a company (or an individual, in rare cases) that takes on all the financial risk for its development and distribution upfront. The programmers, testers, DVD copiers, box designers, attorneys, and others involved in developing, producing, and marketing the software are all paid in advance to create a commercial product that the company then sells to the public. You can purchase commercial software as a boxed product in a retail store or as an online product from the company's website or an online store.

Some software isn't installed on your computer but rather is accessed from the internet or from a secure cloud. This arrangement is sometimes called *web applications* or *Software as a Service* (*SaaS*). Web applications have several advantages over traditionally installed

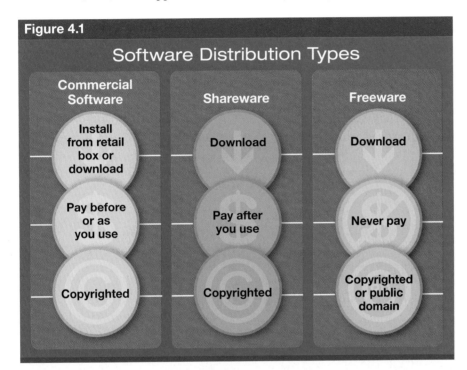

Figure 4.1 Software Distribution Types

Practice
Software Distribution Types

Hotspot

The App-ification of the Software Industry

As more and more people switch to mobile devices as their primary computing tools, apps are becoming an increasingly larger share of the software market. The Apple Store has over 2.1 million apps for the iPhone and iPad, and more are being added every day. App developers have earned more than $25 billion from App Store sales.

A key characteristic of most apps is their specialization: each app performs one task, or a narrow range of tasks. Rather than purchase a large suite of applications, as for a desktop, a mobile user can select a customized combination of apps that accomplish the specific tasks he or she needs to perform, such as playing music clips, browsing a certain social networking site, or checking stock prices. Apps are typically very inexpensive ($5 or less), so consumers can afford to take chances on new apps that promise new capabilities.

This trend toward specialized apps has affected the software industry overall. As consumers become accustomed to using a different app for each task they perform, they start looking at desktop software in new ways. In the future, you may see desktop software become more compact, more specialized, and simpler to install and remove. App popularity is also encouraging programmers to work independently to develop and sell their own apps, rather than work with large software development teams on large desktop applications.

software. For instance, they require little or no local storage space, and you never have to worry about downloading updates, fixes, or new editions. You have an online workspace that is available wherever you are, on whatever device you are using. Also, your data files can be stored in a secure online location, so you don't have to back them up to ensure their safety. Web applications are typically less expensive than an equivalent retail boxed product too, and may even be free, as is the case with Microsoft Office Online applications. However, web apps require that you be connected to the internet to use the software, so they aren't suitable for people with slow or unreliable internet connections. (Web apps are also discussed in Chapter 7, under the heading "Software as a Service.")

Video Installing Applications

Video Applications on the Web

Hands On Uninstalling Windows Applications

Commercial software can either be purchased outright (with a lifetime license to use that particular version) or "rented" using a subscription model. Subscribing to an application has several advantages. The upfront cost is much less, for example, and you get free updates whenever new versions are released. On the other hand, subscribing to an application requires you to keep a credit card on file, so the company can charge you on a monthly or yearly basis for your subscription, and you lose access to the application if you stop paying. Microsoft Office is available as a one-time purchase (Office 2019) or as a subscription (Office 365) (see Figure 4.2).

Because developing commercial software is a big financial risk, the companies involved in producing it want to make sure that everyone who uses the software pays for it. Stealing commercial software by using it without paying is called **software piracy** and costs companies billions of dollars a year. To cut down on piracy, many companies include various kinds of protective measures in their applications.

The most common antipiracy measure is to write the software's setup program so that it requires the user to enter a registration key. A **registration key**, sometimes called an *installation key*, is a string of characters that uniquely identifies the user's purchase of the software. On a retail boxed product, the registration key is typically printed on the disc's sleeve or packaging. For an online purchase, the registration key is delivered in an email.

For additional protection, some software requires activation. If you don't activate the software within a certain number of days, it stops working. **Activation** is a process that generates a unique code based on the hardware in your computer and the registration key you used when installing the software. This unique code is sent online to the company's registration database and effectively locks that registration code to your current hardware. If you try to activate another copy of the software on a different computer, the database may not accept it.

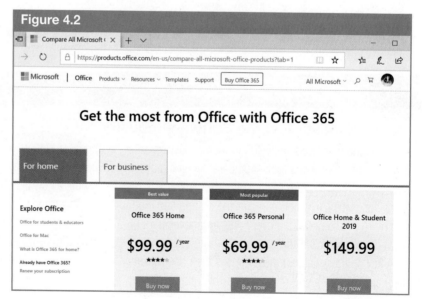

Figure 4.2

You can purchase an Office 365 subscription online.

Most products that require activation allow you to install them on two computers (sometimes more). If you reach the install limit and the product won't activate, you can call the customer support line for the developer and ask for additional activations. These requests are usually granted (within reason).

When you buy commercial software, you aren't really buying the software itself but rather the right to use it in ways that the license agreement permits. The **license agreement** (sometimes called an *end-user license agreement*, or *EULA*) is the legal information that's included in the software packaging or that appears onscreen during setup and asks for your compliance. A license agreement may specify, for instance, the number of computers on which you can legally install the software for your personal use. It may also limit the purposes for which the software may be used. For example, some versions of software are licensed only for use by individuals, students, and nonprofit organizations.

Most PC software is licensed on an individual copy basis, but other license types are also available. For example, Microsoft allows a business to buy a site license for many of its products. A **site license** grants permission to make multiple copies of the software and install it on multiple computers. A site license can allow either per-user or per-seat use. A **per-user site license** grants use by a certain number of users, regardless of the number of computers. A **per-seat site license** grants use for a certain number of computers, regardless of the number of users. See Figure 4.3 for an example showing the difference between the two types of licenses.

> Article
> Software Licensing Arrangements

Many large companies have specialized software needs that can't be met by the mainstream commercial software available. Such companies can choose an off-the-shelf application and then hire programmers to write add-ons and customizations for it, or they can hire developers to create custom software. Although developing custom software costs many thousands of dollars upfront, a company can make up that investment over time through increased worker productivity.

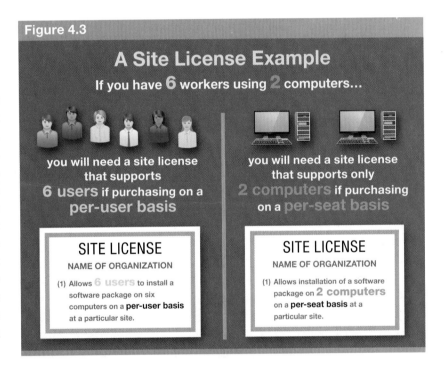

Figure 4.3 A Site License Example

Practical TECH

Purchasing and Downloading Software Online

Buying commercial software online and downloading it to your computer is more convenient than purchasing a DVD at a store and taking it home to install it. To purchase the software online, you can use a credit card or an online payment system (such as PayPal). After you do so, you receive a download link on a web page or in an email (or both). You click that link to download an installer file and then run the installer to set up the application. To complete the installation, you might need a code or registration key; you should receive that from the seller in an email too.

It's important to keep that code somewhere safe, because you will need it if you ever reinstall the software. You might also need it to get a discounted price when buying an upgrade for the software later. You should store the code somewhere other than on your hard drive, such as on a hard copy with your important papers or on a *cloud drive*.

Shareware and Freeware Many small software development companies struggle to get the public's attention for their products. One way that many companies have increased their popularity is to offer their software as **shareware**. With a shareware product, you can try it before you buy it. You can download a trial version for free and use it for a specified number of days or use a limited set of features. If you like the program, you can buy a registration key that will unlock the rest of the features or remove the time limit. Most shareware has a time or feature limitation that lasts until you pay for the product.

Software that's made available at no charge is known as **freeware**. Because freeware is often developed by individuals rather than companies, the quality varies and so do the levels of documentation and support available. The developer of freeware may retain its copyright or give up the copyright and make the program available in the public domain. If a program is in the **public domain**, other people can do what they like with it. In addition, the developer usually makes the **source code** available (called *open source code*) so others can modify and improve the program. LibreOffice is one example of a free, open-source suite of applications.

> Video
> Shareware and Freeware

Application Download Files

Applications that come from OS-specific online stores like the Microsoft Store and the Apple App Store download and install seamlessly with just a few taps or clicks. In the past, this type of delivery was limited to small and simple apps, but now, more and more companies are choosing to release full-featured software in this way.

Desktop applications typically consist of a main executable file plus multiple helper files. Installing these applications requires running a setup utility that unpacks the files from a compressed archive, creates the needed folders, creates shortcuts for easily running the application, and in some cases makes changes to system files such as the Windows Registry.

When you acquire a commercial application by downloading, you typically download a single compressed archive file with an *.exe* extension, which contains everything you need to install that application. When you run this setup file, a setup utility performs all the tasks needed to install the application.

Many smaller software publishers, including most distributing freeware or shareware, make their downloads available in archive packages that are not executable. An archive package is a bundled set of files in a certain compression format. The bundled

Tech Ethics

Paying for Shareware

Some shareware is distributed through an honor system. The developer makes available to the public a full-featured version of the program with no time limits and asks people to send in money if they like and continue to use the product. This distribution method worked well in the early days of the internet. At that time, most of the people who downloaded shareware were tech enthusiasts—among them, many programmers who understood the time and effort put into developing shareware. But today, shareware developers who don't demand payment by limiting the program's features or usage time seldom get paid.

Suppose you download a shareware application for free and digitally sign a license agreement in which you promise to pay for the software if you continue to use it after 14 days. Will you make that payment if you use the software beyond the 14-day limit? Why or why not? Does it matter how much the payment is or what payment methods are available?

files include a setup file. You must open the archive file and extract the files within it to a new folder, open the new folder, and then run the setup file that is inside the new folder. Depending on the OS and the archive format, you might need a third-party file compression utility to open the archive folder and extract the files, or the OS might be able to do this for you. For example, Windows includes built-in support for opening and extracting ZIP files. File extensions for archive files include *.zip*, *.rar*, *.tar*, *.7zip*, *.7z*, *.gzip*, *.gz*, and *.jar*. (These are lowercase because they are commonly used in Linux or UNIX, both of which are case-sensitive and don't commonly use uppercase.) Some Linux archives use the extension *.scexe*. (See Chapter 3 for more information about file types and extensions and file compression utilities.)

Some large application packages can be downloaded as ISO files (Windows) or DMG files (macOS). Such files contain an image of a DVD disc. DMG files on macOS can be directly accessed as if they were drives without any special setup. You download an ISO file, use an optical disc burning utility to save the file on a blank writable disc, and then use the disc to install the application. Advanced users may mount an ISO file to a virtual drive and install it as though they had a physical disc drive. ISOs can also be burned to USB flash drives.

Some Microsoft applications are delivered on DVDs that include archive files with a proprietary format, such as CAB or MSI. During the setup process, the setup utility can extract files from the CAB or MSI archive file as needed.

4.2 Using Business Productivity Software

Watch & Learn
Using Business Productivity Software

No matter what the business or industry, certain common tasks need to be performed. For example, businesses must do the following:
- communicate with customers, suppliers, and coworkers
- store information about customers, products, services, and orders
- schedule meetings and other activities that involve people, locations, and equipment
- market the product or service to potential customers
- make and receive payments and maintain accurate financial records
- analyze current trends to predict future needs

Business productivity software can help with all these activities. You can purchase several applications to perform specific activities, or you can use a suite of related applications to perform multiple tasks. Table 4.1 provides some examples of business productivity software.

A small business (also called a *small office/home office*, or *SOHO*) has different software needs than do large companies. Some applications come in different versions

Table 4.1 Examples of Business Productivity Software

Category	Example(s)	Common Uses
Word processing	Corel Wordperfect Microsoft Word Pages Wordpad	Prepare correspondence (letters, envelopes), labels, reports, and book manuscripts.
Desktop publishing	Adobe InDesign Microsoft Publisher QuarkXPress	Design and produce newsletters, advertisements, brochures, books, and magazines.

continues...

Table 4.1 Examples of Business Productivity Software ...*continued*

Category	Example(s)	Common Uses
Spreadsheet	Numbers Microsoft Excel	Record and organize numerical data and prepare summary charts and reports.
Database management	Filemaker Pro Advanced Microsoft Access	Provide structured data storage for customer, order, and product records, and prepare reports and charts based on the data.
Presentation	Keynote Microsoft PowerPoint	Design and produce text- and graphics-based slides for presentations.
Personal information manager (PIM)	Apple Mail Microsoft Mail and Calendar Microsoft Outlook	Schedule appointments and tasks, and store contact information; may also send and receive email.
Project management	JIRA Microsoft Project ProjectLibre	Schedule and track stages and tasks of complex projects that involve people, locations, and equipment.
Accounting	Intuit QuickBooks	Record and analyze financial data, including bank accounts, transactions, inventory, and payroll.
Page layout creation	Adobe Acrobat Microsoft XPS	Output a document from another application into a print-ready page layout; integrate with other applications as either a file save format or a printing option.
Groupware	Apple iCloud Google Drive Microsoft OneDrive	Provide online storage space for data files; may also provide shared access to cloud-based files and the ability to edit and comment on the files.
Web conferencing	GoToMeeting Skype Webex	Enable people attending an online meeting to collaborate; screen-sharing is built into many video teleconferencing applications.
Remote access	Windows Remote Desktop Windows Remote Assistance	Enable a remote user to take control of a system, to use the system's applications and resources, or to troubleshoot problems on the system via a network or the internet.

customized for one market or another. A version designed for large companies (thousands of users) is often called an *enterprise edition*. A large-scale version may contain extra components that can help professional IT staff administer the software, and may have special pricing programs for large numbers of users.

Word-Processing Software

Word-processing software can be used to create most kinds of printed documents, including letters, labels, newsletters, brochures, reports, and certificates. Word-processing programs are the most widely used of all software applications in business, because nearly every company needs to generate some type of printed information.

Chapter 4 Using Applications to Tackle Tasks

Figure 4.4

Microsoft Word offers a comprehensive set of tools for creating and modifying text and images in word-processing documents. The tools are displayed in a ribbon interface and organized in tabs, groups, and buttons for easy access.

Word-processing programs are often available for more than one operating system. For example, versions of Microsoft Word are available for both the Windows operating system and macOS. Figure 4.4 shows the Windows version of Microsoft Word 365.

You can start a new word-processing document from a blank page, or you can use a previously created and stored model called a **template**. Microsoft Word provides access to a large library of templates that are stored online. Templates are available to help you create newsletters, flyers, reports, greeting cards, certificates, and many more document types.

After creating a document, you can freely type and edit the text by using the keyboard's arrow keys to move the **insertion point** (blinking vertical cursor). You can use the Backspace key to remove text to the left of the insertion point and the Delete key to remove text to the right. To delete an entire block of text, you select the text by dragging across it with the mouse and then pressing Delete.

Perhaps the most valued editing feature in a word-processing program is the **spelling checker**; it checks each word in a document against a word list or dictionary and identifies possible errors. A spelling checker isn't context sensitive, however. This means that it won't flag words that are spelled correctly but used incorrectly—for example, the use of *their* when *there* is correct.

A **grammar checker** checks a document for common errors in grammar, usage, and mechanics. Using a grammar checker is no substitute for the careful review of a knowledgeable editor, but running a grammar check can be useful for identifying problems such as run-on sentences, sentence fragments, double negatives, and misplaced apostrophes.

Word-processing software allows many different types of formatting—that is, manipulation of the text to change its appearance at the word, paragraph, or document level. **Text formatting** is applied to individual characters of text; you can

Hands On
Using a Word Template to Create a Flyer

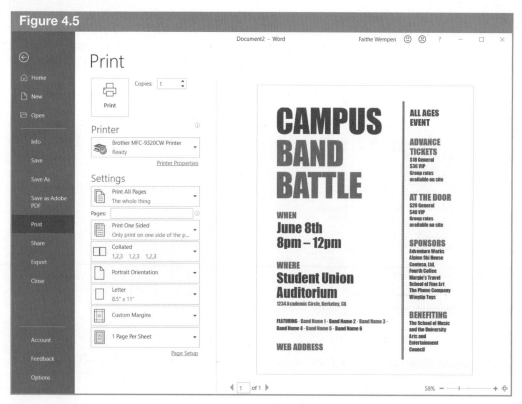

Figure 4.5

In Microsoft Word, the Print Preview feature allows you to see exactly what your document will look like when you print it.

change the font, size, and color of the text. **Paragraph formatting** is applied to entire paragraphs at a time; you can adjust the line spacing, indentation, horizontal alignment between margins, and use of bulleted and numbered lists. **Document formatting** (also called *page formatting* or *section formatting*) is applied to the entire document, page, or section; you can set the paper size, margins, number of columns, and background color. To save time and ensure consistency, you can create and apply styles to a document. A **style** is a named combination of formats; for instance, you can apply a certain font size, font color, and horizontal alignment to all headings in a document.

After creating a document, you can save it for later use and perhaps print a hard copy of it and/or send it to someone by email. In Microsoft Word, when you open the controls for printing, you see a **print preview** of the document. If you see any problems in the document, you can go back and fix them before completing the print command. Figure 4.5 shows the Print controls in Word.

Desktop Publishing Software

Desktop publishing software allows users to create documents with complex page layouts that include text and graphics of various types. At a very basic level, desktop publishing software is similar to word-processing software. Both can be used to combine text and graphics to create many different types of documents. However, desktop publishing software is more sophisticated and more focused on the page layout; it enables many complex layout and typesetting actions. People who design books, magazines, and advertising use desktop publishing software to create professional-quality layouts.

Chapter 4 Using Applications to Tackle Tasks

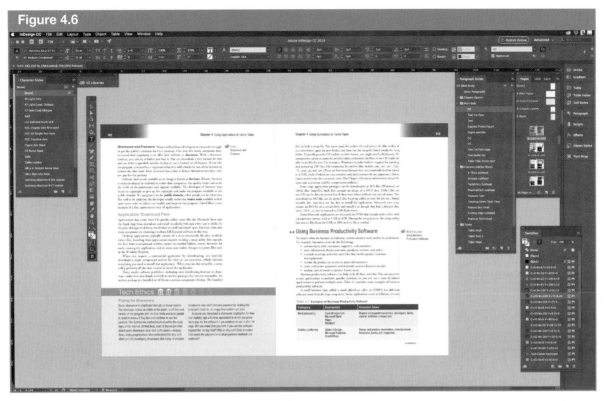

Figure 4.6

High-quality color publications such as textbooks can be designed and produced using desktop publishing software like Adobe InDesign.

Consumer-level desktop publishing software is available (such as Microsoft Publisher), but most desktop publishing software is oriented toward professionals. Programs such as Adobe InDesign and QuarkXPress are expensive and take a lot of time and experience to master. Even so, they are rich in features and can create nearly any special effect you can imagine in print.

A key feature of most desktop publishing software is that all content exists in movable frames. Text isn't placed directly on the background but is contained in one or more frames. This feature makes it easy to create multicolumn layouts and have each column be the exact size and shape you want. To help the designer position the floating frames consistently from page to page, page templates provide nonprinting column and margin guide lines onscreen.

InDesign, a feature-rich program from Adobe Systems, allows the user to create a master page that establishes the formats of repeating elements (such as the page number and the chapter number and title) on all pages of a publication. InDesign also includes content placement features that enable the user to specify how each page will look. Graphics can be cropped and placed precisely on the page. When a graphic is inserted into a publication, InDesign displays tiny rectangles at its edges; dragging the rectangles to the desired position allows resizing of the graphic. This textbook was designed and laid out in InDesign (see Figure 4.6).

Spreadsheet Software

Spreadsheet software provides a means of organizing, calculating, and presenting financial, statistical, and other numerical information. For example, an instructor might use a spreadsheet to calculate student grades. Microsoft Excel is the most

Hands On
Creating an Excel Spreadsheet

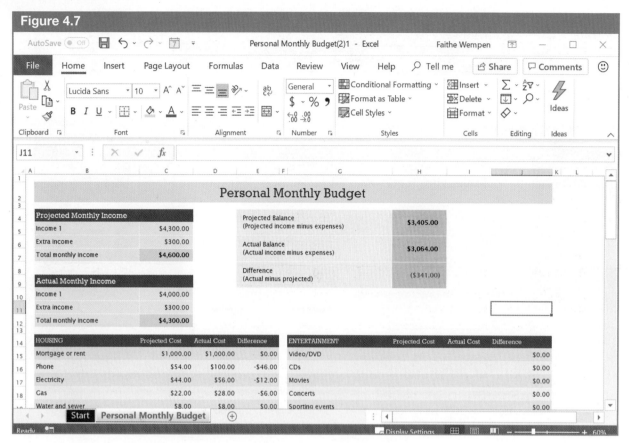

Using a spreadsheet like this one in Excel helps analyze numerical data—in this case, to determine whether personal spending is within a budget.

popular spreadsheet software in use today; it has no major competitors, although a few open source and web-based spreadsheet applications are available.

Businesspeople use spreadsheets to evaluate alternative scenarios when making financial decisions—"what if?" calculations. For example, an event planner might ask, "What if the cost of staffing goes up but the cost of paper supplies goes down? How will our budget be affected?" These types of questions can be answered quickly and accurately using a spreadsheet. A numerical **value** and a mathematical **formula** are entered into the spreadsheet, as shown in Figure 4.7, and then the numerical values are changed to represent different scenarios.

Some types of data manipulation are too complex to set up as formulas. To allow working with these data, the software provides **functions**, which are named, preset actions such as calculations, logical operations, and transformations. For example, the AVERAGE function counts the number of values, sums them, and then divides by the number of values; and the PMT function calculates the payments for a loan based on its terms.

"What if?" calculations are useful in making decisions at home as well. For example, when deciding what type of mortgage loan to apply for, prospective homeowners can compare two or more loans side by side to determine which one saves them the most money or has the lowest monthly payment.

Some common features of spreadsheets include the following:

- **Grid.** Spreadsheets display numbers and text in a matrix (grid) formed of columns and rows. Each intersection, or **cell**, has a unique address that consists of the column and row designations. Columns are usually identified

alphabetically, and rows are numbered. For example, cell address A1 refers to the cell located at the intersection of column A and row 1.
- **Formatting.** Numbers may be formatted in a variety of ways, including decimal point placement (1.00, 0.001), currency value ($, £, ¥), or positive or negative quantity (1.00, −1.00). Cells can be bordered and shaded, and text can be presented in different fonts, sizes, and colors.
- **Formulas.** Mathematical formulas—ranging from addition to standard deviation—can be entered into cells, and they can process information derived from other cells. Formulas use cell addresses, not cell contents. For example, a formula might direct a program to multiply cell F1 by cell A4, and then it will multiply the numerical contents of the two cells. The use of cell addresses means that a spreadsheet can automatically update the result if the value in a cell changes. A spreadsheet formula always begins with an equal symbol (=), like this: =F1*A4.
- **Charting.** A **chart** is a visual representation of data that often makes the data easier to read and understand. Spreadsheet programs allow users to display selected data in line, bar, pie, and other chart forms.
- **Macros.** Most spreadsheets allow the user to create a **macro**, which is a set of commands that automates complex or repetitive actions. For example, a macro can be created to check sales figures to see if they meet quotas and then compile a separate chart for those figures that do not. The macro will automatically perform all the steps required.

Database Management Software

Before computers, organizations typically stored employee, voter, and customer records in file folders housed in metal cabinets, along with thousands of other folders. Locating a particular folder could be time-consuming and frustrating, even if the records were stored in an organized manner.

Many of these manual storage systems have been replaced with electronic databases, which use software to manage data more efficiently. A traditionally organized database is a collection of data organized in one or more data tables. Each data table contains records. A record is a collection of all the data about one specific instance, such as a person, a business, an item, or a transaction. Each individual piece of data in a record is called a *field*, and each field is filled with data related to its field name. For example, in a database that stores information about customers, *FirstName* is one field name and *LastName* is another field name. Each record in a table uses the same field names, and you can sort and filter the database by one or more of those field names. For example, in the same customer database, you can filter it to display only records for which the *City* field shows "Chicago". You can also provide multiple filter criteria, such as only records in which the city is Chicago and the transaction date is less than six months ago.

The software you use to manage a database is called a **database management system (DBMS)**. A large database (sometimes called an **enterprise database**) is stored on a server and managed using professional tools such as Oracle Database and structured query language (SQL). The data can be stored on a local server or on a remote server accessed through a company's data network or a cloud. Multiple users can update and query the database simultaneously using any of a variety of interfaces, both command line and graphical.

Video
How Can Databases Make a Difference in Our Lives?

Personal DBMS applications (such as Microsoft Access and FileMaker Pro Advanced) enable ordinary users to create and manage databases on a small scale. Personal DBMS applications usually have a friendly graphical user interface (GUI) and many tools for easily generating helper objects, such as data entry forms and printable reports. Figure 4.8 shows a data table in Microsoft Access.

Most databases have more than one table, and all the tables are somehow related. Such a database is known as a *relational database*. Figure 4.9 shows a relationships diagram for a database that keeps track of product sales. Notice the tables that have relationships to the Orders table: Customers, Order Details, and Salespeople. When a customer places an order, the salesperson opens an order entry form and chooses his or her own name from the list of salespeople on the form (from the Salespeople table).

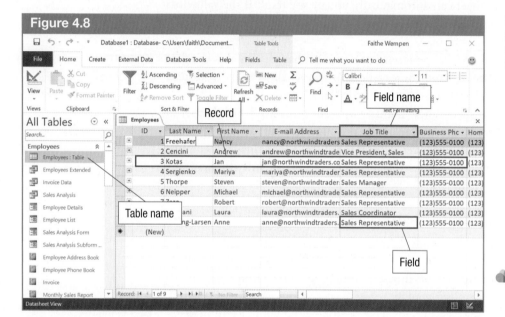

Microsoft Access displays tables in a spreadsheet-like grid for easy browsing.

Practice
A Relational Database Example

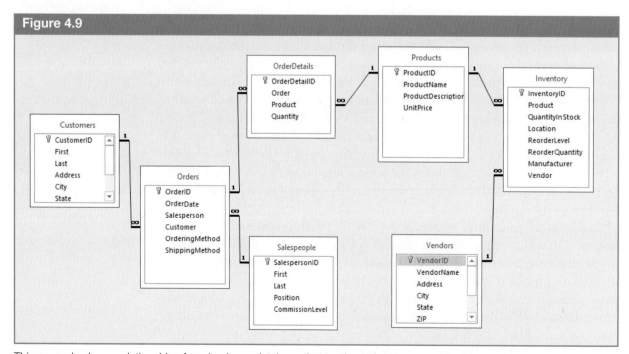

This example shows relationships for a business database that tracks orders.

The salesperson then uses a link to the Customers table to locate the customer's account details and link them to the order. The salesperson selects the items the customer wants from a link from the Products table and specifies a quantity for each product; that data is then added to the Order Details table. As the salesperson adds each item to the order, the database interface checks the inventory level in the Inventory table for each ordered product and alerts the salesperson if any items are out of stock.

Databases are most often created by computer professionals but used by ordinary end users. Whereas the developer who creates a database might interact with the database by typing text commands at a command line prompt on an enterprise-level system, hundreds of clerks and administrative assistants in multiple locations might access the database via a graphical point-and-click interface. Because the people who use the database may not have the same level of expertise as its developer, most database systems enable the developer to create *helper objects* to make the database friendlier and more easily accessed for end-users. For example, helper objects in Microsoft Access can include the following:

- **Forms.** A form is an easy-to-navigate, onscreen document that enables end users to enter, edit, delete, and search for records.
- **Reports.** A report is an attractively laid out, printable document that summarizes raw data with subtotals and other calculations.
- **Queries.** A query is a set of saved sort-and-filter specifications that show only certain records and fields based on specified criteria. Queries can also be used to temporarily join records from multiple tables, so the combined data set can be used for a report, form, or other query.
- **Macros.** A macro is a saved group of commands that can be executed with a single command.

Tables, forms, reports, queries, and macros are all **objects** in an Access database.

Presentation Software

Anyone who has attended group lectures or presentations knows how boring they can be. One way to make a presentation more interesting is to use presentation graphics software. **Presentation graphics software** allows users to create computerized slide shows that combine text, numbers, animation, and graphics, as well as audio and video. Microsoft PowerPoint is the most popular application in this category.

With presentation graphics software, you can create slide shows that combine text, graphics, audio, and video.

A **slide** is an individual page or screen that's created in presentation graphics software. A **slide show** may consist of any number of individual slides. For example, a sales representative might use a slide show to promote a product to customers, highlighting the product's key features. An instructor might use a slide show to accompany a lecture and make it more engaging and informative. A businessperson might use a slide show to deliver information and present strategies at a meeting.

Presentation software accepts content in a variety of external formats, such as text, photos, videos, and music. With the software, users can repurpose existing content in new ways to suit different audiences. Figure 4.10 shows several examples of the types of content you can create and import in PowerPoint.

> **Hands On**
> Exploring PowerPoint

A presentation can be delivered by computer in a real-time environment with a live audience—either in person or remotely over the internet—or it can be recorded to a computer file for later playback. In addition to presenting a slide show on a computer, the user can also output the slides in a variety of ways, including DVD videos, still-image graphics (one graphic per slide), portable document format (PDF) files, and hard-copy handouts.

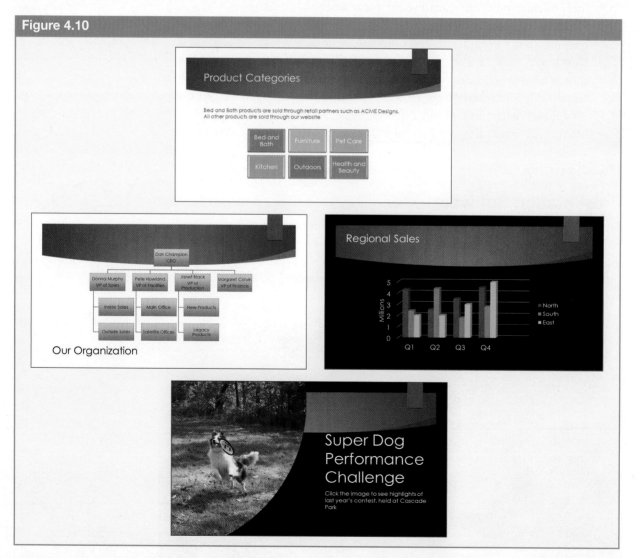

Figure 4.10

With Microsoft PowerPoint, users can create slides with rich content, including appealing backgrounds, effective titles and subheads, compelling bullet points, colorful charts and graphs, clear flowcharts, interesting photos, and even audio, video, web links, and other media elements.

Software Suites

Some software vendors bundle and sell a group of applications as a single package called a **software suite**. Microsoft Office is the best-known software suite; it consists of a word-processing program (Microsoft Word), a spreadsheet program (Microsoft Excel), and a presentation application (Microsoft PowerPoint). Some versions of Microsoft Office contain other programs too, such as a DBMS (Microsoft Access) and an email and personal information manager (Microsoft Outlook). Microsoft sells Office 2019 as a one-time purchase (Office 2019) or as a monthly or yearly subscription (Office 365). The applications are very similar in both purchase arrangements, but Office 365 applications are automatically updated with new features periodically.

Tech Career Explorer
Microsoft Certifications

A software suite is typically less expensive than the combined cost of the applications when purchased separately. Because all the applications are designed with a common interface, the basic features and functions work the same way in every program. This means there's less to learn when you start using an unfamiliar application.

Another advantage of software suites is their ability to share content between applications. For example, both Microsoft Word and Microsoft PowerPoint easily accept charts and spreadsheets from Microsoft Excel. You can freely copy and paste these materials between applications. You can also link or embed content from one application or data file into another. When you **embed** content, the pasted content retains its memory of what application it came from, and if you double-click that embedded content, it reopens in its native application. When you **link** content from one data file to another, the pasted content retains its memory of not only its original application but also its original data file. If the original data file changes, the linked content in another file changes automatically the next time that other file is opened. Embedding and linking are parts of the content-sharing feature in Windows and Office programs known as **object linking and embedding (OLE)**. You don't have to have a software suite to use OLE; it works between any applications that support it (and most do).

Personal Information Management Software

Businesspeople need to keep in contact with a wide variety of people such as customers, suppliers, and coworkers, and they need to organize the contact information for all these individuals, along with their scheduled meetings and phone calls. **Personal information management (PIM) software** fulfills this need by providing an address book, calendar, and to-do list in one convenient interface. Some PIM software also

Practical TECH

Syncing between Phones and Computers

Many smartphones have calendar, email, contacts, and to-do list utilities. If your phone and your computer run operating systems by the same company (for example, if they are both Apple) then you should have no trouble synchronizing PIM data between the two devices without any special setup.

If your phone and computer have different platforms (such as an iPhone and a Windows desktop, or an Android phone and an Apple Mac laptop), then synchronization may be more difficult. If you have an iPhone and a Windows computer, you can use the iTunes application to synchronize your phone and Microsoft Outlook, as described at https://CUT7.ParadigmEducation.com/SyncingMSContacts. If you have a macOS computer and an Android phone, you can use a third-party utility such as Android File Transfer (available at https://CUT7.ParadigmEducation.com/AndroidFileTransfer) or AirDroid (available from the Google Play Store for your phone).

Figure 4.11

Microsoft Outlook includes a full-featured calendar application.

sends and receives email. Microsoft Outlook is best known as an *email application*, but it also offers a full array of PIM components, including Calendar, People, Tasks, and Notes (see Figure 4.11).

Some people choose to use online PIM tools rather than tools that are installed on their individual devices. One popular online PIM tool is Google Calendar, which allows multiple people to share calendars. With an online calendar, contact list, and to-do list, you don't have to worry about the copy on any of your devices ever being out of sync.

Project Management Software

Many businesses regularly engage in planning and designing projects, as well as scheduling and controlling the various activities that occur throughout the term of a project. For example, before a construction firm begins to erect a building, it needs to develop a comprehensive plan for completing the structure. During planning, an architect prepares a detailed building design, or set of blueprints. Schedules are then prepared so that workers, building materials, and other resources are available when needed. Once construction begins, all activities are monitored and controlled to ensure that they are started and completed on schedule.

Before computers were available, projects like this were planned, designed, scheduled, and controlled manually. Today, these tasks are performed electronically using **project management software**. This type of software helps manage complex projects by keeping track of schedules, constraints, and budgets. It can be used for most kinds of projects, including those involving construction, software development, and manufacturing. The most frequently used project management software today is Microsoft Project; it helps users optimize the planning of projects and track schedules and budgets.

Article
Project Management Software

Chapter 4 Using Applications to Tackle Tasks

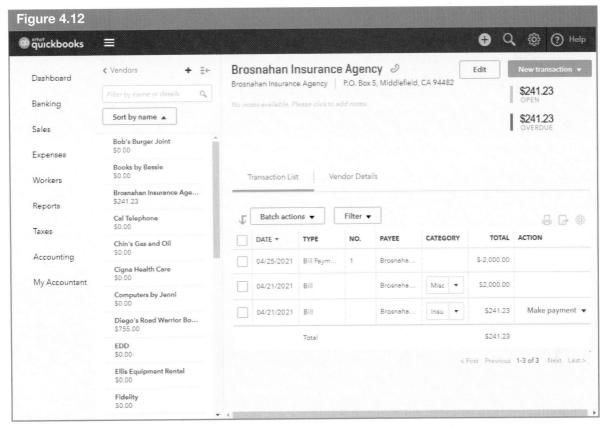

Figure 4.12

Intuit QuickBooks helps small businesses manage their accounting and finances. It is available both as a web application (shown here) and as desktop software.

Accounting Software

Keeping track of a business's financial affairs can be complicated—much too complicated to manage with a simple spreadsheet. **Accounting software** integrates all the financial activities of a company into a single interface. It can track money coming in and going out; monitor bank account transactions and statements (including connecting online to the bank); create purchase orders, estimates, and invoices; and even process employee time cards and payroll.

Large businesses typically have their own enterprise-level software that multiple users can access simultaneously from different computers. Small businesses might use an off-the-shelf product, such as Intuit QuickBooks (shown in Figure 4.12). Accounting software can save money for a small business because a full-time accountant may not be needed. Office workers can handle the bulk of the accounting throughout the year, with an accountant involved only at tax time.

Note-Taking Software

Gathering information often involves collecting multiple types of input. For example, if you are researching a new product, you might have digital ink sketches made at a planning meeting, handwritten notes, typed notes in several different formats, website addresses, and photos from a digital camera. Microsoft OneNote enables users to collect many types of data in a single place. It integrates with the OneNote app on a variety of platforms, so you can take photos and collect information wherever you are. Alternatives to OneNote include Evernote and iCloud Notes.

4.3 Selecting Personal Productivity and Lifestyle Software

Watch & Learn
Selecting Personal Productivity and Lifestyle Software

Personal software is designed for individuals to use for their own enrichment and entertainment. If you want to accomplish a task that isn't related to your work, chances are you can find an application that will help you do it more easily. Popular categories of personal software include lifestyle and hobby, personal finance and tax, legal document preparation, and educational and reference.

Lifestyle and Hobby Software

Lifestyle and hobby applications help you live the way you want to live and make the most of your special interests. Here are just a few examples:
- **Home and garden architecture.** Plan the home you want to build and where to put the furniture. Lay out your garden, specifying the location for each plant.
- **Diet and fitness.** Track your calorie intake, and plan and track your exercise and fitness activities.
- **Home inventory.** Create a database listing everything you own; you can use it to file an insurance claim for a disaster such as a fire.

All the things that these specialized applications do can also be accomplished using general purpose software; you don't necessarily need special programs. For example, you can do home design using a CAD program, track diet and fitness using a spreadsheet, or create a home inventory using a DBMS—if you know how to use these general purpose programs. The specialized programs are designed to be simple, so you can enjoy their benefits without having any previous software experience.

Legal Software

Legal software is designed to help analyze, plan, and prepare a variety of legal documents, including wills and trusts. It can also be used to prepare other legal documents, such as the forms required for real estate purchases and sales, rental contracts, and estate planning. Most packages include standard templates for various legal documents, along with suggestions for preparing them.

This type of program begins by asking the user to select the type of document he or she wants to prepare. To complete a document, the user may need to answer a series of questions or enter information into a form. Once that task has been completed, the software adapts the final document to meet the individual's needs. After a document has been prepared, it can be sent to the appropriate department, agency, or court for processing and registration. It's always a good idea to have an attorney review documents to make certain they are correct and legal in the intended state or local jurisdiction.

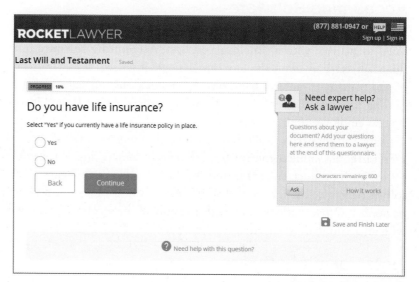

Legal software guides people in creating simple legal documents.

Figure 4.13 Quicken is a popular application for managing personal finances. It includes easy-to-use tools for paying bills, balancing a checkbook, tracking income and expenses, maintaining investment records, and performing other financial tasks.

Personal Finance and Tax Preparation Software

Personal finance software helps users pay bills, balance checkbooks, track income and expenses, maintain investment records, and perform other financial activities. Figure 4.13 shows Quicken, which is a popular personal finance application. This software also enables users to view reports and charts that show how they are spending their money.

Some personal finance programs provide services on the internet and web. Users can go online to learn the status of their investments and insurance coverage, for example. They can also conduct normal banking transactions, such as accessing and printing bank statements that show summaries of monthly transactions. These programs can perform most of the financial activities that used to require mail or telephone contact.

Tax preparation software is designed to aid in analyzing federal and state tax status, as well as preparing and filing tax returns. Most tax preparation programs provide tips that can help identify deductions, possibly resulting in great savings. Some programs include actual state and federal tax forms for entering tax data. Programs that don't include these forms provide instructions for downloading them from the software publisher's website. Completed tax returns can be printed for mailing or filed electronically. Because federal and state tax laws change frequently (as do tax forms), users should be sure to use the software version for the appropriate year or period.

Educational and Reference Software

A computer can be an unparalleled learning tool, especially with the wealth of resources available online. Among the free educational and reference materials on websites are online dictionaries and encyclopedias. You can also purchase educational and reference software—either on a DVD or as a download for local use on your personal computer or as a subscription to an online learning service. You might even take classes at colleges and universities that are delivered entirely through internet-accessed software.

Encyclopedias and Dictionaries Almost everyone has used an encyclopedia or dictionary at one time or another. An encyclopedia is a comprehensive reference work that contains articles on a broad range of subjects. Before computers, encyclopedias and dictionaries were available only in book form. Now, they are available both on the web and on DVDs you can purchase. Figure 4.14 shows a word that's been looked up at Dictionary.com.

Cutting Edge

To Catch a Plagiarist

Think twice about getting a term paper from an essay-writing website or online database. State-of-the-art antiplagiarism software can expose your cheating in a nanosecond.

Plagiarism—the passing off of someone else's writing as your own—has become epidemic on college campuses. According to one study, 36 percent of students admit to having plagiarized a written paper during their college careers. The reasons for the growth of this behavior may be varied. For instance, some college students left high school without being properly educated in citing sources for research papers. Also, today's students, who file-swap music without a care, have a foggier definition of concepts such as copyright. And of course, using the internet makes finding information and then copying and pasting it into a document all too easy.

To combat the misappropriation of the written word, colleges are turning to software such as Turnitin and SafeAssign. These antiplagiarism programs generate a "digital fingerprint" of a document that has been submitted electronically. The document's verbal patterns are cross-checked against a huge database of internet, newspaper, and encyclopedia archives, as well as previously submitted student work. Suspicious sentences and paragraphs are highlighted, and source matches are annotated. The thieving writer is busted in seconds.

Students can also use plagiarism-detection programs as self-screening tools to make sure they present and cite information properly. And even for classes whose professors don't use these programs, their very existence discourages students from copycat writing.

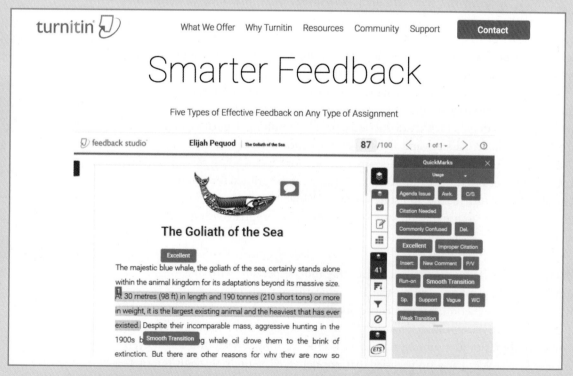

This example from Turnitin shows the overall percentage of nonoriginal work in the submitted paper and identifies specific copied passages and their sources.

Chapter 4 Using Applications to Tackle Tasks

Tutorials Many people learn new skills by using computer-driven tutorials. A **tutorial** is a form of instruction in which the student is guided step by step through the learning process. Tutorials are sometimes referred to as **computer-based training (CBT)** or *web-based training (WBT)*.

Article
Computer-Based Learning

Tutorials are available for almost all subjects, including how to assemble a bicycle, use a word processor, and write a letter. The student and instructor resources for this

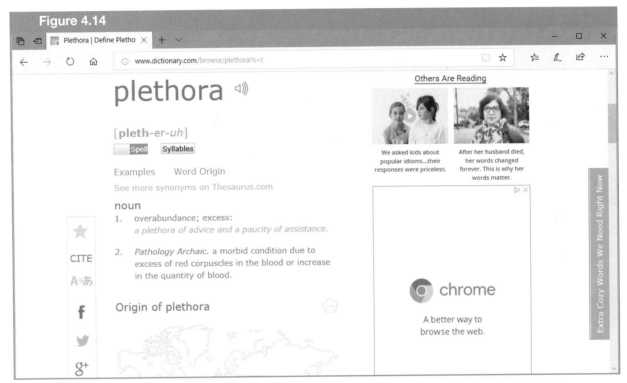

Figure 4.14

Some online dictionaries include features such as audio pronunciations, images, and example sentences.

Tech Ethics

Should You Cite Wikis in Academic Papers?

Wikipedia is the world's largest and most popular online encyclopedia; it contains entries on millions of topics. The term *wiki* refers to an online application that allows people to collaborate by uploading and editing content. In other words, a wiki is a shared project in which anyone may participate.

Some quality safeguards are in place to prevent obvious untruths from being stated in wikis. But wikis are by nature vulnerable to containing misinformation, whether it's placed there accidentally or intentionally. For example, rival scientists might update the wiki page on their competitors' research to indicate that the research is unsuccessful, or a rival business could edit the language on a competitor's page to make the competitor's products seem unreliable.

This potential for misinformation is an important difference between *Wikipedia* (**https://CUT7.Paradigm Education.com/Wikipedia**) and traditional research sources, such as *Encyclopædia Britannica* (**https://CUT7 .ParadigmEducation.com/Britannica**), in which research experts carefully check each article. Many teachers (especially at the college level) don't accept *Wikipedia* as a cited source in academic papers because of its inconsistent quality and accuracy. When you are writing research papers for your own classes, be sure to check with your instructors to find out their policies on citing wikis.

Figure 4.15

Tutorials such as the video activities for this textbook help students understand concepts by providing a combination of text, graphics, animation, and interaction.

book include dozens of tutorials; you may have already been assigned some of them in this class, such as the one in Figure 4.15.

4.4 Exploring Graphics and Multimedia Software

Graphics and **multimedia software** enable both professional and home users to work with graphics, video, and audio. A variety of applications are focused in this area, including painting and drawing, image editing, video and audio editing, web authoring, and computer-aided design (CAD) software.

Watch & Learn
Exploring Graphics and Multimedia Software

Painting and Drawing Software

Painting and drawing programs are available for both professional and home users. The more expensive professional versions typically include more features and greater capabilities than the less expensive personal versions.

Painting software enables users to create and edit bitmap images. As you read in Chapter 2 in the section discussing scanners, a *bitmap* is a grid of pixels, or a map of bits (which is where the name comes from), and each pixel has a color assigned to it. A bitmap image can be created from scratch in a painting program, or it can be imported from a digital camera, digitized with a scanner, captured from a computer screen, or acquired as a file. Another name for a bitmap image is a **raster image**.

Hands On
Trying Out Painting

Adobe Photoshop (shown in Figure 4.16) is a popular and powerful painting program that has many tools for creating new bitmap images and editing photos professionally. Microsoft Windows comes with a very basic 2-D painting program called Paint and a more feature-rich version designed to create three-dimensional (3-D) images called Paint 3D.

Hands On
Trying Out Paint 3D

Drawing software creates and edits images called *vector graphics*. A **vector graphic** is formed using mathematical formulas in the same way that lines and shapes are formed using geometry. Because a particular shape is based on math, it can be resized freely without distorting it. In contrast, a bitmap image can become jagged edged or distorted when resized, especially when it's made larger. Vector graphics also take up much less disk space than bitmap images do, because less data is involved in creating them. The main drawback to vector graphics is that they can't generate photo-realistic

Figure 4.16

Adobe Photoshop is a painting program that's useful for editing digital photos as well as creating new artwork.

images. CorelDRAW and Adobe Illustrator are two full-featured applications that specialize in generating vector graphics. Microsoft Windows doesn't come with a drawing program, but you can experiment with the drawing tools in Microsoft Office applications such as Word, Excel, and PowerPoint.

Photo-Editing Software

Photo-editing software is designed specifically for manipulating photos and has little or no capability for creating original images. For example, the Photos app in Windows 10 transfers pictures from a digital camera to your computer and then organizes them, applies various types of corrections to them, annotates them, and more; you can download this free application from the Microsoft Store. Apple also offers a free photo editing program called iPhoto.

Adobe Photoshop Elements is a basic version of Photoshop that can create new bitmap content, but it's best known for its photo-editing capabilities. Amateur photographers can use this program to fix problems in their photos—for instance, people with red eyes and images that are too dark or too light. Photoshop Elements can also apply interesting special effects to pictures to make them look like watercolor paintings, pencil sketches, photographic negatives, and line drawings.

3-D Modeling and CAD Software

Vector drawing programs can be used to create 3-D models. A **3-D modeling program** is used to generate the graphics for blueprints and technical drawings for items that will be built, as well as 3-D representations of people and objects in games. For example, if you have ever played *The Sims* or *Second Life*, you have seen humanlike avatars developed using 3-D vector graphics.

Each 3-D vector graphic is first drawn as a **wireframe**, which is a surfaceless outline of an object composed of lines. If you were to bend pieces of wire to create the outline of an object, it might look similar to an onscreen wireframe image. Next, a pattern, or **texture**, is added to the surface of the object to make it look more realistic. Finally, light and shadow are added to complete the effect. Figure 4.17 shows a computer-generated chessboard with one of the pieces shown in wireframe mode so you can see how it's constructed.

Technical object modeling with vector graphics is also called **computer-aided design (CAD)**. CAD software enables professionals to create architectural, engineering, product, and scientific designs with a high degree of precision and detail. For example, engineers can use CAD software to design buildings and bridges, and scientists can use it to create graphic designs of plants, animals, and chemical structures. Most CAD software can also create and display designs in three dimensions, so they can be viewed from various angles.

Once a design has been created, changes can easily be made to it until it's finalized. For instance, an aircraft engineer designing a new type of aircraft can use CAD software to test many versions of the design before it's approved and the first prototype is built. Using this process can save companies considerable time and money by eliminating defective designs before beginning production.

Figure 4.17

Vector-based graphics programs store images as a series of geometric shapes.

Practical TECH

Graphics File Formats

The term *bitmap* provides a general description of a type of file, but it is also a specific file format (BMP) with a *.bmp* extension. Some other file formats that are also bitmap graphics, in the larger sense of the word, include portable network graphics (PNG), graphics interchange format (GIF), Joint Photographic Experts Group (JPEG, pronounced "jay-peg"), and tagged image file format (TIFF). Each of these formats has its own benefits and drawbacks.

Some graphics formats support greater color depth than others. For example, a GIF file has a limit of 8 bits per pixel, for a maximum of 256 colors per image. This means that GIF isn't a very good format for photos. The PNG format supports up to 24-bit color. Both GIF and PNG are designed primarily for online use; they don't work well for high-quality print publications, such as magazines.

Both PNG and GIF employ *lossless compression* to make file sizes smaller. In lossless compression, the file loses no quality by being compressed, and when it's decompressed, it retains its original properties. Other graphics formats, however, employ *lossy compression*. When a file is saved in such a format, its quality is permanently reduced to make the file size smaller. JPEG is an example of a format that uses lossy compression. JPEG format's compression is adjustable. The user saving the file can specify an amount of compression and corresponding amount of loss of image quality—whatever is appropriate for the situation at hand. In contrast, the BMP image format does not employ compression of any kind.

TIFF format is designed for high-quality images in print publications. High-quality digital cameras usually save pictures in TIFF format. While TIFF does support some lossless compression, overall its files are larger than the same size images in other formats.

CAD software helps technical professionals design objects for manufacturing and construction.

Full-feature CAD applications, such as AutoCAD, are very expensive. Using them may also require special computer workstations that have more powerful processors and more RAM than standard desktop PCs typically have. Some producers of CAD software offer scaled-down versions of their more expensive software for use by small businesses and for individual and home use.

Tech Career Explorer
CAD across the Board

Animation Software

Film animation used to be a labor-intensive industry, with hundreds of artists drawing cartoon images frame by frame. The same was true of stop-motion photography; to capture live action one frame at a time, an assistant moved a model a fraction of an inch for each new frame. Movie making is changing, however, and more and more films are being made with the aid of **computer-generated imagery (CGI)**. With CGI programs, no pen touches paper to create an animated world. And the quality is so good that the effects created with CGI are more or less indistinguishable from those created using models and other tricks in older movies.

The cost savings in movie making are also significant with CGI. It's far cheaper to use a computer program to draw a dinosaur than it is to create a realistic robotic dinosaur. And instead of building and exploding a model of a spaceship or a building, the same effect can be achieved through computer animation. Even better, the computer animation can be easily manipulated; any image can be redrawn until it's just exactly right.

CGI isn't something that the average user would attempt, because the software is expensive and the learning curve for using it is steep. One popular application that professionals use is LightWave by NewTek. Many different tools for CGI professionals are also available from Autodesk, a company that's a leader in CAD software.

Audio-Editing Software

Computers can store, edit, copy, and reproduce sounds, as you read in Chapter 2 when discussing audio adapters. Sound files vary first and foremost depending on their origin: analog or digital. A **waveform** (often abbreviated *wave* or *WAV*) file is a sound file that has an analog origin—that is, an origin outside a computer system. If you sing into a microphone connected to a computer or record an instrument being played, the sound waves coming from it go through a process called **sampling**, in which they are recorded thousands of times per second, creating a digitized version. When the digitized clip is played back, it sounds to a human ear just like the original live version.

The quality of the recording depends in large part on the number of samples taken per second. A human voice can be captured clearly when sampled at about 11 kilohertz (KHz), or 11,000 digital measurements per second. For this reason, voice equipment such as digital cell phones and voice recording devices usually encode sound at or near this rate. Clear, sharp musical reproductions are recorded at 44 KHz—almost four times the rate used for voice recordings. Creating these recorded sounds as stereo (two-channel) music requires twice as many measurements per second. That's the quality level of most music CDs and DVDs.

The number of bits used to describe each sample is known as the **bit depth**. CD-quality music uses 16 bits per sample, resulting in a file that is about 10 megabytes (MB) in size per one minute of recording. That means that one CD with a 650 MB capacity holds just over an hour's worth of music. Audio CDs use a special file format called **CD audio (CDA)** to store music files. The commercial audio CDs you buy store tracks in this format, and so do those you burn yourself using your computer's CD burner.

A **wave file format** is a noncompressed file type identified with a *.wav* extension, used to produce any kind of sound. Like a graphic file, a wave file can be compressed to make it take up less disk space. The most popular compression format is the Moving Pictures Expert Group Audio Layer III (MP3) format. This format takes a wave file and reduces its size by about 90 percent, leaving behind a high-quality reproduction. This means that a 30 MB wave file will end up as a 3 MB MP3 file. MP3 uses a data compression system similar to that used in JPEG files. Only a highly sensitive ear can hear any difference in the music quality after compression. This dramatic reduction in size, combined with very little loss in quality, has led to an explosion in the use of the MP3 file format. Relatives of the MP3 format include Moving Pictures Expert Group Layer IV (MP4), Moving Pictures Expert Group Layer IV Audio (M4A), and Moving Pictures Expert Group Layer I or Moving Pictures Expert Group Layer II (MPEG or MPG).

The main drawback to using a compressed format for audio files is that not all music players support such formats. Older car and home stereo systems may not support MP3 and other file types; they may require CDs in the traditional CDA format. Other wave compression file formats include Windows Media Audio (WMA) format, which was developed by Microsoft, and the Advanced Audio Coding (AAC) format, which was designed to be a successor to MP3.

In contrast to a waveform file, an all-digital music file isn't analog in origin and it's created, not recorded. This type of music clip is known as **Musical Instrument Digital Interface (MIDI)**, and it's created by digitally simulating the sounds of various musical instruments. A digital piano keyboard can be connected to a computer to record MIDI music, and the software that accepts the input can be programmed to make the piano sound like any of dozens of different instruments. A skilled musical composer can build an entire orchestral performance—woodwind, string, brass, and percussion instruments—with a single digital keyboard by allowing the MIDI software to simulate each instrument as its part is performed.

An advantage of MIDI files is that they are extremely small and compact compared with wave music files. The main drawback of MIDI is that the output doesn't sound

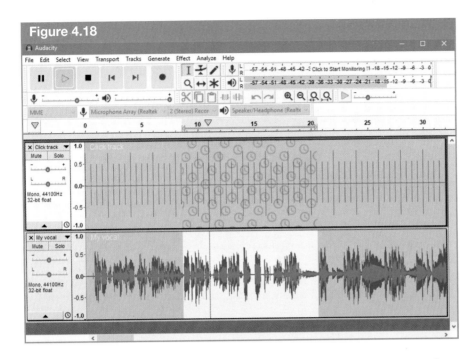

Figure 4.18

Audio-editing programs, such as Audacity, represent sound waves graphically and enable you to crop and adjust them.

exactly like the instruments it's simulating (although with high-quality software and hardware, it can come close).

You can record and edit audio files with a variety of applications: from free hobbyist-quality programs to those designed for professional audio technicians. Figure 4.18 shows a screenshot of Audacity, a free audio-editing program that supports many formats and capabilities. Consumer-level programs, such as Roxio Creator NXT, offer a variety of applications designed for casual users, including the ability to burn CDs and DVDs and perform basic editing.

Video-Editing Software

Video clips are essentially groups of still images, or **frames**, that are played back in a certain order, at a certain rate (the frame rate), and accompanied by an audio soundtrack stored within the same container file. **Video-editing software** enables you to import digital video clips that you have taken with your own video camera or acquired from another source and then modify them in various ways. For instance, you can combine video with audio, superimpose text over the top of the video clip, edit out certain sections, and much more, depending on the video-editing software you are using. Some common video file types include Audio Video Interleave (AVI), Flash Video (FLV), Matroska Multimedia Container (MKV), and MPEG-4 (MP4).

Video-editing software can also compress video clips. The capacity for data compression with movies is in many ways greater than with music or still images. In a digital video clip, most parts of any given frame are identical to those in the previous one. This means that for each frame, it's necessary to store only the portion of the frame that's changed from the previous frame, rather than every pixel. For this reason, compressed movies maintain good quality despite being dramatically reduced in size.

Video-editing software ranges from simple programs intended for amateurs to the powerful programs used by movie-making professionals. Adobe Premiere Pro is a full-featured tool for building movies out of different types of content, including sounds, still images, and raw video footage. Other programs for video editing include Sony Vegas Pro and CyberLink PowerDirector.

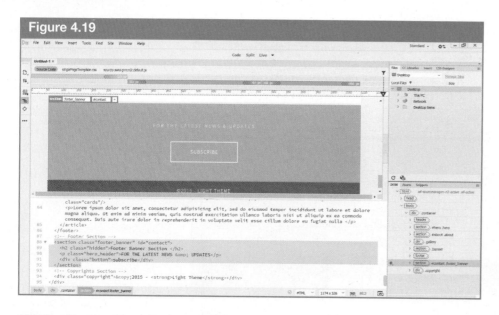

Figure 4.19

Web authoring software such as Adobe Dreamweaver lets users create web pages without having to learn HTML.

Web Authoring Software

Web authoring software (shown in Figure 4.19) helps users develop web pages without having to learn a special language. Software packages such as Adobe DreamWeaver use a **WYSIWYG** (what you see is what you get) approach to web page development. This means that during the development process, you can see the layout and content of the web page within the authoring software.

Basic web creation software focuses on generating code in *Hypertext Markup Language (HTML)*. However, not all of the content found on web pages today can be generated in HTML. More and more sites are including interactive animation and video effects that are created in other formats and then integrated into the web page.

The most popular of these formats are Adobe Flash and Adobe Shockwave. Adobe Flash is a file format in which you can create multimedia content that includes vector graphics and animations. Flash can even be used to create games and rich internet applications (RIAs) that can be played back in a web browser that has the Adobe Flash Player plug-in. However, some browsers do not support Flash playback for security reasons. Flash files are created in a program called Adobe Animate. Adobe Shockwave is a similar file format but has more capabilities. A browser that has the Shockwave Player plug-in installed can play Shockwave content, which can include both animation and user interactivity. Shockwave content is also created using Adobe Animate.

Cutting Edge

Mobile-First Design

Websites have traditionally been developed with desktop and laptop PC screens in mind, and then tweaked to accommodate smartphone users—but that's changing. More and more users access web content from smartphones, and traditional design structures such as vertical navigation panes don't translate very well to that environment.

Mobile-first design is the practice of designing a website with mobile devices in mind. Web designers increasingly create usable page designs that assume a small screen, with variables in the programming that enable the page to expand to fill any browser window width that the user happens to be using. Mobile-first design basics include prioritizing text, not using large graphics, making touch targets sufficiently large for the average-sized finger to hit them accurately on a small screen, and not counting on mouse-focused features like hovering.

4.5 Understanding Gaming Software

Watch & Learn
Understanding Gaming Software

Although many people would have you believe that they keep computers around strictly for serious reasons, the majority of them also play games on their computers. In fact, the Entertainment Software Association estimates that 60 percent of all Americans play video games daily and that the average age of a game player is 34. (See Figure 4.20 for more interesting statistics on game playing in the United States.)

These figures represent a broad section of the population, but then, gaming is a broad subject, including everything from casual Facebook games to the very latest in shoot-em-up monster action. Computer-based versions have been created of almost every traditional card and board game you can imagine, as well as casino games, fantasy-world exploration games, character-based games, city- and civilization-building games, and of course, the very popular first-person shooter games, in which you see what the character sees as you move through a virtual world.

Types of Gaming Software

Gaming software falls into three broad categories depending on the device used to play the game: dedicated gaming devices, PCs, and mobile devices like smartphones.

Applications for Dedicated Gaming Devices Companies that develop video games usually do so for multiple platforms. As you read in Chapter 3, a platform is a type of hardware. Microsoft Windows is one platform (sometimes referred to as *PC*); Apple Macintosh is another. A **game console** is a platform designed specifically for running game software. A game console typically uses a TV as a monitor and hand-held controls such as game pads, steering wheels, and joysticks for input. Motion sensors, cameras, and other physical feedback devices are also available, so your whole body can control input. These devices are also available for PCs, so you can enjoy a full gaming experience with applications developed for delivery on your personal computer.

Game consoles have powerful central processing units (CPUs) and display adapters that are optimized for graphics-intensive games with a lot of motion. Game consoles provide sound support through the TV's speakers, along with support for external speaker systems. Modern game consoles usually have internet connectivity (wired, wireless, or both), so game players can play some games collaboratively online. Users may also be able to download new games online and to download and install game

Figure 4.20

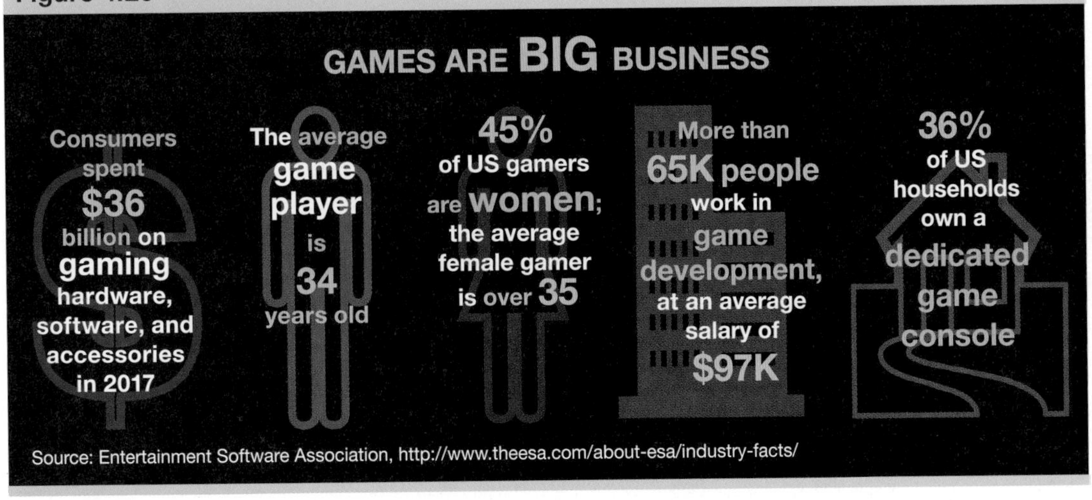

GAMES ARE BIG BUSINESS

Consumers spent **$36 billion** on **gaming hardware, software, and accessories** in 2017

The average **game player** is **34** years old

45% of US gamers are **women**; the average female gamer is over 35

More than **65K people** work in game development, at an average salary of **$97K**

36% of US households own a **dedicated game console**

Source: Entertainment Software Association, http://www.theesa.com/about-esa/industry-facts/

Game consoles typically come with one or more input devices optimized for game playing, such as the game pad (left), steering wheel (middle), and joystick (right), shown here.

updates. Some of the most popular gaming console systems are Microsoft XBox One, Nintendo Switch, and Sony PlayStation 4.

Versions of games designed for a particular game console have system requirements consistent with that console. For example, if you buy a game designed for the Xbox 360, it will play on any Xbox 360—even the model with the minimum amount of RAM and storage.

Applications for PCs With gaming software developed for installation on PCs, you will need to consult the application's system requirements to determine whether the game will run on your particular computer. **System requirements** include the minimum CPU speed, amount of RAM, and amount of disk space required to install and run the application. In some cases, an application might also require a certain brand or model of display adapter, and/or an amount of RAM dedicated to the display adapter. The system requirements are printed on the box the game comes in, if you buy it from a retail store, and can also be found on the manufacturer's website.

For example, Figure 4.21 shows the system requirements provided on the web page for *Diablo III*, a popular game that has some strict requirements for the computers on which it will run. If your computer doesn't meet or exceed the system requirements, the game might not install or run at all, or it might run slowly or with audio or video distortion.

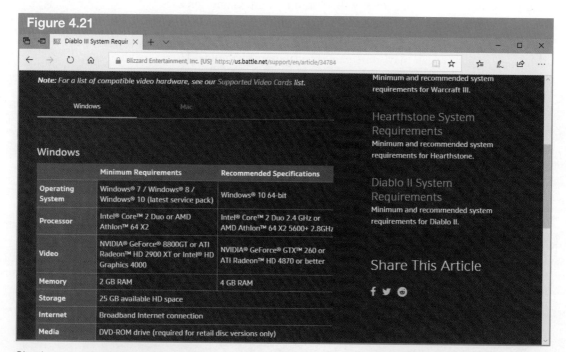

Figure 4.21 Check a game's system requirements before buying it to make sure your computer will run it well.

Game Ratings

Not all games are suitable for all users. Young children shouldn't have access to games with violent or sexual themes, for example, or games that include illegal activity or drug use. Because parents can't possibly prescreen every part of every game before their children play it, they rely on game ratings to determine what games are suitable for children of various ages.

The game rating system summarized in Table 4.2 is administered by the **Entertainment Software Ratings Board (ESRB)**. It classifies each game according to the youngest suitable user. For example, games rated *T* for *Teen* are acceptable for teenagers (ages 13+) and individuals older than that. The descriptions provided are the same ones that Microsoft provides in Windows in the *Family options* feature. (For more information about the ratings system, see https://CUT7.ParadigmEducation.com/ESRBratings.)

Table 4.2 Entertainment Software Ratings Board (ESRB) Game Ratings

Rating	Description
Early Childhood	Content that may be suitable for persons ages 3 and older. Titles in this category contain no material that parents will find inappropriate.
Everyone	Content that may be suitable for persons ages 6 and older. Titles in this category may contain minimal violence, some comic mischief, and/or mild language.
Everyone 10+	Content that may be suitable for persons ages 10 and older. Titles in this category may contain more cartoon, fantasy, or mild violence, as well as mild language and/or minimal suggestive themes.
Teen	Content that may be suitable for persons ages 13 and older. Titles in this category may contain violent content and/or strong language.
Mature	Content that may be suitable for persons ages 17 and older. Titles in this category may contain mature sexual themes, more intense violence, and/or strong language.
Adults Only	Content suitable only for adults—not intended for persons under the age of 18. Titles in this category may include graphic depictions of sex and/or violence.

Cutting Edge

Gaming-Optimized PCs

People who are serious about game playing are often people who don't mind spending more money on a computer to ensure the best possible gaming experience. A whole subclass of desktop computers (and some laptop computers) has been designed with game playing in mind. These computers have the very best of all the components that a game relies on, including CPU, RAM, display adapter, monitor, and sound. Many are also marketed as being "VR-ready," which means they have enough processing power to run demanding virtual reality applications smoothly and interface with virtual reality hardware.

But a gaming computer doesn't stop at having high-quality basic components. Gaming computers are often flashy models designed to express the personalities of their owners, with colored lights, colored components, nontraditional cooling systems (for example, water cooling instead of traditional air cooling), semitransparent or unusually shaped cases, and elaborate logos. One popular line of gaming computers is Alienware, a division of Dell Computers. (Visit https://CUT7.ParadigmEducation.com/Alienware to see their latest offerings and price out your ideal system.)

4.6 Using Communications Software

> **Watch & Learn**
> Using Communications Software

One of the major reasons people use computers is to communicate with others. Software that enables communication over the internet and the web is available for individual, home, and business use. These programs allow users to send and receive email, browse and search the web, engage in group communications and discussions, and participate in web conferencing activities.

Email Applications

Electronic mail (email) is rapidly becoming the main method of communication for many individual, home, and business users. Email involves the transmission and receipt of private electronic messages over a worldwide system of networks and email servers. Messages are sent and received within seconds or minutes, and transmission costs are minimal. Email is a **store-and-forward system**. Messages are forwarded to a mail server and then stored there until the recipient picks them up using either a web interface or an email application.

There are three main types of email accounts: web-based, POP3 and IMAP. Table 4.3 summarizes key points of comparison among these account types.

Web-based email accounts are available through Yahoo Mail, Gmail (Google Mail), and Microsoft's Outlook.com. You don't need any special software to read and compose email with one of these accounts; you simply open a web browser and navigate to the page for that email service. Messages remain on the host company's server, and you can access your inbox from any computer. This type of account is

Table 4.3 Key Characteristics of Email Account Types

Characteristic	Web-Based	POP3	IMAP
Uses email client?	Depends on compatibility between the service and the email client	Yes	Yes
Web interface available?	Yes	Depends on provider	Depends on provider
Message storage location	Mail server	Local PC	Mail server
Key benefit(s)	Can be accessed via any web browser	Provides access to full features of email client; can view received mail even if no internet connection is available	Provides access to full features of email client; can send and receive mail from multiple computers
Key drawback(s)	Web browser interface may not be full featured; account may not work with email client software	Email client should be used to access the account from only one PC, or the mail archive will be incomplete	Email client must be set up separately on each PC

suitable for someone who frequently checks his or her email from different computers and devices. The quality and features of the web-based interface vary depending on the provider. Figure 4.22 shows how an Inbox looks on Outlook.com.

A user who sticks with one computer most of the time may prefer a traditional email account, which is sometimes called a POP3 account. (*POP* stands for *Post Office Protocol*.) This type of email account uses an email client application installed on the user's main PC. An **email client** is an application designed specifically for sending, receiving, storing, and organizing email messages. An email client may have special features that aren't found in a web-based email interface—for instance, linking scheduled meetings to your calendar, flagging messages with high or low importance, and scheduling mail to be delivered after a specified date and time.

After messages are downloaded from the mail server to the email client, they are usually deleted from the mail server and exist only in the local PC's email client. A web-based interface may be available for checking mail while away from the main PC, but mail is organized and stored primarily on that PC. Therefore, this isn't a good type of account for someone who switches between several computers. Although this type of account is commonly known as POP3, POP3 is actually only the protocol for receiving email; there's a separate protocol commonly paired with it for sending email called *SMTP* (which stands for *Simple Mail Transfer Protocol*).

An alternative for someone who uses several computers but wants to use an email client is *IMAP* (which stands for *Internet Message Access Protocol*). An IMAP account keeps all the messages on the mail server and provides on-the-spot access to them via an email client when prompted to do so. The email client doesn't maintain local mail folders; instead, it queries the server for the folders and their contents when it connects to send and receive mail. You can use the same email client software to work with both POP3 and IMAP accounts. Some email client software also enables you to set up web-based email accounts and work with them as if they were POP3 or IMAP, but the client software may have limitations on which services it supports.

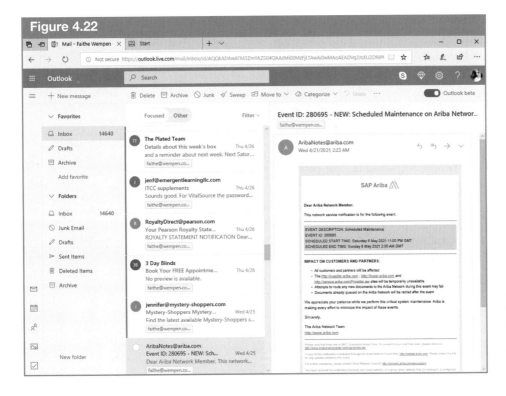

Figure 4.22

Outlook.com provides web-based access to email. Outlook.com's file management tools include options for deleting, archiving, and moving messages.

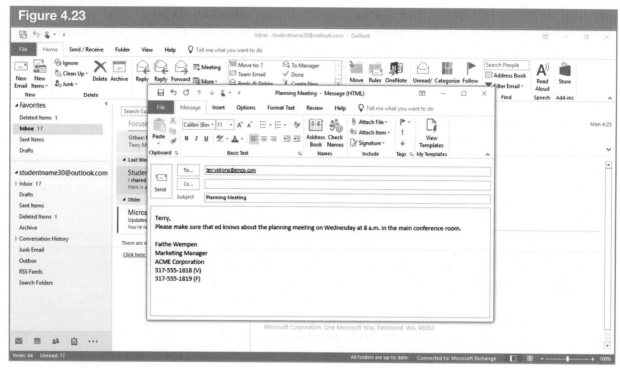

Figure 4.23

Outlook is a powerful program for working with email. The formatting tools available for messages are similar to those of other Microsoft Office applications, such as Word and PowerPoint.

Examples of email clients include Microsoft Outlook, Windows Mail, and Mozilla Thunderbird. Outlook is more than just an email client; it also stores and organizes personal information, such as contacts, appointments, and tasks. Figure 4.23 shows an email message being composed in Outlook.

Web Browser Applications

A web browser is an application that enables users to read web-based content and use web-based applications. Microsoft's Internet Explorer comes with Windows, and Windows 10 also includes another browser, Microsoft Edge. Other popular browsers include Google Chrome, Mozilla Firefox, Safari (the default browser in MacOS), and Opera. Web browsers are covered in detail in Chapter 5.

Text-Based Messaging Applications

Instant messaging (IM) enables people to communicate privately with others using text-based messages. The most popular text messaging method is via cell phone or smartphone, using a communication protocol called Short Message Service (SMS). The term *text message* can refer to any text-based message, but it is most often used specifically to mean an SMS message sent from a phone or tablet.

Some applications enable users to send and receive instant messages within the application. For example, Facebook has a messaging client called Facebook Messenger that you access from within Facebook. Business communication software such as SharePoint allows users to send each other comments within the software—for example, a participant can send a private question to the leader of an online meeting. There are also stand-alone instant messaging systems, such as Discord (often used by people playing online multiplayer games to communicate within the game).

A *chat room* is a group text message, usually hosted on a chat server. Unlike instant messages, which are private and by invitation, many chat rooms are open to anyone using the system. For many years, a popular chat room system was Internet Relay Chat (IRC), with thousands of chat rooms and millions of users. It still exists today, although it is not as popular as it once was. Early proprietary online services, such as AOL, also had their own chat systems.

Instant messaging and chat rooms have declined in popularity because we now have so many other options for meeting and talking with people online. Modern options include professional networking websites, dating applications, and audio and video chat programs like Skype for both business and personal use.

Voice over Internet Protocol Software

Voice over Internet Protocol (VoIP) is the technology that internet phone services use. Rather than rely on traditional phone lines, this kind of service sends and receives phone calls digitally over your internet connection. A VoIP system is driven by a VoIP modem that connects to your internet service box (such as a cable modem) and to a telephone jack in your home or business, serving as a bridge between the two. One example of a VoIP system is Vonage.

VoIP is best known for replacing analog telephone service, but its capabilities are much broader. For example, services like Skype enable video chatting by computer using webcams and the computer's microphones and speakers. Apple's macOS—and iOS 7 and higher—also include video chat capabilities through the FaceTime application, and many other video chat services are also available. Some IM programs, also offer video chatting capabilities, blurring the line between IM and VoIP.

Web Conferencing Software

Web conferencing is an enhanced, group participation version of VoIP. Groups of people can have joint conversations online, including video chat, instant messaging, and application sharing. Participants can show one another presentations, diagrams, documents, and spreadsheets within the web conferencing software, as well as draw on a virtual whiteboard and send one another private instant messages as the meeting progresses. For the audio portion of the meeting, users can either use internet voice capabilities with their computer's microphone or dial in using their telephone.

Because it saves time and travel expenses, many businesses are promoting the use of web conferencing software over face-to-face meetings. Some web conferencing software requires users to download a small application onto their computer—but most is completely web-based, requiring only a browser on each participant's computer. Cisco Webex, Cisco Unified MeetingPlace, and LogMeIn GoToMeeting are some examples of web conferencing software.

Groupware

Groupware, which is also called collaboration software, allows people to share information and collaborate on a project, such as designing a new product or preparing an employee manual. Groupware can be used over a local area network (LAN), wide area network (WAN), or the internet. All group members must use the same groupware program to collaborate on a project.

Microsoft produces a groupware application called *SharePoint* that enables groups of users to collaborate on documents of all types. SharePoint has robust features for managing the sharing of permission levels and creating clear accountability chains

Article
Collaboration Software Trends

that document who reviewed a document, what changes were made, and who approved them. SharePoint is installed on a company's own servers and creates an intranet environment (or a web browser–based internal network). Microsoft Office 365 offers group management capabilities similar to those provided by SharePoint but for smaller groups. Office 365 also manages and stores the group management software in Microsoft's cloud environment, rather than on the company's own server.

Most groupware applications include an address book of members' contact information and an appointment calendar. One desirable feature is a scheduling calendar that allows each member to track the schedules of the other members. Doing so makes it possible to coordinate activities and arrange meetings to discuss project activities and other matters.

Shared calendaring is also a part of Microsoft Exchange, a mail server technology that some large organizations use to manage their email systems, instead of relying on an internet service provider (ISP). On an Exchange server, users can not only send and receive email, but also use many features of Microsoft Outlook, such as meeting scheduling, calendar sharing, and global address book management.

Social Media Apps

If you use social media sites, such as Facebook and Pinterest, you probably access them through your web browser on a desktop or laptop PC. When you access these sites on a mobile device, like a tablet or smartphone, using the web-based interface is sometimes awkward because the sites aren't optimized for small-screen use, slower internet speed, and data usage limitations. To improve the experience on mobile devices, many social media sites provide customized apps. For example, Facebook provides mobile apps for iPhone and Android users. Mobile apps are also available for Snapchat, Instagram, Pinterest, Twitter, and many other sites.

Video
Mobile Apps

Hotspot

Instant Apps

One major modern trend in applications is toward ease of installation. Modern mobile apps install quickly and with minimal user involvement from an online store—and the store itself is accessed through an easy-to-use app. Microsoft Store, Google Play Store, and Apple App Store all make installing modern apps a breeze.

Google's Instant Apps technology is looking to take that simplicity and ease even further. When you enable Instant Apps on your Android device, you can run cloud-based versions of certain apps from your device without having to install them. Not having to install an app means the app doesn't take up storage space on the device, and you don't have to wait for the app to download. News sites like BuzzFeed and shopping assistant apps like Wish are leading the way in this new technology, and many other app providers have instant versions in development.

Google has enabled Instant App links to be embedded in Google search engine results, so you can quickly find instant apps using the Google search engine and don't have to go through Google Play Store. Instant Apps are compatible with all Android versions. The downside is that few apps are available in Instant App versions, and the technology is limited to Android devices (although if it's successful, you can count on Microsoft and Apple having their own versions of it soon).

Chapter Summary

4.1 Distinguishing between Types of Application Software

Application software can be categorized as **individual application software**, **collaboration software**, or **enterprise application software** based on the number of people it serves at once. Application software can also be divided into **desktop applications** and **apps**.

In addition, software can be categorized by how it's developed and sold. **Commercial software** is created by a company (or, rarely, an individual) that takes on all the financial risk for its development and distribution upfront. With a **shareware** product, you can try it for free and then purchase it. Usually the free version has limited features or a limited time frame during which you can use it. Software that's made available at no charge is known as **freeware**.

Commercial software may have **software piracy** prevention (antipiracy) features, such as a **registration key** or **activation** requirement. The terms for use of an application are determined by its **license agreement**. A **site license** grants permission to make multiple copies of the software and install it on multiple computers. Companies that need many copies of an application may choose to buy a **per-user site license** or a **per-seat site license**. Free software that others may modify and redistribute is available in the **public domain**, and the developers make the **source code** available along with the compiled application.

Software can also be distinguished by the way it is installed. To install a desktop application, you run a setup utility, which you can get on a CD or DVD or in a setup file that you download. To install a mobile or modern app, you simply download the app from an online store, such as the Microsoft Store for Windows apps.

4.2 Using Business Productivity Software

Business productivity software provides tools for people to accomplish business tasks, such as writing, performing calculations, and storing information. **Word-processing software** is used to create many types of text-based documents, either from scratch or using a **template**. You position the **insertion point** in the document and then type the desired text. Most word-processing applications have a **spelling checker** and **grammar checker** and enable **text formatting**, **paragraph formatting**, and **document formatting**, as well as the creation and use of **styles**. A **print preview** feature shows the document as it will look when printed. **Desktop publishing software** enables users to create complex page layouts with text and graphics.

Spreadsheet software enables you to organize, calculate, and present numerical data. Numerical entries are called **values**, and the instructions for calculating them are called **formulas**. Formulas can contain **functions**, which are named instructions for performing a calculation (such as AVERAGE and PMT). A spreadsheet consists of **cells** arranged in a grid of rows and columns; each cell can be formatted separately. You can make **charts** based on the values in cells and record **macros** (sets of memorized steps) to save time in performing commonly repeated tasks.

A database is a collection of data organized into one or more data tables. Each table contains records, which are collections of data about specific instances. The parts of a record are called *fields* (for example, *FirstName* and *LastName*). Database software is called a **database management system (DBMS)**. A large database is called an **enterprise database**, and it may be stored and managed using professional tools such as

Oracle Database and structured query language (SQL). A smaller database might be created in a personal program such as Microsoft Access and stored on an individual PC or server. Databases that have multiple tables that are connected are known as *relational databases*; most business databases are relational. In addition to tables, a database can also have saved forms, reports, and queries that can be used to view and modify the data.

Presentation graphics software allows users to create computerized slide shows. A **slide** is an individual page or screen in a presentation. A **slide show** may consist of any number of slides. Slides can incorporate many types of content, including text, photos, charts, and tables.

A **software suite** combines multiple applications in a single package. Microsoft Office, for example, combines Word, PowerPoint, Excel, and other applications. The applications within a suite have a common interface style. You can **embed** or **link** content between applications and documents not only in suites but in any applications that support **object linking and embedding (OLE)**.

Personal information management (PIM) software helps organize your schedule and activities by storing contacts, calendars, lists, and notes in one place. A PIM application may also serve as an email application. **Project management software** provides tools for managing the resources, people, and locations involved in a major business initiative, such as constructing a new building or organizing an event. **Accounting software** helps a company track its accounts payable, accounts receivable, payroll, sales, and so on. Note-taking software (such as Microsoft OneNote) helps bring together different types of content for logical sorting and organizing.

4.3 Selecting Personal Productivity and Lifestyle Software

Personal software is designed for individuals to use for their own enrichment and entertainment. Lifestyle and hobby software helps users plan and track their hobbies such as gardening or fitness. **Legal software** helps users plan and prepare legal documents. **Personal finance software** helps users pay bills and track finances. **Tax preparation software** helps them fill out federal and state tax forms. Encyclopedias and dictionaries provide reference sources. **Tutorials** provide step-by-step learning; this is sometimes called **computer-based training (CBT)**.

4.4 Exploring Graphics and Multimedia Software

Painting and drawing programs both create and modify artwork, but in different formats. **Painting software** works with bitmap images, also called **raster images**. **Drawing software** creates and edits **vector graphics**, which are math-based drawings. **Photo-editing software** is a variant of painting software that is used to correct common problems in photos, such as overexposure and red-eye.

3-D modeling programs and **computer-aided design (CAD)** software also use vector graphics. These programs are used to create sophisticated technical drawings and models in fields such as manufacturing and architecture, and they are also used to create the graphics in many video games. A **wireframe** (outline) is first constructed, and then **texture** and color are applied to the surface. Animation software extends 3-D modeling to create objects that appear to move. Professional animation software is responsible for the **computer-generated imagery (CGI)** in some movies.

Audio-editing software records and modifies **waveform** sounds, which are an analog in origin and **sampled** to create digital versions. The number of bits used to describe each sample is known as the **bit depth**. Uncompressed waveform recordings, such as the **CD audio (CDA) format**, take up a great deal of disk space. Compressed versions, including MP3, Advanced Audio Coding (AAC), and Windows Media Audio (WMA) formats, are much more compact but require a computer (or compression-aware player) to decompress and play back the clips.

A video clip is a group of still-image **frames** that plays back sequentially along with a digital audio track. Video clips can be edited with **video-editing software**.

Web authoring software enables users to develop web pages without having to learn a special language. This software uses a **WYSIWYG** interface to show the page being developed as it will appear in a browser. Web pages are saved in Hypertext Markup Language (HTML), but not all web content is in that format. Animated multimedia content and interactive content are sometimes generated in programs such as Adobe Animate.

4.5 Understanding Gaming Software

Games are available both for desktop and laptop PCs and for **game consoles**. Game consoles are optimized for game playing; they typically use a TV for output and a game pad, joystick, or steering wheel for input. Each game has **system requirements** that detail the minimum hardware specifications required to play that game on a particular platform.

Games have content ratings that advise consumers about the appropriate minimum age for a player. The most widely used rating system is from the **Entertainment Software Ratings Board (ESRB)** and includes ratings such as *Everyone*, *Teen*, and *Mature*.

4.6 Using Communications Software

Electronic mail (email) involves the transmission and receipt of private electronic messages using a **store-and-forward system**. Types of email accounts include web-based, POP3, and IMAP, and each type has both benefits and drawbacks. **Email client** software stores and organizes received and sent messages. A web browser is an application designed to view web pages. Most operating systems include a web browser, and you can also install a third-party web browser.

Instant messaging (IM) enables people to communicate privately with others using text-based messages. The most popular text messaging method is via cell phone or smartphone, using a communication protocol called Short Message Service (SMS). A chat room is a group text message, usually hosted on a chat server.

Groupware, or collaboration software, enables people to share information on various projects. SharePoint is one common groupware product; using it enables companies to create private intranet environments in which employees can collaborate. Office 365 is a Microsoft service that enables business and enterprise customers to collaborate via Microsoft's software cloud, which eliminates the need for a dedicated server for sharing. For companies that operate their own mail systems, Microsoft Exchange can not only manage a company's email but can also offer shared scheduling features.

Social media started out as a web-based system, but today people access social media on mobile devices using applications written specifically for a particular media site such as Facebook or Twitter.

Key Terms

Numbers indicate the pages where terms are first cited with their full definition in the chapter. An alphabetized list of key terms with definitions is included in the end-of-book glossary.

Chapter Glossary

Flash Cards

4.1 Distinguishing between Types of Application Software

application software, 138
individual application software, 138
groupware, 138
enterprise application software, 138
desktop application, 138
app, 138
commercial software, 139
software piracy, 140
registration key, 140
activation, 140
license agreement, 141
site license, 141
per-user site license, 141
per-seat site license, 141
shareware, 142
freeware, 142
public domain, 142
source code, 142

4.2 Using Business Productivity Software

word-processing software, 144
template, 145
insertion point, 145
spelling checker, 145
grammar checker, 145
text formatting, 145
paragraph formatting, 146
document formatting, 146
style, 146
print preview, 146
desktop publishing software, 146
spreadsheet software, 147
value, 148
formula, 148
function, 148
cell, 148
chart, 149
macro, 149
database management system (DBMS), 149
enterprise database, 149
object, 151
presentation graphics software, 151
slide, 152
slide show, 152
software suite, 153
embed, 153
link, 153
object linking and embedding (OLE), 153
personal information management (PIM) software, 153
project management software, 154
accounting software, 155

4.3 Selecting Personal Productivity and Lifestyle Software

legal software, 156
personal finance software, 157
tax preparation software, 157
tutorial, 159
computer-based training (CBT), 159

4.4 Exploring Graphics and Multimedia Software

multimedia software, 160
painting software, 160
raster image, 160
drawing software, 160
vector graphic, 160
photo-editing software, 161
3-D modeling program, 161
wireframe, 161
texture, 162
computer-aided design (CAD), 162
computer-generated imagery (CGI), 163
waveform, 164
sampling, 164

bit depth, 164
CD audio (CDA) format, 164
wave file format, 164
Musical Instrument Digital
    Interface (MIDI), 164
frame, 165
video-editing software, 165
web authoring software, 166
WYSIWYG, 166

4.5 Understanding Gaming Software

game console, 167
system requirements, 168
Entertainment Software Ratings
    Board (ESRB), 169

4.6 Using Communications Software

electronic mail (email), 170
store-and-forward system, 170
email client, 171
groupware, 173

Chapter Exercises

Complete the following exercises to assess your understanding of the material covered in this chapter.

Tech to Come: Brainstorming New Uses

In groups or individually, contemplate the following questions and develop as many answers as you can.

1. Today, a smartphone or tablet app is available for almost any task you can imagine wanting to perform, but what about the tasks of the future? Imagine yourself moving through an average day 10 years from now. What will you be doing, and what kinds of apps will help you do it? Imagine what phone apps might look like then. Will they be voice activated? Will they interact with other devices in your environment?

2. On the desktop platform, multifunction programs like Microsoft Outlook combine several popular activities into a single interface: email, contact management, calendar scheduling, and to-do lists. But on smartphones and tablets, the trend is the opposite; many small, individual apps take care of specific tasks. Which trend will be dominant in the future?

3. Game consoles have traditionally focused only on running games. However, most models now include internet access, and along with it comes access to online audio and video services, such as Netflix, Amazon Video, and Hulu. What additional services do you think the customers of the future will want their game consoles to provide?

- Terms Check
- Knowledge Check
- Key Principles
- Tech Illustrated
  Data Table in
  Microsoft Access
- Chapter Exercises
- Chapter Exam

Tech Literacy: Internet Research and Writing

Conduct internet research to find the information described, and then develop appropriate written responses based on your research. Be sure to document your sources using the MLA format. (See Chapter 1, Tech Literacy: Internet Research and Writing, page 37, to review MLA style guidelines.)

1. Can you accomplish everything you need to do on a computer using only the web? You might be surprised. Web-based applications are available not only for traditional online activities (such as web browsing and email) but also for improving productivity. Productivity applications include word processors, spreadsheets, and databases (such as Microsoft Office, Apple iWork, and Apache OpenOffice) as well as personal information management services (such as Google Calendar). Make a list of everything you do with your computer, and then find a website where you can do each task.

2. What kind of program would help improve your productivity? Visit a store that sells computer software in your area. Select a particular productivity program on display, and read the product description on the package. What platform is the program written for? What's the price? Will the program run on your computer? For what purpose(s) might you be able to use the program?

3. What's hot in application software? Visit your school or public library and look through computer magazines to find an article that describes a new and innovative productivity software program. Write a summary of the program's purpose,

main features, and specifications. Include information about the user interface, amount of internal memory needed to run the program, and amount of disk space required to store the program. Why do you think this program is innovative? What needs does this product fulfill? Are competing products available on the market? Do you think the product will be a commercial success? Why or why not?

4. Is a picture worth a thousand words? Ask your instructor (or another person) for the name of a business or organization in your area that regularly uses presentation software to provide information to clients or to train employees. Find out if you can obtain a copy of a presentation. Watch the slide show, and then write an evaluation of its effectiveness. List the technologies used (hardware and software), and describe the features that impressed you most.

5. The first "killer app" in the history of software development was VisiCalc. This early calculation program is credited with founding the electronic spreadsheet software industry and launching widespread sales of the personal computer. Research the meaning of the term "killer app," and propose some possibilities for the next one. Explain the reasons for your choice.

Tech Issues: Team Problem-Solving

In groups, develop possible solutions to the issues presented.

1. Students are increasingly relying on software for in-class activities such as taking notes, writing reports, and creating graphics for group presentations. Without a laptop or tablet, student learning may be more difficult. At the same time, though, many students abuse the privilege of using software during class to surf the web, write email, chat with friends, or even play games. Write a computer usage policy that establishes rules for allowing students to use computer technology for learning but not for distracting activities, including penalties for breaking the rules.

2. Think about the different types of application software discussed in this chapter, and then determine how you might use some of them to organize activities in your life. Identify three software programs you might use if you were planning a vacation. Why did you pick these programs? What aspects of vacation planning does each program support?

3. Software piracy costs developers billions of dollars each year in lost revenue, and they pass those costs on to consumers in the form of higher prices. However, software companies must balance the aggressiveness of their antipiracy measures against consumers' concerns for convenience and privacy, or risk having products that nobody wants to buy. As a consumer, what balance of higher prices versus convenience and privacy are you comfortable with? Keep in mind that the more lax a company is about piracy, the more each consumer will have to pay for the software to make up for those who get it for free.

Tech Timeline: Predicting Next Steps

The following timeline outlines some of the key milestones in the evolution of encyclopedias from printed books to digital content. Research this topic to predict at least three important milestones that could occur between now and the year 2040, and then add your predictions to the timeline.

1768 The first section of the original *Encyclopædia Britannica* is printed.

1917 The first *World Book Encyclopedia* is published.

1981 The first digital version of *Encyclopædia Britannica* is created for use by LexisNexis subscribers.

1993 Microsoft announces the availability of the first release of *Encarta*, an online encyclopedia on CD-ROM.

1994 *Encyclopædia Britannica* is made available on the internet and on CD-ROM.

1998 The complete World Book Online website is launched.

2001 A multilingual encyclopedia called Wikipedia is launched and supported in an open source model.

2006 Wikipedia announces that the one-millionth user-written article has been published.

2009 Sales of *Encarta* are discontinued and its associated web sites are shut down, partially owing to the rise of its free competitor, Wikipedia.

2012 *Encyclopædia Britannica* is no longer published in print—only online.

2017 Wikipedia reports that it contains over 5.6 million articles in English.

Ethical Dilemmas: Group Discussion and Debate

As a class or within an assigned group, discuss the following ethical dilemma.

When software is downloaded from the internet and installed on a personal computer, it might contain spyware. *Spyware* is software that gathers information through an internet connection without the user's knowledge. Some of the information commonly gathered includes the user's keystrokes, hardware configuration, and internet configuration, as well as data from the user's hard drive and data from cookies. Typically, this information is gathered and sent to the spyware author, who then uses it for advertising purposes or sells the information to another party.

Do you consider this an invasion of privacy? Should it be illegal to include spyware inside another software program? What if the software license agreement includes a disclaimer that says spyware may be installed? Is it harmless if the information is just being gathered for market research? What other problems do you see with this technology? How might internet users be protected from spyware?

Chapter 5

Plugging In to the Internet and All Its Resources

Chapter Goal

To learn how the internet works and how you can use it properly to perform a variety of life-enhancing activities

Learning Objectives

5.1 Describe the overall types of activities made possible by the internet.

5.2 Explain how to connect to the internet, including needed hardware and software and different types of connections.

5.3 Discuss how the internet delivers page information to a web browser from the specified IP address or URL as you navigate the web.

5.4 Describe the language used to create web pages and the basics of publishing to a website.

5.5 Describe how to use basic web browser techniques to view web pages and content.

5.6 Use fundamental search techniques to find information on the web.

5.7 Discuss diverse uses and services on the internet, including social media and services.

5.8 Behave appropriately as a member of the internet community.

Online Resources
The online course includes additional training and assessment resources.

Tracking Down Tech
Observing the Internet in Action

5.1 Exploring Our World's Network: The Internet

Watch & Learn
Exploring Our World's Network: The Internet

The internet is the largest computer network in the world. Its design closely resembles a client/server model, with network groups acting as clients and internet service providers (ISPs) acting as servers. (You will read more about networks and network structures in Chapter 6.) Since the launch of the internet in the early 1970s, this enormous system has expanded to connect more than 4 billion users worldwide (see Figure 5.1).

Many experts within and beyond the information technology (IT) industry consider this vast structure of networked computers and telecommunications systems the most significant technical development of the twentieth century. The internet has the potential to connect every person on Earth to a vast collection of resources, information, and services.

Individuals, organizations, businesses, and governments use the internet to accomplish a number of different activities, which can be divided into these general categories:

- Communications
- Telecommuting and collaboration
- Entertainment and social connections
- Electronic commerce
- Research and reference
- Distance learning

This section provides an overview of these categories. See section 5.7, "Using Other Internet Resources and Services," beginning on page 213, for more details on specific tasks you can accomplish online (on the internet).

Communications

One of the chief functions of the internet is to allow people to communicate with one another quickly and easily. Internet users can take advantage of a number of different communications applications, including email, instant messaging, social networking services, video messaging, and more. Also, users are no longer tethered to the computer to use the internet to communicate. Smartphones and other portable devices can enable you to communicate using various applications (apps) over a wireless connection to the internet or a mobile network. In the case of a mobile (cellular) network, a device such as a tablet would have to be mobile capable, and the user would have to have a mobile data plan for the device. Many restaurants and cafes, including popular chains such as Starbucks and McDonald's,

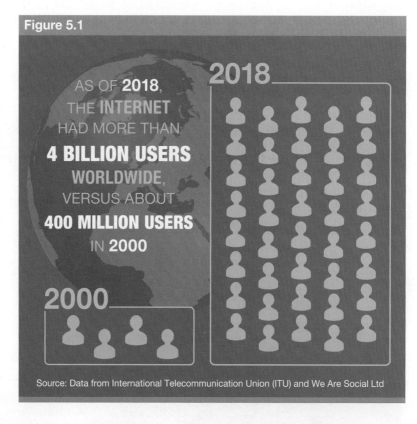

Figure 5.1

AS OF **2018**, THE **INTERNET** HAD MORE THAN **4 BILLION USERS** WORLDWIDE, VERSUS ABOUT **400 MILLION USERS** IN 2000

Source: Data from International Telecommunication Union (ITU) and We Are Social Ltd

offer free wireless service. Having this service lets users connect to the internet with a laptop or other portable, mobile-capable device.

Telecommuting, Collaboration, and the Gig Economy

Millions of workers now perform their work activities at home using a computer and internet connection (see Figure 5.2). This activity is known as **telecommuting** (also called *teleworking*). Some employers have discovered that allowing employees to telecommute offers important advantages—for example, increased worker productivity, savings on travel costs to and from the workplace, and an opportunity to employ individuals who are highly productive but have physical limitations.

Several types of telecommuting arrangements are possible. In some cases, the employer provides the employee with a computer and pays for some or all of the employee's cell phone and internet service. In other cases, the employee supplies his or her own computer equipment and pays for necessary phone and internet service plans. Generally, the employer identifies much of the software the employee will use, such as a particular email program or online collaborative platform (which Chapter 7 covers in greater detail). Many telecommuting employees sign in to the company's email system and also sign in to file and project-sharing systems hosted by the employer. Telecommuting employees might also participate in online video calls or meetings using forms of digital communication that are personal, immediate, and interactive.

Video Real-Time Meetings

The same tools that enable telecommuting also facilitate collaboration among employees working at multiple sites. Employees can work online with team members, as well as outside vendors and contractors, in other cities and countries.

Collaboration tools and apps have also fostered the rise of the *gig economy*, in which a company or an individual hires a worker for a job or task that's usually short-term and temporary in nature. For example, a gig might consist of tasks that can be

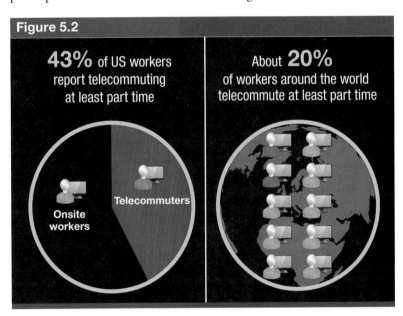

Figure 5.2

43% of US workers report telecommuting at least part time

About 20% of workers around the world telecommute at least part time

Onsite workers / Telecommuters

Practical TECH

Saving Money on Mobile Service

The data plans for mobile phones and other devices can be expensive. You can save money, however, by taking advantage of the options offered by different carriers. For instance, most carriers allow you to share a data plan among multiple devices—perhaps two family phones and a tablet or two sharing one data plan. In such a case, if one user needs a data plan but doesn't use it much, the users of the devices share the monthly data allowance provided by the plan. Most wireless carriers also allow you to purchase additional data as needed to supplement your data plan. Another way to save is to have a small data plan and connect through wireless networks whenever possible. Most smartphones can connect to a home or other wireless network, and through these connections, avoid using the data allowance.

handled remotely, such as designing a website or organizing and analyzing a set of data. Or a gig might be a task handled in person, such as driving a passenger to a location or performing a small home repair. In *crowdwork*, a group of independent individuals collaborate to complete a task. A more comprehensive task is broken up into small, simple tasks known as *micro tasks*, enabling numerous crowdworkers to complete parts of the larger task independently. Both the gig economy and crowdwork provide flexible opportunities to earn income as an independent contractor or freelancer. Trade-offs may include lack of employer benefits and lower overall compensation.

Entertainment and Social Connections

Internet users of all ages take advantage of computers for entertainment purposes. A computer with the right components and accessories can imitate almost any entertainment device. You can play games, listen to music, and even watch movies and TV programs on your computer. You can also message or chat with friends, participate in online social platforms and services (such as online dating sites), place carryout orders or make dinner reservations, and connect socially with others in just about any way imaginable.

Today's dedicated gaming consoles have internal hardware similar to a personal computer (PC) and attach to a TV for display. For gaming, most of these consoles connect to the internet. Mobile phones, tablets, and other portable devices (such as the iPod Touch) can also be used for online gaming and other forms of entertainment. To play online, the user must have a cellular or wireless internet connection for the device.

An enormous number of both paid and free games are available online, including traditional games such as backgammon, checkers, and bridge. Users must buy software to play some games: for example, EVE Online and World of Warcraft. But most (if not all) games can be purchased online and then downloaded or played online.

Playing games with other users online (sometimes called *social gaming*) is just one way to make social connections using the internet. Options for social media and other online social activities are expanding at a fast pace. Advertisers spend a lot of money on trying to reach this growing group of users. According to one research firm, ad spending on social media networks is growing rapidly in the US (see Figure 5.3).

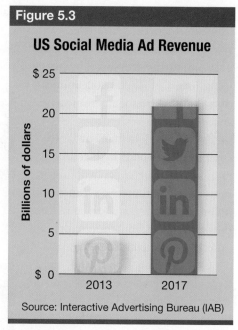

Revenue from social media advertising has grown quickly in the US—a trend that's expected to continue.

Electronic Commerce

Electronic commerce (e-commerce) involves exchanging business information, products, services, and payments online. E-commerce is commonly divided into two categories defined by the target audience: business to consumer (B2C) and business to business (B2B).

Online shopping makes up most of B2C e-commerce. The top retail category is clothing, followed by books, music, videos, auction items, toys, and computer hardware. Each year, retail e-commerce sales continue to grow. By the fourth quarter of 2017, sales had reached $119 billion for that period, or 9.1 percent of all US commerce, according to estimates from the US Department of Commerce.

Many retailers post online catalogs that potential buyers can look through. The buyer can select items of interest and add them to a "shopping cart." This virtual

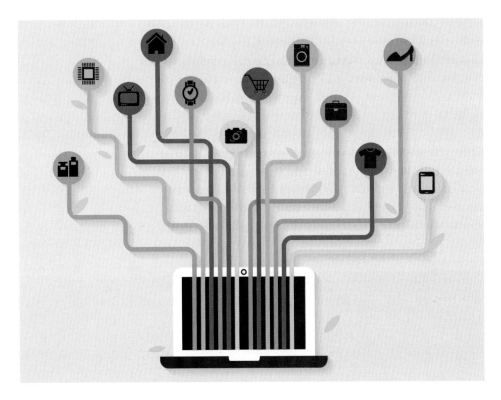

Consumers can purchase nearly any good or service online and have it shipped to a specific location, such as a home or business.

shopping cart functions just like a real shopping cart, allowing the buyer to put items in the cart and remove them later, as desired. When the buyer finishes shopping, he or she proceeds through a virtual checkout. The buyer pays for the purchase by entering a credit card number or using another method of electronic payment. Within a few days, the purchased items arrive at the buyer's "ship to" address.

In addition to selling and advertising products and services to consumers, businesses are using the internet for other purposes: to advertise products, buy goods from manufacturers and wholesalers, order raw materials, recruit employees, file government reports, and perform many other activities. These activities make up the B2B category of e-commerce. New ways for businesses to use the internet are being discovered every day, and current uses are continually being improved to make them even more successful.

Research and Reference

The **World Wide Web (web)** is the part of the internet that enables users to browse and search for information. Because of the web, thousands of opportunities have been created for people interested in research. Aided by the web's increasingly sophisticated search platforms, users can explore a wide range of topics, from anacondas to Zen Buddhism. Students, writers, historians, scientists, and other curious individuals rely on web-based resources to find and share facts and data as part of their pursuit for knowledge.

The numerous resources available on the web include materials from libraries and universities, government agencies, and databases around the world. Researchers can access online books, periodicals (magazines and journals), encyclopedias, photos, videos, and sound files from nearly any location. Users can also perform tasks such as verifying the spelling and use of a word in an online dictionary, answering a health question using medical and science sites, finding a mailing address using a people-search site, looking up property records on a government site, and getting directions to a location using a mapping site. In most cases, users can both read the information online and download it for later use.

Tech Career Explorer
Checking Out TED

Using a search engine within a web browser, the researcher has the tools needed to find practically anything on the web. Most college research projects today begin on the web, rather than in the library, using the student's favorite search engine. Searching for information on the internet is discussed in section 5.6, "Searching for Information on the Internet," on page 210.

Distance Learning

Many colleges and universities offer online courses and study programs in a form of education referred to as **distance learning**. Distance learning involves the electronic transfer of information, course materials, and testing materials between schools and students. A college or university typically delivers distance learning courses (also called *online courses*) to students using a type of online computing platform called a **learning management system (LMS)**.

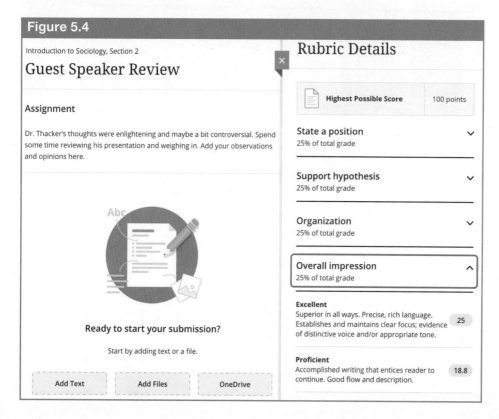

Figure 5.4 Blackboard Learn is an LMS that provides numerous tools for online course delivery, including the ability to post content for both instructors and students.

Practical TECH

Free Online Learning Opportunities

You don't have to be working toward a degree to try online learning. Numerous online courseware sites have partnered with colleges, universities, and organizations such as museums to offer free online courses. One leading site, Coursera, offers more than 2,700 online courses from 150-plus partners. Another site, TED-Ed (**https://CUT7.ParadigmEducation.com/TED-Ed**), offers more than 150,000 free online lessons. Some of these courses are based on TED Talks, which are available at the TED website (**https://CUT7.ParadigmEducation.com/TED**).

The number of free online courses continues to expand and includes offerings from Wikiversity, Udacity, and MIT OpenCourseWare, among others. See the section "Distance Learning Platforms" on page 228 to read even more about these courses, including massive open online courses (MOOCs).

Ready-made LMS platforms are available and provide a range of features. Typical features include pages that provide information about the course, communication tools such as online chat and email, online posting and grading of tests, and learning resources that support course content. Top LMSs include Blackboard Learn and Moodle. With Blackboard (see Figure 5.4), instructors can provide their own course content or use Blackboard-ready content developed by textbook publishers. Other LMSs are Canvas, Edmodo, and Brightspace.

Distance learning has become increasingly popular with students of all ages. It's especially popular with people whose interests aren't met by the course offerings of a standard or nearby college or university. Distance learning is also an attractive option for students whose schedules or careers make it difficult to attend regular classes. Some business schools have distance learning programs and other options for individuals who want to earn a degree while working full time. For example, Duke University's Fuqua School of Business offers blended or hybrid courses in the master's in business administration (MBA) program. Blended or hybrid courses include both distance learning and onsite (at the school) study, providing students with the opportunity to earn a degree on a more flexible schedule while pursuing a career. The Duke MBA program provides online courses with weekend onsite study. Four-year colleges and universities, as well as many community colleges, offer distance learning courses as a low-cost, high-convenience way to get an education.

5.2 Connecting to the Internet

Billions of people throughout the world can connect to the internet. Within each country, high-capacity networks operated by large telecommunications companies form the backbone of the internet. Other organizations, including ISPs, purchase bandwidth from these high-capacity network providers, each of which is also known as a **network service provider (NSP)**. An **internet service provider (ISP)** is an organization that provides user access to the internet, usually charging a subscription fee. An ISP may share data with other ISPs and networks through **internet exchange points (IXPs)**, as well as provide internet connections to customers through a **point of presence (POP)**. High-volume, fiber-optic transmission lines form the internet backbone and move data between the various connection points on what are known as *trunk lines*. In the United States, trunk lines are usually high-speed digital data transmission lines called *tier 1 (T1)* and *tier 3 (T3) lines*. A typical single T3 line to a location can operate from 28 to nearly 30 times faster than a single T1 line. The internet backbone uses even faster fiber-optic transmission lines called *optical carrier (OC) lines*. The OC-768 lines in some sections transmit data at up to 40 gigabits per second. Carriers are continuing to update backbone networks to 100 gigabit speeds in the United States.

Businesses and organizations with high internet bandwidth needs might pay the cost of having a T1 or T3 connection. In some parts of the United States, high-speed fiber-optic connections are available for educational, commercial, and residential users. However, most residential and small business users typically connect using some type of broadband connection, such as digital subscriber line (DSL), cable connection, or satellite internet. Only a small number of users now connect via a traditional telephone landline.

Watch & Learn
Connecting to the Internet

Hands On
Setting Up a New Internet Connection in Your Operating System

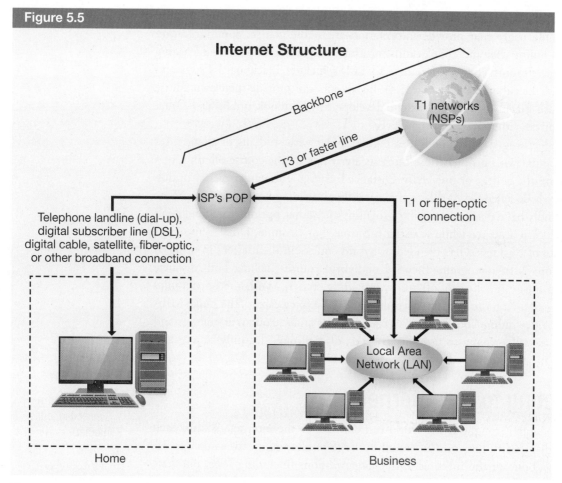

Figure 5.5

The structure of the internet makes it possible for large volumes of data to be delivered to home and business users.

Figure 5.5 illustrates the overall structure of the internet, from the backbone to the connection provided to a user's home or business. Chapter 6 will provide more information about setting up a network.

Practice
The Structure of the Internet

Hardware and Software Requirements

The following equipment and software are required to connect to the internet:
- A computer with an internal wireless or wired network adapter, wireless/wired network card, or USB adapter; a tablet device; or a smartphone
- A modem compatible with your connection type (DSL, cable, and so on)
- An account with an internet service provider (ISP) or value-added network (VAN), including any locally installed equipment required, such as a cable connection or an internet satellite dish
- Wireless and/or wired network equipment, if you want to share the internet connection among multiple devices
- A web browser

As defined earlier, an ISP is a company that provides internet access. Having access usually requires paying a fee but is sometimes free. (Firms that provide free access usually require subscribers to view advertisements when viewing web pages.) A **value-added network (VAN)** is a large ISP that provides specialized service or content, such as reports from the Reuters news service or access to a legal database. Nonetheless, all ISPs and online services can reach the same email users and websites.

The best-known ISPs offer service at a national level. But in the United States, some ISPs offer service only within a particular region based on the limits of the physical network. For example, a large telecommunications company might offer only dial-up or DSL internet service in some ZIP codes that it serves, while offering fiber optic connections elsewhere. Today, most people access the internet via wireless connections when traveling.

Types of Internet Connections

Users can get online using several types of internet connections: dial-up, digital cable, DSL, wireless, fiber-optic, and satellite. All of these connection types except dial-up are considered **broadband**, or high-speed, connections. In 2018, about 65 percent of Americans over age 18 had DSL, digital cable, or satellite service, according to the Pew Research Center.

Table 5.1 compares the connection speeds used to **download** (receive) and **upload** (send) information. Internet transfer speeds are measured according to how many bits of data are transferred per second. Each **Kbps (kilobit per second)** represents 1,000 bits per second, and each **Mbps (megabit per second)** represents 1,000,000 bits per second. A **Gbps (gigabit per second)** is a billion bits per second.

Table 5.1 Comparison of Typical Internet Connection Speeds Offered by ISPs

Connection Hardware	Download Speed*	Upload Speed*
Dial-up access with 56 Kbps modem	40–50 Kbps	28 Kbps
Cable	Up to 105 Mbps	400–600 Kbps
DSL (non-Gfast)	Up to 6 Mbps	Up to 768 Kbps
Wireless and mobile	Varies widely	Varies widely
Fiber-optic cable	Up to 1 Gbps	Up to 1 Gbps
Satellite	Up to 15 Mbps	Up to 3 Mbps

*Download speed measures how quickly one computer can receive a file from another. In most cases, download speed is more important than upload speed. Upload speed measures how quickly a file can be sent from one computer to another.

Connection speeds vary depending on the type of plan purchased from the ISP; more expensive plans generally provide faster download speeds. Of course, having a fast download speed allows data to arrive and display more quickly on your computer and device. Having a fast download speed is also needed to perform activities such as viewing online videos and making calls over the internet; both require high speeds to operate effectively.

Most of the connection types discussed in this section can be shared among several devices by connecting the modem to a local area network (LAN). For example, most users today share a connection at home by setting up a wireless LAN using a wireless router, which connects to a broadband modem, which then connects to the ISP service through a cable. (Some broadband modems now incorporate the wireless router into a single device.) Having a wireless setup makes it possible to connect computers and accessories (such as printers) to the network without wires. It also makes it possible for all of the connected computers and other devices (such as tablets and smartphones) to access the internet connection. Figure 5.6 illustrates how a wireless internet connection operates.

Figure 5.6

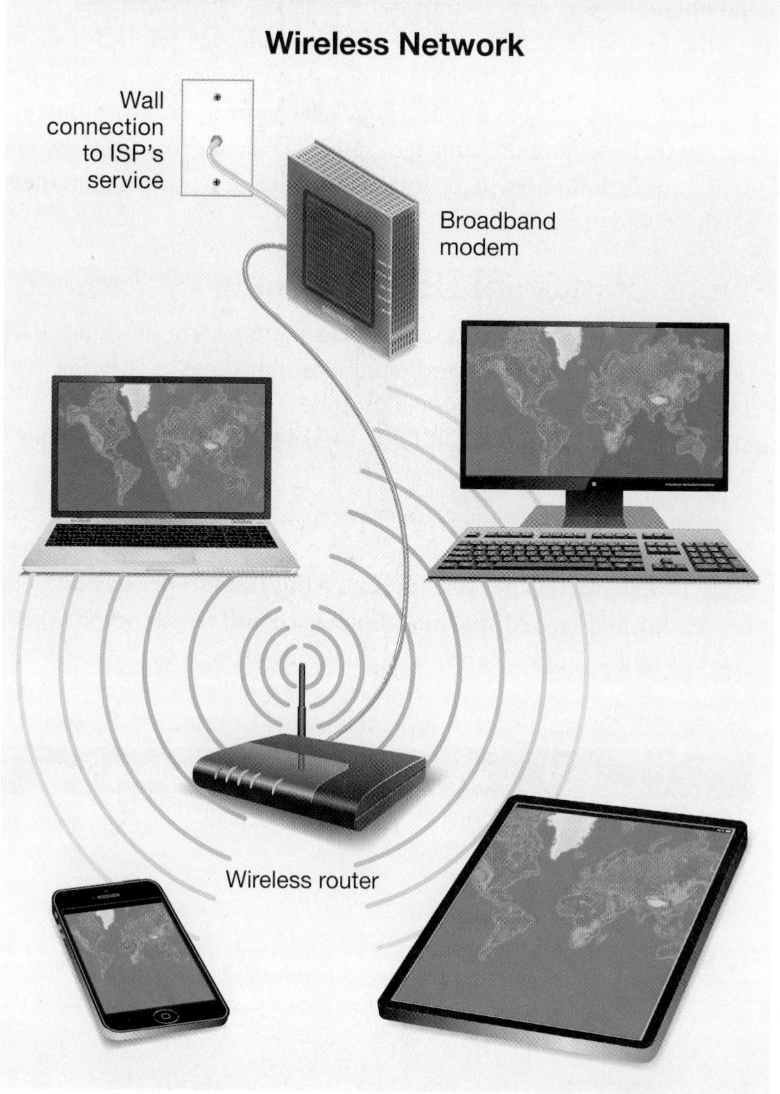

Wireless Network

Having a wireless network enables numerous devices to access a single internet connection.

Dial-Up **Dial-up access** allows connecting to the internet over a standard land-based telephone line (also called a Plain Old Telephone System, or POTS) by using a computer and a modem to dial into an ISP. Dial-up access is a feature typically included with the software provided by the ISP. After installing the software, the user can add a shortcut icon on his or her computer desktop; double-clicking on that icon will launch the connection.

Dial-up access is relatively inexpensive because of the slow connection speeds it provides, as noted earlier. To get faster connection speeds, the vast majority of US internet users utilize a broadband connection (a high-speed internet connection). Broadband connection speeds are better suited to online activities such as viewing videos.

An Integrated Services Digital Network (ISDN) connection is similar to a dial-up connection in that it runs on the public telephone service but offers digital transmission of voice and data. ISDN access enables customers to combine voice and data services on a single connection and offers slightly faster internet connection speeds (128 Kbps for both downloading and uploading) than does dial-up access. While still marketed to businesses, ISDN never gained the adoption rates of other connection types.

Cable The same coaxial cable that provides cable TV service to a home or business can also provide internet access. Cable TV companies such as Comcast and Charter provide a special modem for broadband. Having this type of connection allows simultaneous TV viewing and internet usage. This service is available only in areas in which the provider has installed cable, however, which rules out many rural locations. In addition, with this type of service, the connection speed slows down as more subscribers sign up in a neighborhood or location. Comcast calls its service Xfinity, and Charter calls its service Spectrum.

Digital Subscriber Line **Digital subscriber line (DSL) internet service** is a form of broadband that's delivered over standard telephone lines. It's as fast as service using a cable modem and provides simultaneous web access and telephone use, but it has limited availability. DSL service is usually available only to users within three miles of the telephone carrier's digital subscriber line access multiplexer (DSLAM). A digital subscriber line is dedicated to one household and isn't shared with neighbors.

You can buy DSL service in a bundle with phone service from your phone company, or you can buy a naked DSL service, a connection without the accompanying phone service. DSL has become less popular in the United States as consumers have shifted to faster broadband technologies.

Wireless One fast-growing segment of internet service involves wireless broadband connections. Tens of millions of wireless access points, or **Wi-Fi hotspots**, have sprung up worldwide, and the number is expected to continue to grow. Using these hotspots, you can access the internet in public places and even aboard airplanes using a laptop computer or mobile device (such as a tablet or smartphone).

Just like you can create a wireless network at home, the owner of a coffee shop or other public location can create a Wi-Fi hotspot by sharing a broadband internet connection through a wireless network. In some cases, the owner may protect the password for the wireless network so that only authorized customers can use it. In many other cases, though, using a Wi-Fi hotspot is free for customers and designated users. Numerous mobile apps, such as the one shown in Figure 5.7, enable you to search for location and password information for free hotspots. Some wireless service providers and private operators charge a fee to give users access to their hotspots.

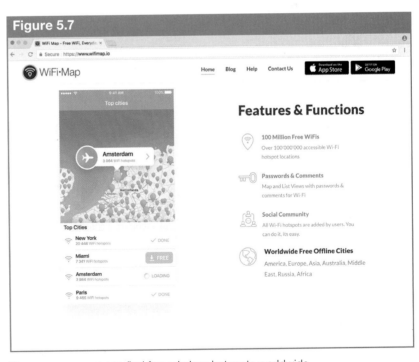

Figure 5.7

You can use an app to find free wireless hotspots worldwide.

Wireless connections to the internet can be slower than wired connections. Even so, they provide a great deal of flexibility and portability, because you aren't required to plug into a connection in a wall.

To guarantee that you will have access to hotspots while traveling, check the services offered by your mobile phone provider. Some providers offer access to a network of hotspots, either bundled as part of a plan or as an add-on service. Some providers also let you purchase short-term, prepaid access to enable your devices to connect to Wi-Fi hotspots. For example, AT&T offers prepaid DataConnect Pass plans in more than 100 countries, as well as global plans and both daily and weekly plans. AT&T also offers On the Spot one-time access, which lets you buy a session at one of its more than 30,000 hotspots for several dollars or less, depending on the location and session time. You can purchase this access even if you aren't an AT&T customer.

Wireless ISPs (also known as Wi-Fi ISPs) provide another form of wireless internet access—one not many people know about. A Wi-Fi ISP uses directional antennas to direct wireless radio signals from one location to another within a particular area. Directing these signals creates a large hotspot, in effect, which provides continuous service within that area. Some cities have experimented with building this type of network to offer free wireless internet access to the people that live and work in the area. Small wireless ISPs have also emerged in rural areas to offer a choice of services to customers who may not have access to other types of broadband. Service may be poor, however, in areas with lots of hills or trees.

Video
Wireless Access Point

Fiber Optic To make up for declining business in landline telephone service, large telecommunications providers in the United States and other countries are offering other services. For example, AT&T and Verizon are delivering digital phone, TV, and internet service to homes using fiber-optic cable.

A fiber-optic cable bundles many strands of thin glass or plastic cable within an external sheath. In terms of service, fiber optics offers the advantages of freedom from

Hands On
Get Ready with Airplane Mode

Hotspot

Understanding Airplane Mode

Making a call using a mobile phone during a commercial airline flight is prohibited in the US by the Federal Aviation Administration (FAA). However, the FAA does allow passengers to use some personal electronic devices (PEDs) during all phases of a flight, including takeoff and landing. Most major airlines offer Wi-Fi service, so now you can use your ebook reader or tablet throughout a flight. (Because of the sizes of laptop computers and certain devices, you must still stow them during takeoff and landing.) As of early 2014, additional rule changes with regard to the use of cell phones were also under consideration. Each carrier also may set its own specific policies for use of personal electronic devices, as well as setting fees (if any) for online wireless services.

There's one catch for using your device in flight—you must put it in airplane mode or otherwise disable cellular service in the device's settings. (Depending on the device and manufacturer, you may see another name for airplane mode, such as *flight mode*, *offline mode*, or *stand-alone mode*.) Using airplane mode prevents the device  from sending and receiving calls, text messages, and other forms of data. Enabling airplane mode may also turn off additional signaling features that could interfere with the plane's avionics.

While in airplane mode, devices generally consume less power. This means that using airplane mode will help preserve battery life, which can be an advantage on a long flight.

electromagnetic interference and reduced signal loss; it also provides the greatest bandwidths over the longest distances. (Network bandwidth is discussed in more detail in Chapter 6.) A fiber-optic connection not only offers the speediest internet downloads, but it can also carry digital TV and phone (voice) signals. This means that fiber optics allows the bundling of internet, TV, and phone services. Some providers offer promotional discounts and free or discounted equipment to new bundle subscribers.

The biggest limitation of fiber optics is availability. Fiber-optic lines are expensive to install, so major providers have been slow to roll out *fiber-to-the-home (FTTH) service*. The competition for providing fiber-optic cable is growing, however. Well-funded companies that haven't previously provided internet or other content delivery services see the potential of fiber optics and are getting into the business. For example, Google Fiber has launched a digital internet and TV service in several US cities, with more cities to follow. Google Fiber's hardware eliminates the need to have a separate modem and Wi-Fi router. Wireless network capability is built into Google's Network+ Box gigabit router. Google's TV Box can connect wirelessly to the signal from the Network+ Box and deliver HDTV.

Google Fiber uses a powerful Wi-Fi router to deliver crystal clear HDTV.

Another feature included with many fiber-optic TV bundles and some cable and satellite TV providers allows playing live TV or streaming other shows to a mobile device. After installing a free app on the device and entering the user name and password for the internet account, you can watch content wherever a wireless connection is available.

A technology called Gfast works as a hybrid DSL and fiber-optic internet connection. This technology eliminates the need for full fiber-to-the-home service. With Gfast, the provider runs a fiber to a location relatively close to homes where it can connect with the older copper phone lines running into the homes. Depending on the specific deployment, this arrangement is called *fiber-to-the-node (FTTN)* or *fiber-to-the-distribution-point (FTTdp)* service. Gfast can help customers in more areas see

Practical TECH

Shopping for an ISP

When shopping for an ISP, consider more than just what type of connection is available in your area. With the exception of dial-up access, ISPs offer most types of connections through a variety of service plans based on download speeds, number of email addresses, and other features. Some plans even bundle services by combining cable TV or telephone with internet for a single monthly fee.

You can find out what providers serve your area, what types of connections are available, and how much different plans cost by doing some research. Go to the library to use its internet-connected computers or take your laptop or tablet to a place with free wireless service. To learn about service options, go to an ISP's website and look for a link or button for setting up a new service. You will likely have to enter your phone number or address to see whether service is provided in your area, as well as what plans and features the ISP offers and what they cost. If telephone or cable service is already available at your location, it's likely that your existing carrier offers internet service too. If that's the case, then start with that ISP's website.

faster internet connections. Note that DSL service advertised with speeds well in excess of those listed in Table 5.1 is Gfast hybrid technology, not standard DSL.

Satellite Like having satellite TV, having a satellite internet connection involves installing a dish at the service location. For a basic plan, the dish might be as small as a satellite TV dish, but for a plan with high download speeds, a somewhat larger dish will probably be required. Advances in the technology have reduced the sizes of satellite dishes over time, however.

Installing a satellite internet dish requires hiring a trained technician. The technician will know how to point the dish in the precise southern direction and angle needed to communicate with the satellite carrying the service. The satellite ISP might lease space on the satellite or own the satellite outright. For example, leading satellite ISP HughesNet uses EchoStar XIX, which it calls the world's highest-capacity broadband satellite, to deliver its Gen5 satellite internet service. ViaSat Internet is the other major satellite ISP offering services to residential and business customers in the United States.

A satellite dish must be installed where there's a clear line of sight to the satellite. At some locations, this requires installing the dish away from the building, whether a home or office, and mounting it on a metal pole. Coaxial cable runs from the dish to the building and connects to a satellite internet modem. (The cable is buried if the dish is installed away from the building.) The modem is then connected to the wired or wireless network router to provide a broadband connection.

Downloading data can be quick via satellite, but uploading is slower. (This is also true for most other types of broadband connections.) Some satellite ISPs allow bundling of TV, internet calling, and other services. The costs of satellite plans are also similar to those of other types of broadband services. Costs will vary depending not only on download speed but also on the amount of data downloaded. Because satellite capacity is limited, satellite ISPs charge users more for downloading large amounts of data (similar to most data plans for mobile devices). For example, a plan might limit you to downloading 15 gigabytes (GB) of data per month, which might not be enough if you stream a lot of video. Satellite ISPs usually offer ways to get around data limitations. For example, you might be able to purchase additional data allowances as needed or to download more data during low-traffic periods (such as 2 a.m. to 8 a.m.). Satellite internet service has long been offered in rural areas with no other forms of broadband connection available, a great advantage.

Mobile Hotspots If you need to connect to the internet frequently while on the go and don't want to constantly look for Wi-Fi hotspots, you can bring your own hotspot with you. Major carriers such as Verizon offer mobile hotspot devices, including devices that integrate mobile hotspot functionality with other features in your car. To use the device, you have to purchase a mobile data plan for it. You tell the device to connect with the mobile (cellular) network, and once it's connected, it wirelessly shares the internet connection. The number of users or devices that can connect at any time varies depending on the device and provider. (See the section "Using a Cell Phone as a Hotspot," in Chapter 6 on page 278, for instructions on setting up your phone as a hotspot.) As providers deploy next-generation 5G cellular networks and more 5G devices become available, using a mobile hotspot to go online should become even faster and more convenient.

A hotspot device enables multiple devices and users to connect to the internet wirelessly from any location. A mobile data plan is required and service is provided by the mobile network.

If you need only a single connection while on the go, you may be able to use your smartphone as a hotspot—a setup called **tethering**. Many models of smartphones are equipped to act as mobile modems. To use this function, you must pay for a monthly data plan. Once you have activated the proper plan, you can turn on the hotspot feature by changing a setting on your phone. The phone will display a request for a sign-in password to make the connection. The wireless networking hardware on your computer or tablet should discover the new wireless network (the phone hotspot). When it does, you will select that connection and enter the password to sign on. Your computer or tablet will then be connected to the internet using your mobile phone provider's network. Any data that you download will be charged against your data plan.

5.3 Navigating the Internet

Once you have connected your computer or tablet to the internet, you can start **surfing**, or navigating between pages and locations on the web using the browser. To navigate the web effectively, you should also know about Internet Protocol (IP) addresses and uniform resource locators (URLs) and how they are used to identify and locate resources on the internet.

Watch & Learn
Navigating the Internet

Web Browsers

A **web browser**, or browser program, displays web pages on the screen of a computer or mobile device. The Microsoft Windows operating system comes with the Microsoft Edge and/or Internet Explorer browser built in (see Figure 5.8). Apple's macOS and iOS operating systems come with the browser Safari built in. Google's Chrome browser works on Android and Chrome OS devices. Google also offers versions of Chrome that you can install and use instead of the default browsers on the Windows, macOS, and iOS operating systems. A number of other browsers are also available for computers and other devices, such as Mozilla's popular Firefox browser and Opera Software's Opera browser.

Article
Browser Evolution

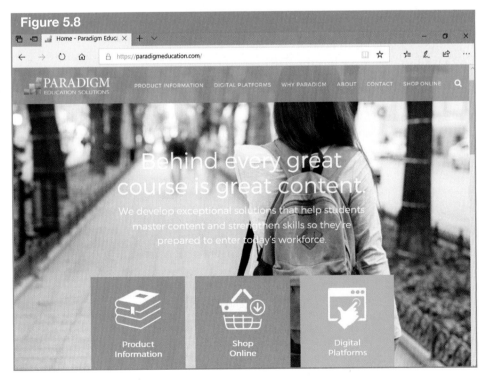

Figure 5.8

Use a browser such as Microsoft Edge to display a web page.

As a part of the internet, website content must be built and accessed using consistent technology. Because of this, most browsers have similar features, including the following:
- The ability to interpret and display HTML code (the language of web pages).
- Support for compiled programming languages (such as Java) and scripting languages (such as JavaScript, Perl, PHP, and Ruby). The mini-programs or scripts written in these languages extend the web browser's capabilities—for example, adding the ability to stream video. In some cases, an additional software component—either a plug-in or a runtime environment—must be installed within the browser to enable it to run the mini-programs or scripts.
- An easy-to-use interface that allows navigating backward and forward, tracking favorite websites, and more. You will read more about these features later in the chapter.

IP Addresses and URLs

Web browsers locate specific material on the internet using an **Internet Protocol (IP) address**. An IP address works like an internet phone number. Every device connected to the web can be located using its IP address.

Currently, a large majority of IP addresses follow the **IPv4** format and use a four-group series of numbers separated by periods, such as 207.171.181.16. Using the IPv4 format, only 4.3 billion addresses can be created, and some of these addresses must be reserved for special purposes on the web. Because of the recent explosion in the number of devices connecting to the internet, a new IP format has been developed, **IPv6**. An IPv6 address uses eight groups of four **hexadecimal digits** (that is, numbers with a base of 16), and the groups are separated by colons (:), as in 2001:0db8:85a3:0000:0000:8a2e:0370:7334. (The IPv6 format can be abbreviated or stated in other ways. For example, you could remove the leading zeros within each group between colons, so that :0000: becomes just :0: or :0370: is just :370:.) Each IPv6 address is also called a **128-bit address**, which means it allows for 2^{128} addresses. That is:

    340 undecillion, or
    340,000,000,000,000,000,000,000,000,000,000,000,000, or
    340 billion billion billion billion.

IPv6 is slowly replacing IPv4; the two protocols aren't compatible.

Because remembering IP addresses would be difficult, every website host has a corresponding web address called a **uniform resource locator (URL)**. URLs generally use descriptive names rather than numbers, so they are easier to remember. For example, to go to the home page of Amazon, you type the URL *http://www.amazon.com*, not the IP address. The **Domain Name Service (DNS)** is an internet utility that uses Domain Name System servers (also called *DNS servers* or *name servers*) to map the URL to the host's underlying IP address.

Practical TECH

Is Your Computer System IPv6 Ready?

Most new computer systems are set up to use the IPv6 protocol, so users don't have to do anything to start browsing to sites using IPv6. To see what your computer's IPv4 and IPv6 addresses are and to test your computer's IPv6 readiness, you can run a test on a site such as https://CUT7.ParadigmEducation.com/test-ipv6. For some older computers, it's possible to add the IPv6 protocol for the internet connection being used. If you encounter this rare situation and have trouble browsing to specific websites, search the web for instructions about adding the IPv6 protocol to your computer's operating system. Making this change might solve the problem.

The Path of a URL

A URL consists of an address that indicates where the information can be found on the internet. It contains several parts that are separated by forward slashes (/), a colon (:), and dots (.). The first part of a URL identifies the communications protocol to be used. One protocol is Hypertext Transfer Protocol (HTTP); it's designated by *http://* (as in the Amazon URL mentioned previously). The other protocol is Hypertext Transfer Protocol Secure (HTTPS); it's designated by *https://* and indicates a secure connection.

Immediately after the protocol is the overall **domain name**, divided into various parts. First may come format information, such as *www* for World Wide Web pages. Following the format information is the **second-level domain**, which identifies the person, organization, server, or topic (such as Amazon) responsible for the web page. The **domain suffix**, or *top-level domain (TLD)*, comes next (*.com* in the Amazon example). After that, there may be additional folder path information (with folder names separated by forward slashes), as well as the name of a specific web page file or object. To go directly to a specific folder or file, you must type the full path and file name portions of the URL. Figure 5.9 illustrates the various parts of a URL.

The domain suffix, or TLD, identifies the type of organization or the country of origin. In the Amazon example, the domain suffix *.com* stands for *commercial organization*. A web-based enterprise is often referred to as a **dot-com company**, because the company's domain name ends with *.com*. Table 5.2 lists other domain suffixes that are in common use.

In 1998, the US Department of Commerce created the Internet Corporation for Assigned Names and Numbers (ICANN) and assigned it the task of expanding the list of existing domain suffixes. In 2013, ICANN approved and began rolling out the latest group of new suffixes (see Table 5.3). These suffixes serve very specific purposes and may be reapproved for use on a limited basis. You may not see them used a lot for national organizations, but you may see local organizations using them.

Video
URL Structure

Figure 5.9 Parts of a URL

Protocol	Domain name	Path	File name
http://	www.nasa.gov/	audience/forstudents/	index.html

Table 5.2 Common Domain Suffixes Used in URLs

Domain Suffix	Type of Institution or Organization	Example
.com	commercial organization	Ford, Intel
.edu	educational institution	Harvard University, Washington University
.gov	government agency	NASA, IRS
.int	international treaty organization, internet database	NATO
.mil	military agency	US Navy
.net	administrative site for the internet or ISPs	EarthLink
.org	nonprofit or private organization or society	Red Cross

Table 5.3 New Domain Suffixes

Domain Suffix	Type of Institution or Organization
.aero	airline groups
.biz	business groups
.club	social groups
.coop	business cooperatives
.guru	experts on a subject
.info	general use
.llc	limited liability companies
.museum	museums
.name	personal websites
.pro	professionals

Some URLs also include a two-letter country abbreviation. Table 5.4 lists the country abbreviations.

Typing a URL into a browser and pressing Enter or Return sends a request to the internet (see Figure 5.10). Routers identify the request and forward it to DNS servers that translate the URL—including a root DNS server, which identifies the location of the top-level domain server. The top-level DNS server uses the URL to determine which page, file, or object is being requested. Upon finding the item, the server sends it back to the originating computer via the user's ISP server using the HTTP or HTTPS protocol, so the browser can display the delivered page on the screen of the computer.

Table 5.4 Country Abbreviations Used in URLs

af	Afghanistan	fr	France	nz	New Zealand	ch	Switzerland
au	Australia	de	Germany	no	Norway	tw	Taiwan
at	Austria	il	Israel	pl	Poland	uk	United Kingdom
be	Belgium	it	Italy	pt	Portugal	us	United States
br	Brazil	jp	Japan	ru	Russia	yu	Yugoslavia
ca	Canada	kr	South Korea	za	South Africa	zw	Zimbabwe
dk	Denmark	mx	Mexico	es	Spain		
fi	Finland	nl	Netherlands	se	Sweden		

Figure 5.10

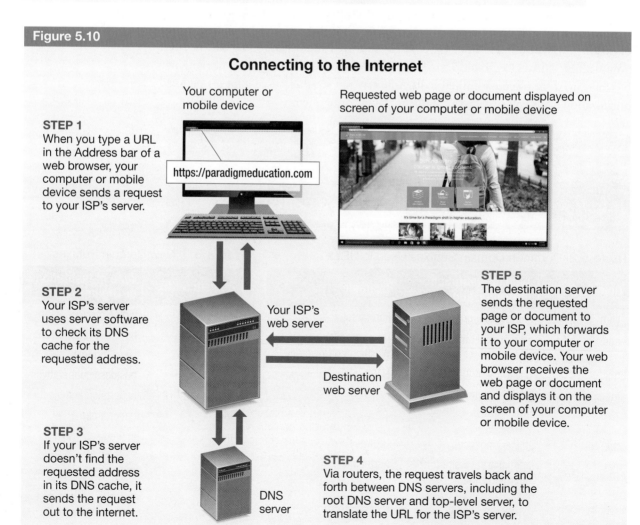

Connecting to the Internet

STEP 1 When you type a URL in the Address bar of a web browser, your computer or mobile device sends a request to your ISP's server.

STEP 2 Your ISP's server uses server software to check its DNS cache for the requested address.

STEP 3 If your ISP's server doesn't find the requested address in its DNS cache, it sends the request out to the internet.

STEP 4 Via routers, the request travels back and forth between DNS servers, including the root DNS server and top-level server, to translate the URL for the ISP's server.

STEP 5 The destination server sends the requested page or document to your ISP, which forwards it to your computer or mobile device. Your web browser receives the web page or document and displays it on the screen of your computer or mobile device.

Packets

A file isn't sent over the internet as a single file. Instead, messaging software breaks the file into units called **packets** and then sends the packets over separate paths to a final internet destination. The packets are reassembled when they reach that destination. What paths individual packets take depends on which routers are available. This process of breaking a message into packets, directing the packets over available routers to their final internet destination, and then reassembling them is called **packet switching**.

Figure 5.11 shows the journey of an internet message sent from a computer in Seattle and received on a computer in Miami. The message is broken into three packets, and each packet follows a different route to the final destination. If a packet arrives at a node (that is, a router or other communications-switching equipment) that is not the final destination, the network passes it along to the next node. At the destination, the computer receives the packets and reassembles the file. If any packet is missing, the receiving computer requests that it be sent again. This re-requesting of information explains why web pages sometimes appear incomplete and some portions take longer than others to fully load.

The idea of dividing files into packets originated during the Cold War (1945–1991). During that era, the internet was thought of as a system for maintaining communication among the military and other government agencies in the event of a

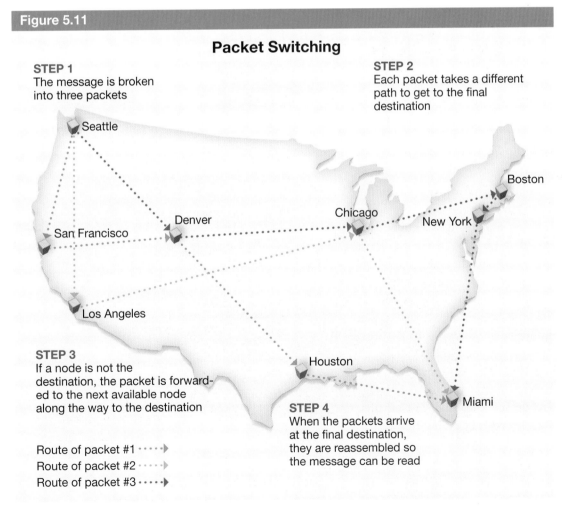

Figure 5.11

Messages sent over the internet are broken into separate units called *packets*. The packets are directed by routers and reassembled when they arrive at the final destination.

nuclear war. To prevent breakdowns, the system was designed to keep working even if part of it was destroyed or not functioning. A packet sent from New York to Los Angeles, for example, might attempt to travel through Denver or Dallas. But if both of those systems are busy or unavailable, the packet could go up to Toronto and then on to Los Angeles. This design feature, called **dynamic routing**, is part of what makes the internet work well even with a heavy load of traffic.

IPv6 enables routers to handle packets more efficiently than IPv4. Because of this and many other differences between the two protocols, the internet will have to operate using the two protocols as side-by-side networks until IPv6 has been fully implemented. Special translation gateways and some other translation technologies enable limited traffic to flow between the IPv4 and IPv6 networks.

5.4 Understanding Web Page Markup Languages

Watch & Learn
Understanding Web Page Markup Languages

Text files created with a word-processing program follow a particular file format that the program can display. Similarly, web page files follow a particular format that a web browser can display. Web page files are different, however, because special software isn't needed to create them. You can use practically any word-processing or plain text editing program that you want. However, to describe the content for a web page so that the browser can display it, you must use a specific language.

HTML and CSS

Hypertext Markup Language (HTML) has long been used to create web pages. HTML is a tagging or **markup language**, which includes a set of specifications that describe elements that appear on a page—for example, headings, paragraphs, backgrounds, and lists. Like other programming languages, HTML has evolved over time to include new features and specifications; the current standard is HTML5.

Along with HTML, you can use **cascading style sheets (CSS)** to make the design for a web page or site more consistent and easier to update. CSS is a separate formatting language. Most modern websites also have functions that have been added using various scripting languages. (Chapter 11 will cover web development in more detail, because creating a complex website requires significant programming skills.)

HTML gives web developers a lot of freedom in determining the appearance and design of web pages. Within an HTML file, tags are inserted to define various page elements, such as the language type, body, headings, paragraph text, and so on. A **tag** is enclosed in angle brackets, and most tags must be used in pairs: an opening tag and a closing tag. (The closing tag includes a forward slash.) As shown in Figure 5.12, the opening tag *<html>* should appear at

Figure 5.12

```
<!doctype html>
<html>
<head>
        <title>Page title</title>
</head>
<body>
        <header>
                <h1>Page title</h1>
        </header>
        <nav>
                <!-- Navigation -->
        </nav>
        <section id="intro">
                <!-- Introduction -->
        </section>
        <section>
                <!-- Main content area -->
        </section>
        <aside>
                <!-- Sidebar -->
        </aside>
        <footer>
                <!-- Footer -->
        </footer>
</body>
</html>
```

Opening tag defining the language <html>

Closing tag defining the language </html>

HTML tags are enclosed within angle brackets (< >) and are usually used in pairs. An opening tag and closing tag define the text between them as a particular type of element for the web browser to display.

the top of the document (below the <!doctype> tag). The closing tag </html> should appear at the bottom of the document to indicate that all of the text in the document is HTML code. The tag pair <body> and </body> define what should appear in the main browser window, <title> and </title> indicate the page title, and so on.

A web page also typically functions as a **hypertext document**, presenting information that's been enhanced with **hyperlinks** (also called *web links*) to other websites and pages. The user clicks a link to access additional information on another web page. Links most commonly appear as underlined text, but they can also take the forms of buttons, linked photos or drawings, navigation bars or menus, and so on.

XML

Whereas HTML defines the format of a web page, **Extensible Markup Language (XML)** organizes and standardizes the structure of data so computers can communicate with each other directly. XML is more flexible than HTML, because it's really a **metalanguage**, a language for describing other languages. XML allows developers to design custom languages that work with limitless types of documents and file formats.

Web developers frequently use XML to manipulate and present information for large online databases. Many commonly used applications, such as Microsoft Excel and Access, can now import and export XML data. This compatibility makes XML the ideal tool for facilitating data sharing between the web and other platforms.

Website Publishing Basics

You can create a simple website with just a little knowledge of HTML. (As noted earlier, Chapter 11 will provide more information about the scripting languages used for web programming.) Many ISPs include storage for a basic website and supply a predetermined URL based on the information in your account name. You can upload your web page to that location, often using a method called File Transfer Protocol (FTP), and have your site ready to go. (FTP is explored further later in this chapter.) However, if you want to have your own domain name, the process requires a few more steps:

1. **Obtain your domain name.** You can check to see if a particular domain name (for example, *mydomainname.com*) is available by using name registration services such as GoDaddy (at https://CUT7.ParadigmEducation.com/GoDaddy) and Network Solutions (at https://CUT7.ParadigmEducation.com/NetworkSolutions). If the name you want is available, you can register it by paying a fee. Registration services also enable you to renew your domain name, and in some cases, they allow you to bid on names already owned by others.

2. **Obtain hosting.** The internet hosting service operates the web server that will store your website's content and deliver it to the web. The monthly or annual fee you pay will vary depending on the level of hosting services you need; you can usually get started for a low monthly cost. (Of course, a large organization might have its own servers to self-host the website.) GoDaddy and Network Solutions are among the many national hosting companies; you might also find local hosting companies that you can work with.

3. **Create your site content.** You can create your own HTML and other content from scratch, or you can use fee-based or free online tools such as Google Web Designer, Joomla, WordPress, and Drupal. (Keep in mind, however, that the features you build in may require a higher level of web hosting to support them.)

Tech Career Explorer
Trying Out HTML

Article
Using XML to Share Information

Video
Creating a Simple Web Page

If you include graphics files, use the file formats JPEG (Joint Photographic Experts Group), PNG (Portable Network Graphics), and GIF (Graphics Interchange Format). These files are smaller and preferred because they download more quickly. Similarly, for any video or audio you include, use an appropriately compressed file type, such as MP4 for video and MP3 for audio.

4. **Upload your content if needed.** Some hosting services include built-in tools for transferring files using FTP or another method. In other cases, you will need FTP software to complete the transfer.

Also keep in mind two more points. First, most hosting services require the home page for a website to have a specific file name (such as *index.htm*). Following this practice ensures that when a user navigates to your domain name (URL), the correct page will display. Second, some users take advantage of popular blogging websites (such as WordPress.com and Blogger) to get content creation tools and hosting services in one package. A URL may also be included in that package (although typically, the URL will include the name of the blogging website). WordPress.com will let you use your custom domain URL if you upgrade to a monthly hosting plan.

5.5 Viewing Web Pages

A **web page** is a single document that's viewable on the World Wide Web. A **website** consists of one or more web pages devoted to a particular topic, organization, person, or the like, and generally stored on a single domain. When you navigate to a website, the first page displayed is usually the site's home page. Like the table of contents in a book, the **home page** provides an overview of the information and features presented in the website.

This section covers some of the basic actions for using a web browser program to view, find, and mark online information. It also introduces the basic elements incorporated within today's multimedia sites.

Watch & Learn
Viewing Web Pages

Hands On
Taking a Spin Online

Basic Browsing Actions

Various web browsers look a little different and offer some varying features. Even so, the basic browsing actions work the same no matter which browser you are using. Look at Figure 5.13 to see where to find the bars, buttons, and links used to perform these actions:

- **Go to a URL.** Select any text in the Address bar, type a new URL, and then press Enter or Return.
- **Go back, go forward, and reload.** Click the Back button to return to the previous page and click the Forward button to return to the previous page. Click the Reload button to reload a page that didn't display properly. While a page is reloading, the Reload button changes into the Stop button. Click this button to stop the reloading (which you may want to do if it's taking too long, for example).
- **Add a new page tab.** Click the New tab button, type the URL for the new web page, and then press Enter or Return.
- **Browse by following a hyperlink.** Click the hyperlinked text, image, or button in the body of the page, on the navigation bar, in a drop-down list, or elsewhere on the page.

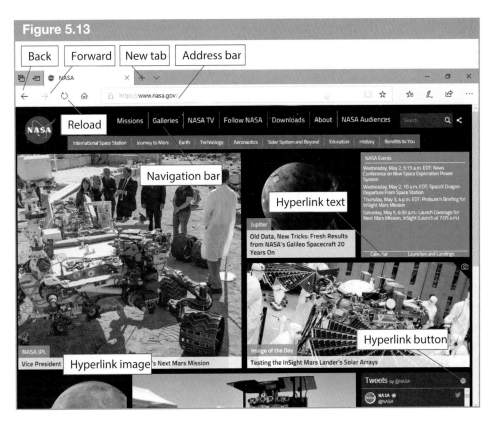

Figure 5.13 Use these tools to browse the web.

Audio, Video, and Animation Elements

As the web grows more sophisticated, multimedia elements such as video and sound are being incorporated into web pages both to grasp viewers' attention and to expand the range of activities a web page can perform. Various applets and scripts make adding the elements possible, as do plug-ins such as Apple QuickTime, Adobe Acrobat Reader, and Adobe Shockwave player.

Java Applets and Scripts **Java** is a programming language that website designers frequently use to produce interactive web applications. It was created for use on the internet and is similar to the C and C++ programming languages. Applets are small Java programs that web browsers run. (*Applet* is the term for a miniature program.) Like macros, Java applets provide the ability to program online games and highly interactive interfaces.

Other scripting languages, such as JavaScript and Ajax, enable web developers to add web page functions, such as user login tools, shopping carts, dynamically loading content (such as articles that update periodically), and more. (Numerous scripting languages are available, including Perl, PHP, and Ruby; they will be discussed again in Chapter 11.)

Cookies A **cookie** is a very small file that a website places on a user's hard drive when he or she visits the site. Often, these files are harmless and even helpful. For example, one type of cookie may be used to verify whether you are currently logged in to your subscription account on a site such as an online newspaper. Sometimes, however, websites use cookies to track the surfing habits of users without their knowledge. One website might place a file on your computer that indicates what sites you have visited

Tech Career Explorer
IT for Space Travelers

Practice
Web Browsing Features

Article
Plug-ins, Applets, ActiveX Controls, and Scripting Languages

and then allow related sites to read this information and track your actions. You can adjust the security settings on your browser to warn you when cookie files are being accessed or to prevent their operation altogether.

Plug-Ins Sometimes, a website will ask for approval to add a plug-in to your web browser. A **plug-in** is a mini-program that extends the capabilities of web browsers in a variety of ways, usually by improving graphic, sound, and video elements. A missing plug-in often causes viewing errors on a web page, such as a message that a video can't play because the appropriate plug-in hasn't been installed.

Most plug-ins don't carry hidden or destructive features that can cause problems. However, as a general rule, it's a good idea to perform a web search and verify that a plug-in is safe and legitimate before giving permission to load it onto your computer. If you think the website recommending the plug-in download is trustworthy, click the Refresh button on the browser. Doing so will reload the page and redisplay the prompt to install the plug-in. Follow the prompts to install the plug-in for your browser.

One widely known plug-in is Adobe Shockwave Player. Sites using Shockwave can take longer to load but offer higher-resolution graphics, superior interaction, and streaming audio. Adobe Flash Player (which is being phased out) and Apple QuickTime are two other popular plug-ins that let users experience animation, audio, and video.

Ads

Advertising on a company's or individual's website produces income. For some companies and individuals, this ad revenue supplements the primary revenue stream. For example, the giant online retailer Amazon not only promotes the products it sells, but also sells ad space to companies that want to feature their products or services on Amazon's prime web "real estate." For other companies and individuals, income from ads posted on their websites provides the majority of their revenue. A social media site such as Facebook, a news media site such as CNN.com, and an individual blogger's website may all rely heavily on ad-generated revenue.

Website advertisers typically pay for the advertising on a per-click basis, but other payment models are followed as well. A website may also present ads in many different formats, including ad banners and pop-up windows. Users should be aware that some less-than-legitimate content can masquerade as ads, including blind links and hijackers.

Banners A **banner ad** invites the viewer to click it to display a new page or site selling a product or service. Banners were originally rectangles that appeared across the tops and bottoms of web pages, but today, they can take any shape and appear in any location on a page. Many banners include bright colors, animation, and video to attract attention, as shown in Figure 5.14.

Banners can provide helpful information to people interested in the product being advertised. Also, banners often provide special pricing when the user clicks the banner to go to the linked product site.

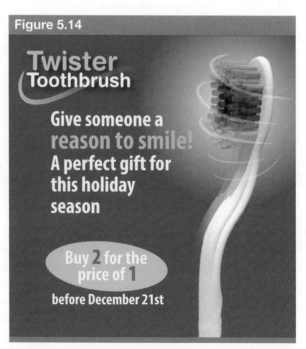

Figure 5.14

Websites sell advertising space in the form of banner ads.

Pop-Ups An online ad may also appear as a **pop-up ad**, as shown in Figure 5.15. Named for its tendency to appear unexpectedly in the middle or along the side of the screen, a pop-up ad (or *pop-up*) typically hides a main part of the web page. When the window pops up, that part of the page may also darken or flicker.

On most devices, you can click a Close (X) button in the pop-up to remove it from the screen. However, some pop-ups are extremely persistent—reappearing immediately after you close them or even lacking a Close button or any other means for shutting them. On a Windows computer, you might be able to close a pop-up window that doesn't have a Close button by pressing the keyboard shortcut Alt + F4. Most web browser software includes **pop-up blocker** features that you can activate to avoid this sort of nuisance. (Chapter 9 will explain more about web privacy and security, including pop-up blocking.)

Figure 5.15

A pop-up ad covers part of the web page and prompts the user to buy a product or subscribe to information.

Blind Links A link sometimes misrepresents its true destination, taking you to an unexpected page when you click on it. For example, a link such as Next Page might actually take you to an advertising web page. Called a **blind link**, this type of deceptive device appears only on websites that aren't trustworthy, such as some free-host sites. Free-host websites don't charge a fee for hosting web pages. However, they do require that the pages display banners and other forms of advertising chosen by the companies that host the pages.

Hijackers Web hijackers can disrupt your browsing experience and expose you to unwelcome ads. A **hijacker** is usually an extension or plug-in that's installed with your web browser. It functions by taking you to pages you didn't select—generally, pages filled with advertisements.

Hijackers typically install on your computer without your awareness and sometimes attempt to make you pay a fee to remove them. You can often get rid of hijackers by working with the security settings in your web browser and by deleting or uninstalling web browser extensions. In extreme cases, you may need to reinstall your computer's operating system.

A new form of hijacker replaces ads that the publisher sold with ads from the company that created the hijacker. Each time a user clicks one of these ads on the publisher's website, the hijacker company—not the publisher—receives revenue.

Web Page Traps Some websites can change your browser's settings permanently or attempt to prevent viewers from leaving by continually popping up more windows and disabling the Back button. To avoid this so-called **web page trap**, change your web browser's settings to increase the level of security. Having higher security settings may cause the browser to prompt you whenever it

Hotspot

Wireless, USA

5G wireless technology will soon replace 4G technology on cellular networks throughout the US, bringing faster speeds and better responsiveness to mobile devices. 5G upload and download speeds are comparable to those of other types of broadband internet connections, meaning 5G could serve as the first wireless "internet" network covering the country. At least one mobile provider plans to offer in-home fixed 5G installations, with 5G serving as the main broadband internet connection.

comes across suspicious behavior, as shown in Figure 5.16. Unfortunately, having higher security settings may cause the browser to prompt you constantly, even when you are visiting legitimate websites.

If you fall into one of these traps and are using Windows, press the key sequence Control + Alt + Delete to open the Task Manager. (Depending on your Windows version, you may have to press this sequence twice.) From the Task Manager, you can choose to reboot the computer or close the web browser.

Private Internets

Security and privacy online, which you'll read more about in Chapter 9, have become an increasing concern for users of all types. A variety of technologies and methods for working online securely continues to emerge, addressing concerns of both individuals and groups of professionals. A private internet-type network can help professionals with similar communications needs collaborate more effectively. For example, the US Department of Energy's Energy Sciences Network, also known as ESnet (see Figure 5.17), provides 100 Gbps transfer speeds to handle the intensive data needs for researchers at the country's National Laboratory system and other research facilities.

Individual users and groups can take advantage of the Tor Browser, which works in conjunction with another type of private internet, the Tor Project. The Tor Browser

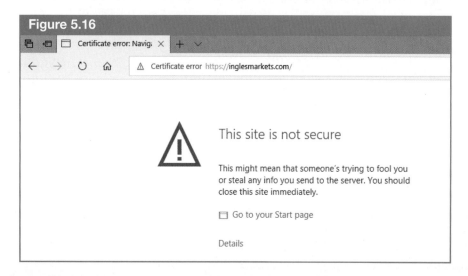

Figure 5.16

This type of warning indicates a problem with the site's online security certificate, which verifies the site's authenticity to users. The certificate might have expired, for example.

Cutting Edge

Introducing Web 3.0, the Semantic Web

The term *Web 2.0* was coined in 2004 by Tim O'Reilly at a conference sponsored by the company he founded, O'Reilly Media. Interaction and collaboration were beginning to occur online, most notably through websites such as Facebook and Twitter. Now that Web 2.0 is part of the internet's "present," experts are increasingly using the term *Web 3.0* (sometimes called the *Semantic Web*) to describe the internet's future. Possibilities for Web 3.0 developments may include enhanced security and privacy, decentralization, machine-generated (rather than human-sourced) information, three-dimensional (3-D) simulations and augmented reality (including the widespread use of sensors), greatly expanded use of high-quality video, and much more.

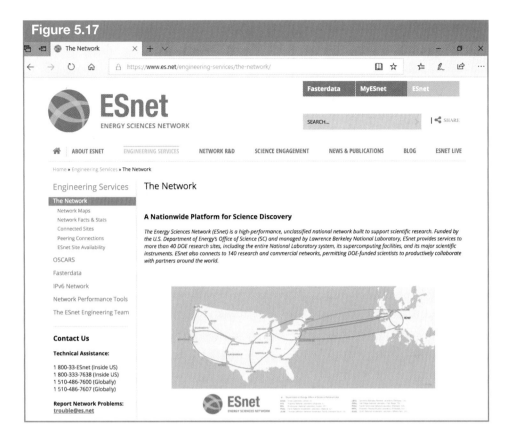

Figure 5.17

The Energy Sciences Network, also known as ESnet, is a fast private network handling the intensive data needs for researchers at the US National Laboratory system and other research facilities.

is available as a free download on the Tor Project website. When you use the Tor Browser, your communications travel through private tunnels that are overlaid on the public internet and facilitated by volunteer-operated servers worldwide. Tor allows you to navigate the *dark web*, which is a part of the internet made up of websites that hide their IP addresses. The dark web is part of the *deep web*, which consists of parts of the World Wide Web that are not indexed by search engines.

With Tor, you can access websites using the .onion domain, which is not part of standard DNS naming and typically represents anonymous services. Tor not only directly enhances your privacy, but also prevents websites from tracking your online activities. Similarly, the private network Freenet aims to help users avoid internet censorship, and I2P focuses on anonymous website hosting.

Tools like Tor can play a role in unethical activities. Unfortunately, anonymous services can consist of black markets selling illegal goods, hacking groups, scams, and worse. The infamous Silk Road black market was operated as a Tor hidden service. Because users could navigate anonymously on Silk Road, it became known for drug sales and other illegal trade. It was ultimately shut down by the Federal Bureau of Investigation, and its founder was convicted of several charges and sentenced to life in prison.

According to the Tor Project, many people use Tor for positive reasons. For example, a user in a country that blocks some web content might use Tor to search for information about sensitive subject matter, such as particular health conditions. Journalists use Tor to receive information from sources and whistleblowers or to distribute stories to media beyond the state-controlled outlets in their home countries. Law enforcement officials can use Tor to protect the identities of persons submitting crime tips. Human rights workers use Tor to protect themselves when reporting abuses and corruption.

5.6 Searching for Information on the Internet

More than 90 percent of adult internet users search the internet to tap into the global library of limitless data that's available on practically any topic. Some experts now estimate that the World Wide Web holds trillions of pages of information. Finding any particular item among so much information requires the proper search tools and techniques.

Users can search for and retrieve information from web pages by using a search engine in a web browser. A **search engine** is essentially a website or service you use to locate information. You type search criteria or **keywords** in the site's search text box or in your browser's address or location bar. The leading search engines in the US include Google, Yahoo, and Bing.

Suppose you want to find information about the Battle of Vicksburg for a history class report. You will begin by typing the search criteria—in this case, the words *Battle of Vicksburg*—in the search text box and then pressing Enter or clicking the Search button. Doing so will cause a list of hyperlinked articles to appear in the browser. Clicking on an article's title will make the article display and allow you to read it. Most search tools provide a basic search page that contains a search text box with a search command button beside or below it (see Figure 5.18).

Search Engine Choices

The search engine that was set up as the default when you installed your web browser may not be the one you prefer to use. However, every web browser program enables you to change the default search engine. In addition, most browsers enable you to add another search engine of your choice.

Not all search engines offer the same features, and some search engines perform certain types of searches better than others. Search results differ depending on the search algorithm (step-by-step search method) and software tools used to generate

Watch & Learn
Searching for Information on the Internet

Hands On
Adding a Search Engine to Your Browser

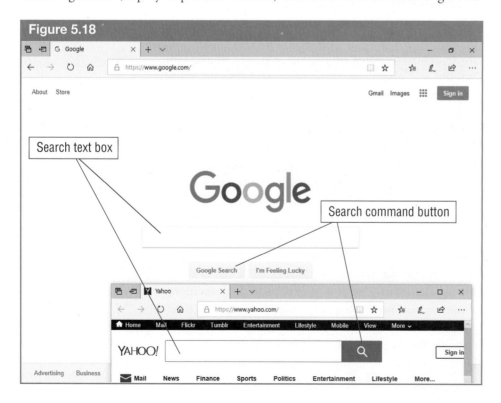

Figure 5.18

Google and Yahoo are two frequently used search engines. Both provide a basic search page with a search text box and a search command button.

the results. Search results also may differ depending on the number of web pages an engine indexes (catalogs) and on the methods it makes available for refining a search. Also consider that some search engines accept fees for placing websites at the top of search results lists. This means that the owners of the first few websites in a results list are likely to have paid for that premier position, and the websites may not contain the information that's most relevant for you.

In addition to working with a single search engine, you can use a metasearch engine. A metasearch engine sends your search request to multiple search engines at once, and may also search other libraries and directories or paid-placement services. Metasearch engines include Search.com, Dogpile, and Excite.com.

Default Search Engine

You can change the default search engine in your browser at any time. For example, you might prefer to use Google for most searches but switch to the Wikipedia Visual Search engine when you are looking for content with images for reference. Most browsers enable you to switch between search engines easily, but keep in mind that the new search engine you choose will remain the active or default engine until you choose another one.

Hands On
Choosing the Active or Default Search Engine in Your Browser

Advanced Search Techniques

To get the most targeted results from any internet search, you must enter the right keyword or keywords in the search engine's search text box. Using too many keywords will produce hundreds or even thousands of search results, which you will need to wade through to find what you are looking for. Using vague, obsolete, or incorrectly spelled terms will further reduce the chances of conducting a successful search. Think carefully about what combinations of words most likely appear in the material you need. Some search engines, such as Google, let you work with an advanced search page that prompts you to enter details about the information you want, as shown in Figure 5.19.

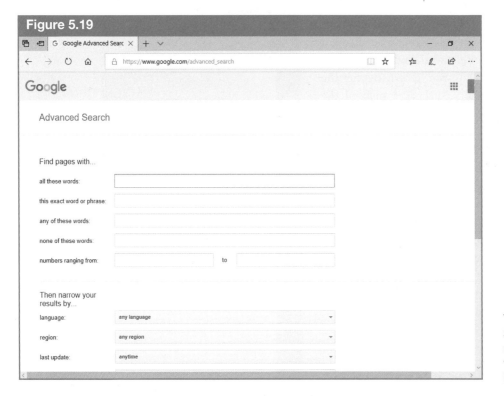

Figure 5.19

The Google Advanced Search page prompts users to provide details that will return more specific information.

The first and often overlooked method for getting more accurate search results involves using quotation marks. If you enter the phrase *four score and seven years ago*, you will get different results than if you enclose the phrase in quotation marks, for example *"four score and seven years ago"*. For a well-known phrase such as this, the results might be similar due to the advanced intelligence of the algorithms used by today's search engines. However, for a less common phrase or a special item such as a name, enclosing the phrase or name in quotation marks will produce more targeted results.

Other advanced searching methods use a logic statement containing one or more **search operators** to refine searches. Three common search operators are AND, OR, and NOT:

- AND (or the plus sign, +) connects search terms and returns search results that contain references to all of the terms used. For example, asking the search engine to search for *dogs AND cats* (or *dogs + cats*) will return only sites containing references to both dogs and cats.
- OR returns results that contain references to any of the search terms. Asking the search engine to search for *dogs OR cats* will return sites that have references to either dogs or cats or both. OR is usually the default logic option on a search engine.
- NOT is used to exclude a keyword in a search. Searching for *dogs NOT cats* will produce only sites that refer to dogs but don't mention cats.

You can use search operators along with phrases contained in quotation marks to conduct even more detailed searches.

Hands On
Reinforcing Your Browsing and Search Skills

Searching with an Intelligent Personal Assistant

In Chapter 1, you read about intelligent personal assistant software that you use to interact with voice-controlled smart speakers. When you ask Siri, Google Assistant, or Alexa a question using a smart speaker, the software converts the question to a web query and then delivers the answer verbally.

Intelligent personal assistant software can also be used with other devices and operating systems. For example, Siri is included with iOS for iPhone and iPad, and with some versions of macOS. You can activate Siri using a built-in or added microphone, and then ask a question (such as "Where is the Eiffel Tower?"). Siri searches the web and displays information (such as a map of the tower's location, a list of its open hours, and its address) on the device's screen.

Google Assistant ships with Google Home speakers and Pixelbook laptops. It is also available as an app that can be added to a variety of devices, such as Android phones and wearables. Alexa comes on voice-controlled smart speakers from Amazon and other third-party devices.

Cortana is an intelligent personal assistant included with the Windows 10 operating system. This assistant is built into the Windows taskbar as a search box and a microphone button. You can type a query in the search box or, if you have a built-in or added microphone, you can click the microphone button and say your query. Like Siri, Cortana displays the results onscreen. For example, Figure 5.20 shows the results of the query "Grand Canyon National Park".

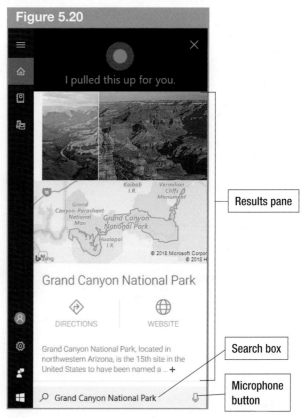

Figure 5.20

In some cases, intelligent personal assistants may provide lists of web results onscreen in response to a question. If that happens, click a result in the list to open the system's web browser and display the linked information. When you search in Microsoft Edge using the Address bar or a search engine or go to a particular type of website such as a restaurant's website, a Cortana icon may appear in the Address bar. If that happens, click the icon to display a results pane with links to matching information or details or coupons for the displayed site.

Bookmarking and Favorites

If you frequently use the web for research, you can mark and organize web pages to make it easier to find them again later. Some web browsers (such as Chrome and Firefox) call a marked page a **bookmark**, while others (such as Microsoft Edge and Safari) call it a *favorite*.

Once you have navigated to the page to mark, choose the button or command for marking favorites in your browser; this happens to be a star icon in the upper-right corner of both Chrome and Microsoft Edge. If needed, click the button for adding favorites, edit the name to use for the site, choose or create the folder to store the bookmark in, and then click Save, Add, or Done. Using folders enables you to organize the sites you visit by subject matter or type, as shown in the folders listed in Figure 5.21. To use a bookmark or favorite, simply display the list or pane that holds them in your browser, expand a folder if needed, and click the bookmarked web location to visit.

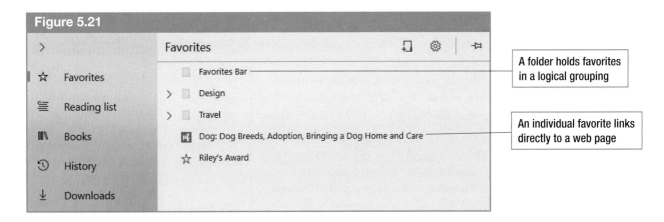

Figure 5.21

5.7 Using Other Internet Resources and Services

Watch & Learn
Using Other Internet Resources and Services

The internet originated as a platform for communication and information sharing. As its infrastructure and underlying programming technologies have evolved, the internet has become a platform for many activities beyond communication and research. This section provides a detailed look at the broad variety of applications and activities that you can take advantage of when online.

Communicating with electronic devices might sound cold and impersonal, but in reality, the internet continues to evolve as a primary channel for staying in touch with others. Less than a dozen years ago, you could use the internet to communicate in only a handful of ways, but now, new digital platforms for connecting and sharing with others online emerge almost daily.

Electronic Mail

Electronic mail (email) remains one of the top activities performed on the internet. In fact, there were more than 6.65 billion active email accounts in 2018 (see Figure 5.22). You can use email to create, send, receive, save, and forward messages in electronic form. Email is a fast, convenient, and inexpensive way to communicate.

A typical internet account includes at least one email address, and some ISPs allow you to create multiple email addresses per account. Other web-based email services (such as Gmail, Yahoo Mail, and Outlook.com) allow you to set up an email address independent of a particular ISP. Gmail, Yahoo, and Outlook.com accounts are free, but paying a small monthly fee will typically let you upgrade your account with features such as additional storage space.

You also can choose to work with and manage email in two different ways, depending on your ISP's capabilities and your own preferences. One way is to use an email application that's installed on your computer, such as Microsoft Outlook or Mozilla Thunderbird (a free alternative). Both Windows and macOS include built-in applications for reading and sending email. After installing the software (if necessary), you set up your email account information so the software can connect with your ISP to send and receive messages. Email applications typically download all messages to your computer or device. However, you can choose to have messages stored online, as well, so that you can access them from other computers or devices.

The other way to work with and manage email is to use **webmail**. This method has become more common as users have increasingly wanted to access email via a smartphone or tablet in addition to a computer. With webmail, you use your web browser to navigate to your email account on the website of your ISP or web-based mail service. Depending on your service and device, you may have to navigate to a particular URL (such as *mail.yahoo.com*) to access your sign-in page (see Figure 5.23). After you sign in with your account information, you can send, receive, and view messages online.

Even though Gmail, Outlook.com, and Yahoo Mail are all webmail services, they also enable you to access your email through an email application such as Outlook. For Gmail and Yahoo, the

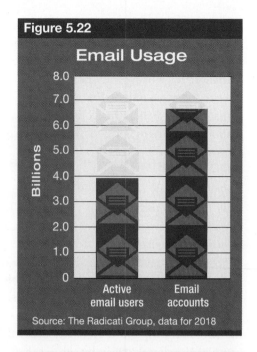

Figure 5.22

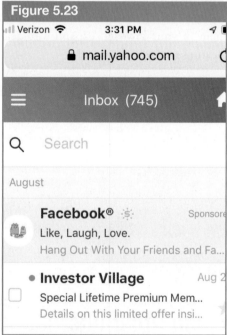

Figure 5.23

You can view webmail using your mobile device browser.

Practical TECH

Mail on Mobile

If you use a tablet or smartphone for email, you can probably download an email app to make the process simpler than logging on through the device's web browser. For example, you can download apps for Gmail and Yahoo Mail. To find apps specific to other email providers, search the app store for your device type. For some other types of accounts, such as Outlook.com, you enter the email account information in the device's settings and use the device's mail app to work with your email. iOS and Android devices all have built-in mail apps.

Chapter 5 Plugging In to the Internet and All Its Resources

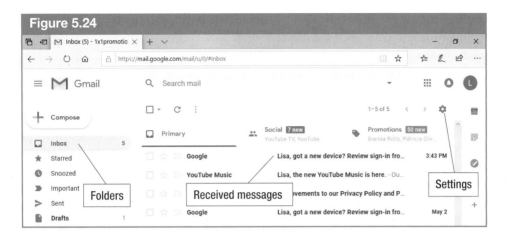

Figure 5.24

Google's Gmail is a free email service that includes many valuable features.

process for connecting to your email application typically means making sure that the POP email protocol is enabled in your online account settings and then setting up the account information in your email program.

Figure 5.24 shows common features of both email applications and web-based email as they appear in Gmail. These features allow you to perform the following actions:
- organize messages in folders
- assign priorities to the messages you create
- sort the messages in a folder
- change settings
- mark messages as "Read" or "Unread"
- print messages and save attachments

Whether you are using an email application or webmail, the process for sending an email message generally works the same. You click the command or button for composing a new message, provide the recipient's email address, type a subject in the *Subject* text box, type a message, and then click the Send button. In addition to sending messages that contain text, you can attach various types of files to email messages, including reports, spreadsheets, photos, and video files. Recipients can then open the attached files to view or save them.

In most email programs and webmail systems, you insert an attachment by clicking an Attach button (which often has a paper clip icon). Some email programs allow you to insert other types of information in a message as well. For example, if you click the Insert tab in an Outlook message window (see Figure 5.25), you will find options for inserting or attaching a calendar item, virtual business card, table, and more.

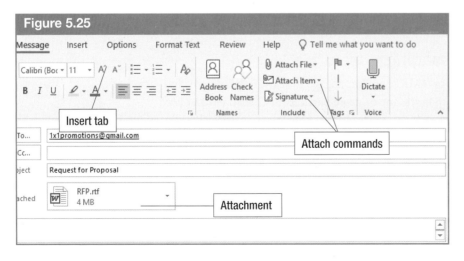

Figure 5.25

In the Outlook message window, click the Attach File button to select a file to send with a message or use the options available on the Insert tab to include other types of content.

For efficiency, you may want to compress or "zip" large file attachments or multiple files. A **compressed file** or **zipped file** is smaller than the original file, which means it takes less time to send and download (copy from the host's computer to the recipient's computer). The recipient can then open or "unzip" the received file to view its contents.

Be aware that because malicious software can be hidden within certain types of files, some ISPs and IT departments block certain types of file attachments—including zipped files, image files, and executable files. It's a good idea to check with the recipient before sending a very large file or a file in a special format. If sending and receiving such files causes problems, you can use the FTP method (covered later in this chapter) to transfer the information.

Also be aware that most ISPs enable you to have some control over spam (unwanted junk emails). You can often change the overall level of spam filtering in your account settings or within your email program, and block specific senders. You also can mark messages as spam, so that messages from that sender will automatically be directed to a Spam or Junk folder (also called the **spam trap**) in the future. The only drawback to this screening feature is that legitimate messages sometimes end up in the spam trap. Be sure to check this folder routinely for messages you might want.

Social Media, Sharing, and Networking

One genre of web services enables people to create personal online spaces and interact socially. **Social networking services**—such as Facebook, Twitter, Pinterest, LinkedIn, and Reddit—have become popular, boasting hundreds of millions of users. These free sites give anyone the opportunity to open an account and share content.

Article
Internet Social Networks

For example, Facebook users can personalize their Timelines with profile information, photos, music and video, and status updates. Animal rescue groups use their Facebook pages to raise funds and place animals in adoptive homes.

Twitter operates differently, allowing people to share 280-character messages called *tweets* with anyone who cares to follow and read their messages. Tweets can include photos, videos, and shared information, as well as text. Celebrities and journalists use their Twitter accounts to connect with millions of fans and followers.

Some sites bring together content from a variety of sources, either organizing it by tags or letting contributors rank or vote on its value. When tagging, the publisher or user applies a keyword (or *tag*) to identify a particular type of content.

New types of social media sites emerge nearly daily to address nearly any interest. Table 5.5 lists a variety of popular and emerging services of various types. To use some social media services, such as Instagram, you have to download an app for your mobile device. Other services simply work within your web browser.

Video
How Are Businesses Using Social Networking?

Chat and Instant Messaging

Using **chat** and **instant messaging (IM)**, users can engage in real-time dialog, or live, instantaneous online conversations, with one or more participants. Chat and IM used to be thought of as two different services, with chats taking place in special online locations called *chat rooms* and IM usually taking place between two people in a back-and-forth dialog, like a typed phone conversation. Many users now call IM interactions *chats*; for example, some sales and help websites give you the option of initiating a chat (IM) session with a customer service representative.

ICQ ("I seek you") offers web-based chat room environments, where users can discuss a variety of topics, such as climate change and stocks. IM used to require

Table 5.5 Popular Social Media Services Beyond Facebook and Twitter

Service	Description
Pinterest	Pinterest lets you bookmark (or *pin*) items from around the web on a virtual corkboard or scrapbook; you can also follow other users' boards.
Instagram	Instagram lets you capture photos from your mobile device, apply fun filters, and then share the photos on Instagram and at other social sites, such as Facebook, Twitter, and Tumblr.
tumblr. / reddit	Some social sites aggregate (or collect) a particular type of content and allow users to follow and interact. Tumblr aggregates more than 150 million blogs, and Reddit acts as a social news service.
LinkedIn / LinkedIn Learning	LinkedIn is a professional social network on which users can create an online résumé, search for jobs, and share information about business topics. Users can upgrade to a Premium plan to take advantage of advanced features for job searching, reputation management, finding business leads, and finding talent. LinkedIn offers other fee- and subscription-based services, such as advertising and LinkedIn Learning online courses.
DeviantArt / flickr	Some social sites let you share and distribute content you have created. Digital artists and others can share and sell their creations on DeviantArt. Using digital photocentric services such as Flickr, you can upload and share images, have printed items made, and distribute images for a fee or under Creative Commons licensing.

having a specific application. However, most IM services have been moved to a web-based platform or rolled into another service, such as Skype, Facebook, or Slack. Internet Relay Chat (IRC) is a form of instant messaging (even though it includes the word *chat* in its name) that requires specific software, such as mIRC (https://CUT7.ParadigmEducation.com/mIRC). In addition, Google Hangouts incorporates text chat along with other messaging features for communicating with another user or a group.

Most online and social services provide some form of chat or messaging. Gmail enables you to chat with your online contacts, and Facebook enables you to exchange **private messages (PMs)**, sometimes called *direct messages (DMs)*, which are visible only to you and others invited to the conversation. With these types of social services, you can initiate a chat or message when you see one of your contacts or friends online. Type a comment and then press Enter or click Send to send it to the other person, who can respond to you in the same way. The messages remain visible in the message area as you continue the typed conversation, much like text messaging using a smartphone. You also can use social messaging apps such as WhatsApp, Whisper, and ASKfm to message more privately with friends.

The corporate world is increasingly implementing forms of chat as a business communications tool. For instance, the team communication platform Slack lets users communicate by DMs or channels. Employees accustomed to instant messaging in their personal lives also find it convenient to communicate with coworkers and clients using this medium.

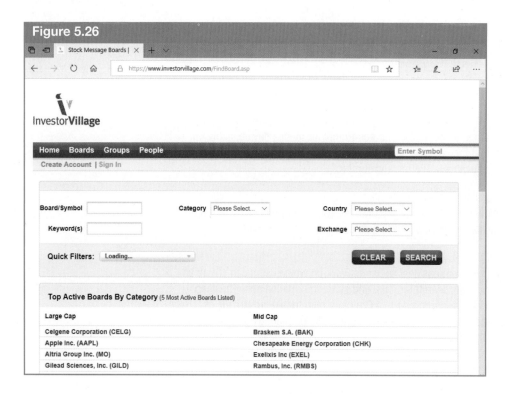

Figure 5.26 Users go to message boards, like the ones available at InvestorVillage, to read and post messages about topics of interest.

Message Boards

A **message board**—often called a *discussion forum* or simply a *forum*—presents an electronically stored list of messages that anyone with access to the board can read and respond to. Like a classroom or dormitory bulletin board, a message board allows users to post messages, read existing messages, and delete messages.

Similar to chat rooms, most message boards center on particular topics. Yahoo Finance provides a message board, or list of conversations, for almost every stock so that investors can discuss a company's performance and stock price. InvestorVillage is a stand-alone website that also offers a variety of boards about specific stocks and general investing topics (see Figure 5.26). In the InvestorVillage message boards, you can read and respond to posts and typically assign ratings to comments by others. Because the messages remain in a message board (unless deleted by a moderator), you can typically search a board to find information posted by a particular user or information about a particular keyword. This searchability feature makes many message boards useful bodies of information that users can access for future reference or research.

Blogs

A **blog** is a frequently updated online journal or chronological log offering personal thoughts and web links. (It was originally called a *weblog*—a combination of the words *web* and *log*.) The content and style of blogs vary as widely as the people who maintain them (who are called *bloggers*). In general, blogs serve as personal diaries or guides to others with similar interests. Collectively, the world of blogs is known as the **blogosphere**. Users can publish blogs at websites such as Blogger and WordPress.com. (Professional writers might choose to publish on Medium, a paid platform.)

Many corporate websites now include one or more blogs to communicate with employees, customers, and partners. Bloggers add a personal, informal tone to company communications and provide information that's more timely and relevant than the content traditionally provided in a glossy marketing brochure.

Online Shopping

Online shopping involves using a web-connected computer or mobile device to locate, examine, purchase, and pay for products. Although online shopping offers many advantages for both consumers and businesses, you should take care to be a savvy consumer, which includes knowing some guidelines for shopping online.

Video
What Computer Jobs Has the Web Produced?

Online shopping is increasingly replacing shopping in stores and malls. For the consumer, online shopping offers several distinct advantages over traditional shopping methods:

- convenience
- greater selection
- easy comparison shopping
- potential sales tax savings

Some US states do not require an online retailer to collect sales tax when the retailer has no physical location within the state; some states do. Other states, such as North Carolina, require taxpayers to estimate an amount of sales tax due for online sales even when retailers did not collect sales tax. The federal government has not passed any legislation regarding sales taxes for internet sales, though proposals have been made to standardize sales tax collection practices.

Many "bricks-and-clicks" businesses encourage consumers to shop online because it saves employee time, thus reducing staff needs and costs for the company. For example, some major airlines offer special discounts to travelers who purchase their tickets online, and most are eliminating paper tickets altogether. In addition, with an airline ticket purchased online—called an e-ticket—a consumer can often check in for a flight and print a boarding pass online up to 24 hours prior to departure. See Figure 5.27.

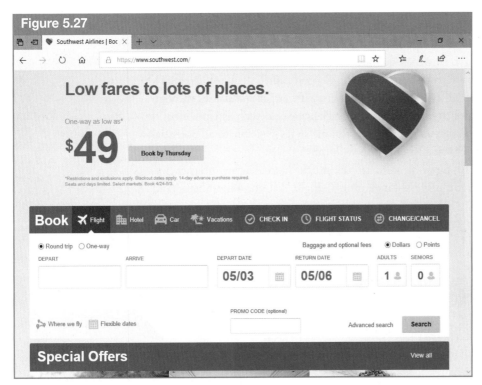

Figure 5.27

Most airlines, including Southwest Airlines, enable consumers to purchase e-tickets and check in to flights online, as well as print boarding passes.

News and Weather

Many users are turning away from print and TV sources of news and weather and relying instead on mobile and standard web versions of their favorite information sources. For example, CNN, ABC News, and other national networks offer free news online, 24/7. Some print-based publications, such as the newspapers owned by the Gannett Company, attract readers by offering limited online articles to individuals who don't subscribe to the print edition and unlimited online articles to print subscribers. Other publications, such as the *New York Times* and the *Wall Street Journal*, offer digital (online-only) subscriptions.

Other more topical information sites are offshoots of popular subscription TV channels, such as ESPN and The Weather Channel. Some sites have more video content than other news sources and may offer localized content as well. For example, at many weather news sites, such as Intellicast.com, you can get local forecast information by entering your zip code. So you can read and access vital information while on the go, most news and weather sites (and many of the informational sites covered in the next section) also have mobile apps available.

Business, Career, and Finance

While plenty of entertainment can be found on the web, plenty of business resources are available as well. If you are looking for business news or strategy information, you can consult online editions of such popular magazines as *Fast Company*, *Businessweek*, *Entrepreneur*, and *Forbes*. You can also consult web-only business news sources, such as *Business Insider*. You will find content covering a variety of business segments and niches in specialty online publications such as *Wired*, and you will find informative publications and newsletters at the website of the trade or professional association for the industry you work in.

Most news and business sites offer a "Career" or "Job Postings" section, and some sites are dedicated to helping people find jobs (or to helping businesses that are hiring find employees). The top job search sites include Indeed, Monster, CareerBuilder, Glassdoor, and Simply Hired. Some job search sites are specific to professions, such as Dice for IT professionals. Job-hunters of all types can also connect with target companies through LinkedIn. There are even sites to help people find freelance work, such as Upwork and Guru. Most of these sites enable you to post your résumé so it's available to all potential employers, and provide career statistics and help with career development.

If you are interested in news specific to the stock market, the internet has what you need. Many major players—including Google, Yahoo, Bloomberg, and

Tech Ethics

Keeping Clean in Company Communications

Electronic communications in the workplace often involve legal privacy issues. Companies generally keep backups of communication in the event of a data loss, but some industries require that all forms of digital communication be archived for legal reasons. For example, the financial industry requires that all email and instant message content be kept for several years. Doing so is necessary to ensure it will be available as evidence in the event of an investigation by the Securities and Exchange Commission or another legal body.

Employees should keep the content of email and instant messages in the workplace professional and courteous, because their employers will have digital records of who said what and when.

MarketWatch—provide daily stock market and company news, along with stock quotes, charting, and more. Some sites also allow you to track a portfolio of stocks, and most provide information on topics such as personal finance and commodities.

Government and Portals

Any website with the domain *.gov* is a government website. Many .gov websites are operated by federal agencies. For example, you can go to the top of the US government by visiting WhiteHouse.gov. Other .gov websites are operated by governments at the state, county, and local or municipal levels. These sites not only offer information about government departments and services, but they also publish important announcements, including those required by public meeting laws.

A special type of website called a **portal** acts as a gateway for accessing a variety of information. A portal serves as a "launching pad" for users to navigate categorized web pages within the same website or across multiple websites. The US government's USA.gov website (https://CUT7.ParadigmEducation.com/USA.gov) is an example of a portal (see Figure 5.28). The majority of the material on the USA.gov site is located on the USA.gov servers, but links also take users to individual federal, state, and local government agency websites. The home page for USA.gov provides an overview of the information contained within the site.

FTP and Online Collaboration

As mentioned earlier in the chapter, **File Transfer Protocol (FTP)** enables users to transfer files to and from remote computers. FTP was the original method used for transferring files on the internet, and it remains an option for transmitting files that can't be emailed because of their large size or because they are blocked by the recipient's system.

Although most web browsers can be used to connect with an FTP server, heavy FTP users usually prefer to use some type of FTP client software, such as the free

Tech Career Explorer
Telecommuting Techs

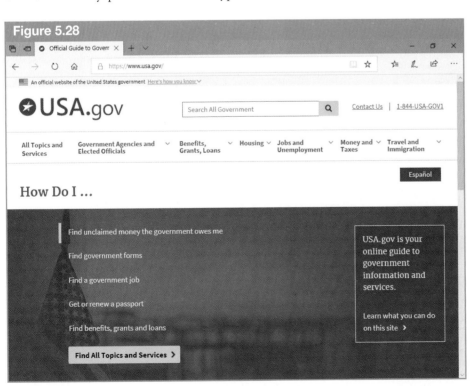

Figure 5.28

The official website of the US government, USA.gov, is a portal that offers access to information about all aspects of the federal government. The USA.gov home page provides links to information on millions of government web pages.

program FileZilla (shown in Figure 5.29). Most FTP servers are secure, so a user name and password are needed to sign on, but some site operators allow users to log on anonymously.

Additional types of FTP provide greater security and encryption. File Transfer Protocol over SSL (FTPS) (also known as FTPES, FTP-SSL, or FTP Secure), adds another layer to the base FTP protocol. Secure File Transfer Protocol (SFTP) (also known as SSH File Transfer Protocol) is a completely different protocol.

An FTP server is set up to use a particular type of FTP. When you use the FTP client software to connect to the FTP server or site, you must specify whether the connection uses regular FTP or one of the more secure types, as required by the server. For example, FileZilla lets you establish regular FTP connections as well as FTPS and SFTP connections.

Typically, you can download files from the FTP site to your computer and upload files from your computer to the FTP site. FTP allows any kind of file to be transferred. For example, a student can download lecture outlines that an instructor has made available on a school's FTP site, or an engineer can download and view blueprints that an architect has placed on the firm's FTP site. And of course, telecommuting employees and contract workers can submit completed projects and documents via FTP and download resource documents.

While FTP used to be the main way to transfer documents for telecommuting and other forms of remote work, more modern online workspaces have evolved over the years. Online workspaces have developed both in conjunction with web-based email services and as standalone services implemented by corporations and other organizations. For example, when you have an Outlook.com account (or any Microsoft account), you also have an online storage area known as your OneDrive. You can upload, download, and even share documents from your OneDrive with other users. Corporate-level collaborative platforms include SharePoint, which allows online file sharing and real-time file collaboration. (Chapter 7 explores online sharing and collaboration features in greater detail.)

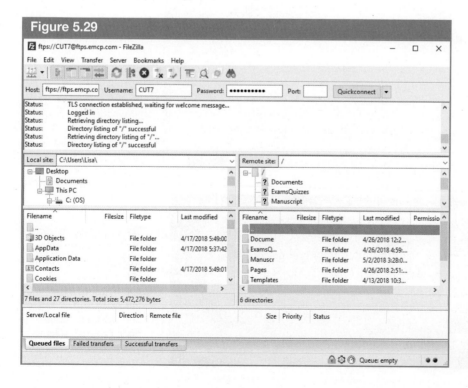

Figure 5.29

The authors of this textbook used FileZilla to upload all of the manuscript and image files to the publisher's FTP site.

Peer-to-Peer File Sharing

Peer-to-peer (P2P) file sharing has caused a lot of controversy on the internet. P2P enables users to download material directly from other users' hard drives, rather than from files located on web servers. Napster, the famous pioneer of peer-to-peer file sharing, operated by maintaining a list of files made available for sharing by subscribers to the system. For example, someone would let Napster know that he had 50 music files on his hard disk that he would be willing to share. Other users could then use Napster to locate these files and request that they be sent to their computers. Newer P2P systems remove the central server entirely and allow user computers with the fastest connections to provide the search function and keep track of which computers have shared a file. P2P is a powerful idea that allows every computer to function as both a server and a client.

Napster was launched in 1999, and in less than a year, it had more than 20 million users. At its peak, Napster had more than 70 million users worldwide and was being used to download music files by almost 70 percent of US college students. Unfortunately, many of the files being shared in this way were copyrighted material. Napster was forced to shut down in 2001 after two court injunctions ordered it to stop doing business. After filing for bankruptcy and being acquired by Roxio, Napster was relaunched as a music download service in 2003. Eventually, it was incorporated into Rhapsody, one of many services chasing market leader iTunes, and was then rebranded as Napster once again.

Today's P2P technologies allow sharing any types of files, including games, movies, and software programs. Since Napster, the biggest file sharing technology has been BitTorrent. Industry research firms estimate that BitTorrent usage accounts for 35 percent of all internet traffic. Although some individuals use BitTorrent to illegally find and download copyrighted files, others use the software to share their own files, which is legal. Also, software and media companies such as TimeWarner are interested in the distribution technology as another sales channel.

Using P2P technology to harness the individual contents of millions of computers around the world represents a vast opportunity for communications. However, with additional access come additional security risks.

Voice over Internet Protocol

Through an application of **internet telephony** (digital communications using different IP standards) technology called **Voice over Internet Protocol (VoIP)**, two or more people with good-quality connections can use the internet to make telephone-style audio and video calls around the world. The process digitizes the voices and videos, breaking them down into packets that can be transmitted anywhere, just like any other form of digital data.

VoIP can be used in three different ways: from device to device, via internet-ready phones, and via an analog telephone adapter (ATA). Through popular device-to-device services, such as Skype, the user at each end of the connection downloads and installs software that uses the sound features in the computer, mobile device, or Skype-ready TV. Adding video requires having a webcam. The software enables the users to connect through a "call" over the internet. Then, it translates the words spoken into the microphone by the sender into sounds that are heard through speakers by the recipient, with video transmitting simultaneously.

Two-camera models of the iPhone, iPad, and iPod Touch, as well as some recent MacBook models, have the ability to make FaceTime audio and video calls over a Wi-Fi network. These devices all come loaded with the FaceTime software.

Having a Gmail account enables you to use a feature called Google Voice to call telephones of other Gmail account holders. These individuals can make free calls within the United States and Canada.

Although less common now than in the past, internet-ready phones and devices provide another option for users. This type of phone plugs directly into an internet connection and performs the same translation without separate software. An ATA device takes the analog signal from a traditional phone and converts it into digital data that can be transmitted over the internet.

Using VoIP service eliminates the need to have long-distance telephone service, which sometimes involves paying extra charges. To take advantage of this savings, some users eliminate traditional telephone lines from their homes and make all calls using VoIP service along with a mobile phone. Vonage pioneered VoIP service, but many high-speed ISPs now offer VoIP calling plans.

You can make audio and video calls over the web using your computer or mobile device.

Audio, Video, and Podcasts

To watch a video or listen to a song stored on the internet, users previously had to download the file to a computer and use a dedicated application to open it or download the file to a digital media player, such as an iPod.

Now, you can either download the file or stream it online.

Downloading The most popular music download service, Apple's iTunes, works with a variety of devices, including Windows-based computers. Most music download websites charge a per-track or per-album download fee. You "own" the music but can sync or copy it to only a limited number of devices.

Cutting Edge

Accelerating Innovation

The development of advanced, high-speed internet applications and technologies is being facilitated by the Internet2 research platform. Internet2 is a consortium of more than 200 universities that are working in partnership with industry and government. It enables large US research universities to collaborate and share huge amounts of complex scientific information at amazing speeds. The goal is to someday transfer these capabilities to the broader Internet community.

Internet2 provides a testing ground for universities to work together and develop advanced Internet technologies, such as telemedicine, digital libraries, and virtual laboratories. An example of this collaboration was the Informedia Digital Video Library (IDVL) project. IDVL uses a combination of speech recognition, image understanding, and natural language processing technology to automatically transcribe, partition, and index video segments. Doing so enables intelligent searching and navigation, along with selective retrieval of information.

Internet2 operates the Internet2 Network backbone, which is capable of supporting speeds of more than 100 gigabits per second (Gbps). To learn more about Internet2, visit the Internet2 website at **https://CUT7.ParadigmEducation.com/Internet2**.

For some time, the most widely used music file format for downloads was Moving Pictures Expert Group Audio Layer III—also known as MPEG Audio Layer-3 format or, more commonly, **MP3 format**. The MP3 compression format reduces the size of CD-quality sound files by a factor of 10 to 14. It reduces file sizes by removing recorded sounds that the human ear can't perceive. The MP3 format creates files that are much smaller and easier to download.

Compression and file size are an issue of some controversy for audio fanatics. iTunes files use the **Advanced Audio Coding (AAC) format**, which is thought to provide higher audio quality with comparable compression. Windows has another format: **Windows Media Audio (WMA) format**. Other enthusiasts prefer the **Free Lossless Audio Codec (FLAC) format**. This open source method provides efficient, **lossless compression**, which means none of the original sound information is eliminated.

Using a portable music player is a convenient way to listen to songs downloaded from the internet.

iTunes and Windows Media Player, as well as other digital music management programs, enable you to **rip** (copy) songs from a CD and choose the file format and compression level you prefer. Once you have downloaded or ripped a music file to your computer's hard disk, you can transfer the song to a portable digital music player or smartphone for mobile playback.

You can view downloaded video files using an application such as Movies & TV, Microsoft Windows Media Player, or VLC media player, an open source application. Many formats are used for digital movies, such as **MPEG (Moving Picture Experts Group) format** and the newer Moving Pictures Expert Group Layer IV (**MP4 format**) and **WMV (Windows Media Video) format**. Because video files are very large, downloading movies or long clips takes a long time unless you have a high-speed internet connection.

Some online audio and video files are called *podcasts*. In some ways, a **podcast** is like a typical downloadable audio or video file (or even a PDF or ebook file), but it's different in the sense that each download is often part of an ongoing series. Podcasts may be offered as daily or weekly episodes or as parts of a larger, comprehensive broadcast. Although most users typically listen to podcasts on portable devices, you can also play them back on a computer using applicable software for the podcast file format.

Article
Streaming Media

Streaming An alternative to downloading a song or video from the internet is a technique called *streaming*. **Streaming** sends a continuous stream of data to the receiving computer's web browser, which plays the audio or video. Old data is erased as new data arrives, which means no complete copy of the material downloads. Streaming protects the owners of copyrighted materials to some degree, because it eliminates the possibility of copying and sharing.

The entertainment and rich media (streaming and interactive media) that are offered on the web have become increasingly full of features and easier to use. Sites such as YouTube and Pandora have become the internet standard in video and music streaming.

YouTube enables people to upload, view, share, and comment about videos. The videos range from political commentaries to music videos to product reviews, plus some clips of random silliness. YouTube streams many videos at low resolutions to speed download and viewing time, but some are high-definition (HD) quality. YouTube is free to users, because the service is supported by advertising.

Pandora is a free, ad-supported online radio service that personalizes the variety of songs it plays for listeners. It selects songs for you by analyzing what songs you have listened to previously. In 2013, Apple launched a similar service, iTunes Radio, offering hundreds of online music stations. Spotify is another popular music service offering free and subscription options.

Other services offer fee-based video and music streaming or offer some free content along with the option to pay for additional content. For example, the Napster music streaming service allows you to play music via a web browser, a mobile device app, or certain voice-controlled smart speakers. Hulu offers some free online TV episodes and other content (such as documentaries), but it also lets you upgrade to Hulu Plus's more advanced content for a fee. And of course, the monthly subscription service Netflix enables you to stream TV and movie content to a variety of devices. Google Play also enables you to purchase movies and music.

It's also possible to view TV shows, music videos, movies, and other types of videos from a variety of websites. Many news networks and newspapers offer video on their websites. You can click on a story and then view a short video newscast that gives you all of the details.

Downloading versus Streaming So which is better: downloading or streaming audio and video content? The answer depends on your preferences. If you like to control the cost on a case-by-case basis and "own" the content, then you should rip or purchase and download your media. But if you love to sample new music and movies, you will prefer the variety available through streaming subscription services.

Purchasing media allows you more control over access to the content. Every online service has usage agreements with particular artists and content providers. If any of those agreements changes, the service may have to remove certain songs and videos from its catalog. Should that happen, subscribers would no longer have access to that content.

Audio Books and Ebooks

Online books represent the latest "battleground" in terms of competition among digital entertainment services. Audible and other services provide downloadable audio books that users can play on digital music players. Amazon's Kindle reader spawned the first large group of **ebook** consumers. Amazon's Kindle eBooks store is the leading ebook retailer, holding a significant share of the market. Even if you don't have a Kindle, you can download a free Kindle app for another device and then purchase ebooks and read them using the app. You can also buy downloadable ebooks at the Apple Books store, the Barnes & Noble NOOK Books store, and the Google Play service, along with other ebook stores and publishers.

While publishing hard copies of a book can be expensive, publishing a digital book is much less costly. With the right software and expertise, virtually anyone can publish an ebook. You can learn the ins and outs of the different file formats required by various ebook retailers, or you can produce and promote your ebook using a service such as BookBaby. These services produce and promote your ebook for a flat fee and, in some

Hotspot

Replacing TV with Telephone?

According to Nielsen, a global information and measurement company, adults over the age of 18 are increasingly using their smartphones to consume media rather than watching television. In addition, younger adults spend less time watching traditional TV than do older adults.

Be aware that streaming media uses a lot of data. Know the data limit on your phone plan, and watch the amount of data you are using. You can reduce the amount of data you use through your phone service by connecting to a wireless network whenever possible.

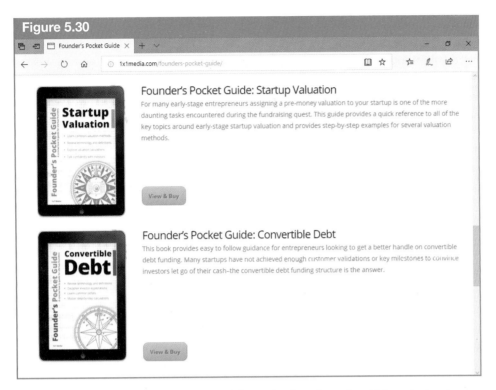

Figure 5.30

Ebooks are available for sale from digital booksellers and publishers, such as independent publisher 1x1 Media.

cases, a commission on royalties received from the digital retailer. Although ebooks typically sell for less than print books, ebook publishing can offer a great money-making opportunity for individuals and independent publishers (see Figure 5.30).

iBooks Author from Apple enables creation of a new generation of ebooks. Using this free app, you can create interactive ebooks for iPad, iPhone, and macOS. iBooks Author makes it possible to include not only text and images but also videos, galleries, interactive diagrams, 3-D objects, and more.

Health and Science

The web now also teems with medical information. Your personal medical providers and local medical facilities all likely have websites with basic information about location, hours, contact numbers, and service offerings. Some sites also offer health news and information that you might find useful, such as details about upcoming wellness programs. National health resources are also online, such as WebMD and MayoClinic.org. Sites such as these provide both current medical news and information about specific diseases and conditions, treatments, and medications.

If you are scientifically minded, you might enjoy exploring the range of resources available on the web. You can check out the online versions of popular publications, such as *Science* (Sciencemag.org), *Science Daily* (Sciencedaily.com), and *Popular Science* (Popsci.com). You can also check out the science offerings of media outlets such as Discovery Channel and National Geographic. At their websites, you can search to find serious scientific research and publications or fun science in the form of home science experiments.

Online Reference Tools

If you are a writer or researcher, you may feel overwhelmed by the huge number and variety of resources on the web. To make your search more efficient, use one of the reference tools available online, such as Refdesk.com. Refdesk.com, which calls itself the "Fact Checker for the Internet," serves as a dashboard to other core resources for researchers. It includes direct connections to all of the most popular search sites. In addition, it provides thought provokers such as "Fact of the Day," "Word of the Day," and "This Day in History"; links to news and weather sites; games and other diversions; translation tools; links to online encyclopedias and more specific information sources; and writer's resources, such as online dictionaries, quotations, and style and writing guides.

Other reference sites address specific subject matter or provide specific types of help. For example, HowStuffWorks offer instructions and information about a lot of tasks and topics. By searching, you can find sites offering useful information about everything from cleaning to legal advice. Of course, the websites of many schools and universities offer reference information about their programs. A search might turn up school-approved writing guidelines or research results of a study conducted by a particular department.

Wikipedia represents another valuable reference resource: a wiki. A **wiki** works like an online encyclopedia, but it allows anyone to contribute information. Users can post new articles about subjects, or they can edit or contribute to existing articles, add "Notes" or "See also" entries, or respond to questions or disputed information in an article. The idea behind a wiki is that having many users' contributions will improve the quality of the information in the article. (The concept of having large numbers of people contribute paid or unpaid work to create content or complete another task is called *crowdsourcing*.) Another popular wiki, wikiHow, is devoted to teaching users how to do anything. You also may see user-created "Help" and "How-to" wikis for open source software.

One potential drawback about using wikis applies to many online reference resources: Make sure the resource is credible and has a good reputation. In other words, know your sources.

Gaming and Gambling

Online gaming and gambling have grown into popular internet pastimes for many people. Users can play by themselves or compete with other players, often in very real online worlds. For example, virtual reality (VR) involves a computer simulation of an imagined but convincing environment or set of surroundings. Games enhance the VR effect by enabling each player to select a virtual body, called an **avatar**, which serves as his or her point of view in the game world.

Millions of players pay monthly fees to play *World of Warcraft*, the most popular online game in the world. Many users also spend significant time gaming on mobile devices; in fact, apps for games are some of the most frequently downloaded mobile apps. Users also enjoy gaming through other platforms, such as Facebook.

Online casinos are a unique and controversial form of entertainment. Users can log on and gamble online in a virtual casino. Although online casinos are prohibited by law in many areas, they are difficult to police because they may be located in any part of the world. The experience may seem like playing a game, but any losses are real and will be billed to the user's credit card.

Distance Learning Platforms

As discussed earlier in this chapter, online or distance learning platforms serve the needs of degree-seeking and non-degree-seeking learners around the world. These platforms have developed in response to the rising costs of advanced education and

Article
Earning a Degree Online

the ongoing need for skills development and training. Some universities provide fee-based online learning services, such as online or hybrid (part online, part on campus) courses. MBA programs are offered online by a number of major business schools.

As noted earlier, you can also take advantage of free online courses from websites such as Coursera. A **massive open online course (MOOC)** provides free and open access and frequently offers the best content from top schools and partners. MOOCs typically include online video lectures with accompanying project assignments and tests. Some provide a completion certificate (either free or for a fee for some Coursera courses), which can be used as an employment credential. In addition to Coursera, other MOOC providers include Udacity and edX, whose partners include the Massachusetts Institute of Technology (MIT), Harvard University, and the University of California, Berkeley.

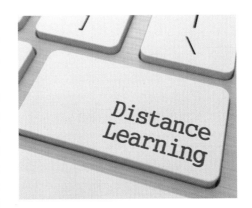

Some universities have adopted MOOCs to offer online degree programs. Using this successful technology has allowed schools to offer low-cost educational opportunities to a broader range of students than they can normally serve. While these programs aren't free, they are offered at a reduced cost. Students therefore have the chance to earn a college degree online for a reduced cost, compared with an on-campus degree. In 2013, the Georgia Institute of Technology (Georgia Tech) became the first top-ranked university to offer an online master's degree program. The cost of earning the degree online was less than one-fourth what it would be for earning the degree on campus. Twice as many students enrolled in the online program than in the standard program.

Research about the success of distance learning programs is ongoing. However, even the early data suggest that students stay more motivated when taking online courses than regular courses. Offerings for distance learning and online degrees will most likely continue to expand in the next decade.

Web Demonstrations, Presentations, and Meetings

Not only educators are using the web as a platform for information delivery. In addition, businesses and organizations are using the web increasingly to share information and provide education. For example, many companies create media-rich and interactive product demos to show potential buyers product features and options.

Microsoft's PowerPoint presentation graphics program enables you to share a presentation online with users that you invite. A tool will prompt you to invite the users and then lead you through starting the online presentation. So, while you are holding a conference call, for example, your colleagues can view the presentation you have prepared as you walk them through it, slide by slide, over the phone.

Other online meeting services offer a more robust experience, including audio and video (of both participants and information), whiteboard-style brainstorming, and more. Team members around the country or the world can participate in live meetings via computer or mobile device. These services go beyond what can be done with PowerPoint and similar one-way presentation programs, because they allow all of the participants to interact with the online content. The leaders in these services include Webex and GoToMeeting.

Sometimes, organizations use online meeting services to host a webinar, or web-based seminar. Conducting a webinar allows communicating directly with customers and others. A service such as AnyMeeting allows users to conduct online webinars and meetings with up to 1,000 participants. (Chapter 7 discusses other forms of online collaboration and teamwork.)

5.8 Respecting the Internet Community

Internet users around the world form a community, and like members of any community, they exhibit the entire range of behaviors possible—from considerate and creative to insulting and damaging. Unfortunately, the anonymous nature of internet interaction tends to bring out the worst in some people. The fear of embarrassment or shame that sometimes governs behavior in face-to-face encounters fades away when people interact on the internet. This means that some individuals act very differently than they would if they were in a public forum. These individuals ruin the internet experience for others but experience few consequences themselves.

Guidelines for good net behavior have been developed to encourage people to be considerate and productive. Providing moderated environments is another way to manage inappropriate behavior. In addition, a number of technical and legal issues influence the direction and development of the internet. Companies and individuals insist that standard protocols and increased transmission bandwidths be provided, consumers worry about the privacy and security of internet communications and transactions, and copyright holders want stronger protection for their intellectual property.

> **Watch & Learn**
> Respecting the Internet Community

Netiquette

Netiquette (*net* and *etiquette*) is a collection of guidelines that define good net behavior. Netiquette is based on the idea that people should treat others as they would like others to treat them.

Some points of netiquette address problem behaviors such as **flaming**, which is the internet equivalent of insulting someone face to face. By taking advantage of the anonymity that's offered online, some people are as rude as possible, to the point that they drive others away. So-called flame wars are instances of flaming that are traded back and forth, often among multiple parties.

Other points of netiquette deal with internet conventions that users should learn to avoid offending others unintentionally. For example, new email users commonly type messages in all capital letters without realizing that, by convention, using all caps is equivalent to shouting. Without meaning to, these email writers may make people

Practical TECH

The Internet Is Forever

The fact that Facebook and Twitter feeds seem to scroll off and Snapchat snaps seem to disappear immediately may tempt you to act out or reveal too much information online. You should remember, though, that the servers hosting your data may archive it for a long time, and other users may capture and keep your material. For example, it's possible for a recipient to make a screenshot of a Snapchat snap and post it on another service, such as Facebook or Twitter. In short, don't put anything online that you wouldn't want a future employer or anyone else to see. When posting "selfies," avoid photos that show you nude, behaving badly, and so on.

uncomfortable or even angry. Knowing the common rules of netiquette listed in Figure 5.31 can help you avoid this and other unintentional offenses.

Having some specific suggestions for composing and sending email and other types of digital messages (including text messages and posts to social media) may help you better understand netiquette. Because composing and sending email is so quick and easy, you should be careful to avoid mistakes that you might regret later. Keep in mind these important points when writing and sending email messages and making posts:

- In most cases, an email message can't be retrieved after it's been sent.
- A permanent copy of any email message probably exists somewhere on the internet.
- Others can easily forward or copy your email messages and posts.

For these reasons, it's a good idea to avoid sending or posting any message that you have written in anger or haste. Save the message and then look at it again later, after you have cooled down. If you still feel the message should be sent or posted, you will still have that option. However, in most cases, you will realize that you would have regretted sending or posting the original message.

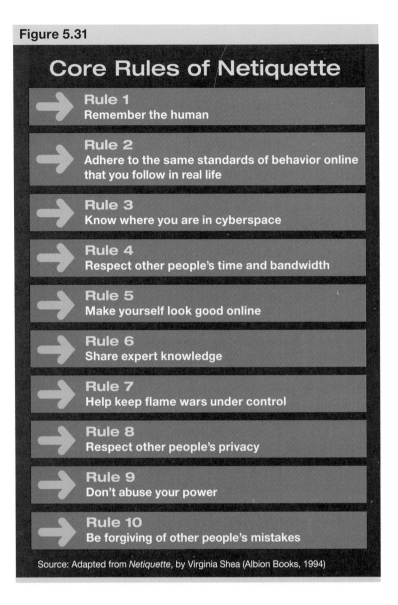

Figure 5.31

Core Rules of Netiquette

- Rule 1: Remember the human
- Rule 2: Adhere to the same standards of behavior online that you follow in real life
- Rule 3: Know where you are in cyberspace
- Rule 4: Respect other people's time and bandwidth
- Rule 5: Make yourself look good online
- Rule 6: Share expert knowledge
- Rule 7: Help keep flame wars under control
- Rule 8: Respect other people's privacy
- Rule 9: Don't abuse your power
- Rule 10: Be forgiving of other people's mistakes

Source: Adapted from *Netiquette*, by Virginia Shea (Albion Books, 1994)

Moderated Environments

Many people who want to avoid the seedy side of the internet interact with others in moderated environments. Many chat rooms, message boards, and mailing lists have a **moderator**, an individual with the power to filter messages and ban people who break the rules. Rule violations can range from making insults to not staying on topic. A moderator running a chat room on travel, for example, might warn or ban people for discussing their favorite movies at length. Usually, a moderator has complete power over the situation and can discipline people any way he or she sees fit. If a moderator is too harsh, however, people might switch to another group.

Net Neutrality

A common principle among networks is that of *network neutrality*, often shortened to **net neutrality**. Net neutrality is a doctrine or code of fairness that states that all internet traffic will be treated with equal priority. In other words, one packet won't be favored or ignored over another, no matter who the sender or receiver might be. Nor will data be discriminated against owing to its content or the fact that it might be streaming content from a particular provider.

Article
The Battle for Net Neutrality

Following this policy has several benefits. One key benefit is that the system's job of transmitting data is simplified. No data is more important than other data; all data is the same in value. This means that the system simply passes on whatever data it gets to the intended destination without judging the content. The policy of net neutrality also provides for greater innovation. When a new website is put up on the web, it's treated by the internal structure of the internet with the same respect as the most popular site on the planet. By leveling the playing field in this way, the policy allows new sites to grow rapidly. The FCC has ruled in favor of net neutrality on some occasions, and against it at other times. The future of this doctrine remains in question pending legislative and court action.

Privacy Issues

Privacy is a major concern for many internet users, particularly with email communications and e-commerce transactions. Most users know that email messages can be intercepted and read by others. In the workplace, employees' email messages may be read by their supervisors, and current law gives employers the right to do that. Monitoring employees' email is becoming more common as businesses discover that their workers are spending time surfing the web for personal reasons, instead of doing their work. (For more on employee monitoring, security, and similar privacy topics, see Chapter 9.)

Copyright Infringement

Many of the materials found on the web are copyrighted, and copying and using them without permission is illegal. Most websites include a copyright notice that states general guidelines for how the site's content may be used. Nevertheless, users frequently ignore copyright laws, and some violators end up in court, such as those who used P2P services to obtain copyrighted music files.

Because most copyright laws were written with printed materials in mind, the US Congress passed a law in 1998 that addressed the major issues related to protecting digital content on the internet (which can include text, videos, music, and many other file formats). The **Digital Millennium Copyright Act of 1998 (DMCA)** generally prohibits people from disabling software encryption programs and other safeguards that copyright holders have put in place to control access to their works. Entertainment companies have tried to protect their movies by including security codes, but hackers quickly developed programs capable of cracking them. A key provision of the Digital Millennium Copyright Act makes the use and distribution of security-cracking codes illegal, and the penalty involves civil damages ranging from $200 to $2,500. Repeat offenders face criminal penalties of up to $1 million in fines and 10 years in prison.

Chapter Summary

5.1 Exploring Our World's Network: The Internet

The internet is used for communication, working remotely (**telecommuting**) and collaboration. Computers can be used for social connections and entertainment purposes, including playing games, listening to music, and viewing movies and videos. Electronic commerce (e-commerce) refers to the internet exchange of business information, products, services, and payments. The **World Wide Web (web)** is the part of the internet that enables users to browse and search for information. The internet is a vast source of information to support both research and **distance learning**.

5.2 Connecting to the Internet

To connect to the internet, the user must have an account with an **internet service provider (ISP)**, which is a company that provides internet access. The ISP will give the user a user name and password. Several types of internet service are available. The older type of **dial-up access** is being replaced by **broadband** connections; they include cable, **digital subscriber line (DSL)**, wireless, fiber-optic, and satellite connections. Many users can share an internet connection over a local area network (LAN), which is often wireless, or from a wireless **Wi-Fi hotspot**.

Connection speeds vary depending on the type of plan purchased from the ISP, and different connection speeds may be offered for **downloading** (receiving) and **uploading** (sending) information. Internet transfer speeds are measured according to how many bits of data are transferred per second. Each **Kbps (kilobit per second)** represents 1,000 bits per second, and each **Mbps (megabit per second)** represents 1,000,000 bits per second. A **Gbps (gigabit per second)** is a billion bits per second.

5.3 Navigating the Internet

Surfing the internet means gaining access to and moving about the web using a browser. **Web browsers** locate material on the internet using **Internet Protocol (IP) addresses**. Every IP address also has a corresponding web address called a **uniform resource locator (URL)**, which is a path name that describes where the material can be found. The first part of a URL identifies the communications protocol to be used (for example, *http://*). The protocol is followed by the overall **domain name**, divided into parts that may identify the format (for example, *www* for World Wide Web) and the party responsible for the web page (such as *nasa* for National Aeronautics and Space Administration). The **domain suffix**, or *top-level domain (TLD)* (for example, *.gov*) comes next. After that, there may be additional folder path information. The internet breaks files into many pieces of data called **packets** and sends them out over separate routes in a process called **packet switching**.

5.4 Understanding Web Page Markup Languages

Web pages are usually created using **Hypertext Markup Language (HTML)**. HTML is a tagging or **markup language**, which includes a set of specifications that describe elements that appear on a page. A *tag* is enclosed in angle brackets, and tags must be used in pairs: an opening tag and a closing tag.

Extensible Markup Language (XML) is a new and improved web language that allows computers to manage data more effectively. A **hyperlink** is any element on the screen that's coded to transport viewers to another page or site.

5.5 Viewing Web Pages

A **web page** is a single document that's viewable on the World Wide Web. A **website** is a collection of web pages about a particular topic. You use a web browser program to view and navigate between web pages and websites. The **home page** appears initially when you go to a website. You can enter the URL for a page in the browser Address bar and then press Enter to load the page. Website designers frequently program using **Java** and a variety of other scripting languages. A **plug-in** is a mini-program that extends the capabilities of web browsers in a variety of ways. Companies advertise on websites using **banner ads** and **pop-up ads**.

5.6 Searching for Information on the Internet

Search engines help you look for information using **keywords** (search terms). Advanced searching requires the use of logic statements known as **search operators**; using these operators will refine searches and improve results. You can mark and organize web pages to make it easier to find them again later. Some web browsers call a marked page a **bookmark**, while others call it a *favorite*. You also can perform voice searches with intelligent personal assistants on both a voice-controlled smart speaker or a desktop or mobile operating system.

5.7 Using Other Internet Resources and Services

Most ISP accounts provide free email, but users also have the option of choosing a web-based service such as Gmail, Yahoo Mail, or Outlook.com. Another option is to use **webmail**, rather than email software, to access email with a variety of devices, including mobile devices. When using webmail, you may want to compress or "zip" large file attachments or multiple files. A **compressed file** or **zipped file** is smaller than the original file, which means it takes less time to send and download.

Dozens of **social networking services** enable users to connect and interact online for social and professional purposes. Facebook enables you to exchange **private messages (PMs)**, sometimes called *direct messages (DMs)*, which are visible only to you and others invited to the conversation. Users can also set up **blogs** to write articles about various topics and share them online.

File Transfer Protocol (FTP) enables users to transfer files to and from remote computers. **Peer-to-peer (P2P) file sharing** enables users to download material directly from other users' hard drives, rather than from files located on web servers.

Voice over Internet Protocol (VoIP) technology provides free video and voice calling service over the internet, rather than the telephone network. Tiny video cameras called *webcams* allow video to be included in the calls.

Music and video files can be downloaded from websites to computer devices, usually for a fee. A widely used music file format for downloads is Moving Pictures Expert Group Layer III (**MP3 format**), which reduces the size of sound files by removing recorded sounds that the human ear can't perceive. The **MP4 format** is a newer format used for digital movies. A **podcast** is similar to a downloadable audio or video file except that each download is often part of an ongoing series.

Rather than download a piece of music or a video, users can access it using **streaming**. Streaming online entertainment to a variety of devices represents a fast-growing use of the internet. Similarly, many users are purchasing and downloading **ebooks**, which can be read on computers, mobile devices, and even special readers such as Kindles. A **wiki** works like an online encyclopedia, but it allows anyone to contribute information. Users also enjoy a variety of gaming and gambling opportunities on the web. Students can take advantage of **massive open online courses (MOOCs)** to learn new skills or even get college degrees. Online presentation and meeting services improve businesses' productivity through webinars.

5.8 Respecting the Internet Community

Guidelines for good online behavior, called **netiquette**, have been developed to encourage people to interact politely and productively. **Flaming** is one of the most frequently encountered examples of rude internet behavior. Providing a **moderator** is another way of managing inappropriate behavior. Other concerns for internet use include net neutrality, privacy, and copyright infringement. **Net neutrality** is the policy that all internet traffic will be treated with equal priority. Privacy is a concern particularly with email communications and e-commerce transactions. Copyright infringement occurs frequently on the internet. The **Digital Millennium Copyright Act of 1998 (DMCA)** generally prohibits people from breaking software encryption and other safeguards.

Key Terms

Numbers indicate the pages where terms are first cited with their full definition in the chapter. An alphabetized list of key terms with definitions is included in the end-of-book glossary.

Chapter Glossary

Flash Cards

5.1 Exploring Our World's Network: The Internet

telecommuting, 185
World Wide Web (web), 187
distance learning, 188
learning management system (LMS), 188

5.2 Connecting to the Internet

network service provider (NSP), 189
internet service provider (ISP), 189
internet exchange point (IXP), 189
point of presence (POP), 189
value-added network (VAN), 190
broadband, 191
download, 191
upload, 191
Kbps (kilobit per second), 191
Mbps (megabit per second), 191
Gbps (gigabit per second), 191
dial-up access, 192
digital subscriber line (DSL) internet service, 193
Wi-Fi hotspot, 193
tethering, 197

5.3 Navigating the Internet

surfing, 197
web browser, 197
Internet Protocol (IP) address, 198
IPv4, 198
IPv6, 198
hexadecimal digit, 198
128-bit address, 198
uniform resource locator (URL), 198
Domain Name Service (DNS), 198
domain name, 199
second-level domain, 199
domain suffix, 199
dot-com company, 199
packet, 201
packet switching, 201
dynamic routing, 202

5.4 Understanding Web Page Markup Languages

Hypertext Markup Language (HTML), 202
markup language, 202
cascading style sheets (CSS), 202
tag, 202
hypertext document, 203
hyperlink, 203
Extensible Markup Language (XML), 203
metalanguage, 203

5.5 Viewing Web Pages

web page, 204
website, 204
home page, 204
Java, 205
cookie, 205
plug-in, 206
banner ad, 206
pop-up ad, 207
pop-up blocker, 207
blind link, 207
hijacker, 207
web page trap, 207

5.6 Searching for Information on the Internet

search engine, 210
keyword, 210
search operator, 212
bookmark, 213

5.7 Using Other Internet Resources and Services

webmail, 214
compressed file, 216
zipped file, 216
spam trap, 216
social networking service, 216
chat, 216
instant messaging (IM), 216
private message (PM), 217
message board, 218
blog, 218
blogosphere, 218
portal, 221
File Transfer Protocol (FTP), 221
peer-to-peer (P2P) file sharing, 223
internet telephony, 223
Voice over Internet Protocol (VoIP), 223
MP3 format, 225
Advanced Audio Coding (AAC) format, 225
Windows Media Audio (WMA) format, 225
Free Lossless Audio Codec (FLAC) format, 225
lossless compression, 225
rip, 225
MPEG (Moving Picture Experts Group) format, 225
MP4 format, 225
WMV (Windows Media Video) format, 225
podcast, 225
streaming, 225
ebook, 226
wiki, 228
avatar, 228
massive open online course (MOOC), 229

5.8 Respecting the Internet Community

netiquette, 230
flaming, 230
moderator, 231
net neutrality, 231
Digital Millennium Copyright Act of 1998 (DMCA), 232

Chapter Exercises

Complete the following exercises to assess your understanding of the material covered in this chapter.

- Terms Check
- Knowledge Check
- Key Principles
- Tech Illustrated Packet Switching
- Tech Illustrated URL Structure
- Chapter Exercises
- Chapter Exam

Tech to Come: Brainstorming New Uses

In groups or individually, contemplate the following questions and develop as many answers as you can.

1. Smartphones, household appliances, vending machines, road systems, and buildings are just a few of the things that can be connected to the internet for control and monitoring purposes. How will developing new uses for the internet change its nature and how we think about it? What other new applications might be possible if roads can tell us how much traffic there is, if cars can tell how much fuel they have left, and if refrigerators can tell us there's no more juice?

2. Chat and instant messaging (IM) technology have grown from forms of social media to tools for business communication. Financial services and retail websites offer customers live "chat" with customer service representatives, and colleagues within organizations can chat to meet informally whenever the need arises. How will the use of chat and IM likely evolve now that video messaging and calling have become even more popular? What additional benefits will this technology provide for businesses, both internally and externally?

Tech Literacy: Internet Research and Writing

Conduct internet research to find the information described, and then develop appropriate written responses based on your research. Be sure to document your sources using the MLA format. (See Chapter 1, Tech Literacy: Internet Research and Writing, page 37, to review MLA style guidelines.)

1. Find at least four providers that offer free or low-cost website hosting, such as Google Sites (https://CUT7.ParadigmEducation.com/GoogleSites) and Weebly (https://CUT7.ParadigmEducation.com/Weebly). Identify and compare the features they support. Which providers have wizards that automatically build websites? Which allow form-based uploading? Which allow FTP client connections? Which, if any, come with downloadable web page editors? Visit some sites developed using these free hosts. Do they load quickly? Do they have an excessive number of advertisements? Write a report comparing and contrasting these free host providers, and identify the one you think is the best.

2. Join LinkedIn and develop a profile that presents your résumé to the world. Keep in mind that your profile should impress potential employers. Invite your instructor to view your profile.

3. Find out what ISPs are available in your area. Which one would best meet your needs? If you don't have access to a computer at home, assume that you are researching this information for your school. For each ISP, identify the type of service it offers (cable, DSL, satellite, and so on), speeds, other account features, and data limitations (if any). Also find out information about the costs of the services. Then create tables, charts, and other graphics that compare the ISPs in terms of services offered and costs in a presentation or word processing program.

Present the information to your class using the slide show or document you created.

4. Research the pros and cons of using a wireless router to connect several home or office computers to the internet. Consider factors such as the approximate costs of equipment and installation, as well as system speeds, convenience, and security. Support your answers with specific data from the web. A good place to start looking for information about this equipment is Newegg.com (**https://CUT7.ParadigmEducation.com/Newegg**).

5. Find a web page that uses animation. Game advertisements from companies such as Disney, Sony, and Microsoft tend to use animation frequently. At the web page, were you asked you to download a plug-in? Are the resulting graphics superior? How much longer does it take to load a page with Shockwave graphics compared with a page with simple HTML? (Be sure to mention the type of internet connection you are using.)

Tech Issues: Team Problem-Solving

In groups, develop possible solutions to the issues presented.

1. Should federal and state governments invest heavily in providing high-speed internet connections to schools and libraries? Why or why not? If the government pays for internet service, should it have a say in how schools and libraries use the internet? After the technology has been perfected, should high schools and colleges require students to use electronic versions of textbooks downloaded from the internet? What would be the advantages and disadvantages of requiring this?

2. Many people use the web to operate their own businesses or to telecommute. If you were to consider working from home, what advantages and disadvantages would you consider? Would more risk be involved? More freedom? Less work or more work? What would be your greatest concern? Support your answers by conducting web research on this topic.

3. Experts claim that the "shelf life" of knowledge is only two to three years in many fields, including areas as diverse as medicine, technology, engineering, and history. How can distance learning and web-based tools help address this issue? Given this problem, should college diplomas be stamped with an "expiration date"? Why or why not?

Tech Timeline: Predicting Next Steps

The following timeline outlines some of the major events in the history of cybercrime. Research this topic to predict at least three important milestones that could occur between now and the year 2040, and then add your predictions to the timeline.

1984 The press learns of several high-profile incidences of criminals breaking security systems and uses the term *hacker* to describe these criminals.

1986 The Electronic Communications Privacy Act and the Computer Fraud and Abuse Act both pass Congress.

1988 Robert Morris, a college student, releases a worm that brings much of the internet to a halt.

1990 On January 15, AT&T's long-distance telephone switching system crashes, disrupting 70 million phone calls.

1990 In May, Operation Sundevil commences. Sundevil was the code name for the government's sweeping effort to crack down on cybercrime.

2000 In January hackers shut down Yahoo.com, Amazon.com, CNN.com, and eBay.com, among others, for one hour.

2000 In March, a 13-year-old hacker breaks into a government security system that tracks US Air Force planes worldwide, damaging a "secret" system.

2001 A cyberwar flares up between Chinese and American hackers after a US Navy plane collides with a Chinese fighter aircraft, killing the Chinese pilot. Each group of hackers attacks thousands of sites in the other nation.

2002 The Federal Bureau of Investigation (FBI) arrests three men who gained unauthorized access to credit reports and caused consumer losses of more than $2.7 million.

2003 The Department of Justice, FBI, and Federal Trade Commission conduct a major cybercrime sweep called *Operation E-Con* that results in 130 arrests and $17 million in property seizures related to internet auction scams, bogus investments, credit card fraud, and identity theft.

2004 The fastest-growing internet scam is now "phishing," in which criminals pretend to represent a legitimate web site or institution and request updates to individuals' financial records.

2005 Hackers favor the use of keystroke-logging technologies to steal sensitive information.

2005 Hackers crack Microsoft's antipiracy system within 24 hours after its launch.

2006 Jeanson James Ancheta pleads guilty to four felony charges for creating and selling so-called botnets: thousands of computers infected with malicious code and turned into zombie computers to commit crimes.

2009 An 18-year-old hacker going by the name GMZ cracks a Twitter staffer's password and gains access to high-profile celebrity Twitter accounts.

2010 The Stuxnet worm is deployed to disable equipment at Iran's nuclear facilities.

2013 In July, US Army soldier Chelsea Manning is convicted of violating the Espionage Act by releasing digital copies of classified documents to the public.

2013 In August, US National Security Agency contractor Edward Snowden flees the United States after copying and releasing classified documents and is granted asylum in Russia.

2015 Ross Ulbricht, creator and owner of Silk Road, an online black market for illegal goods, is sentenced to life in prison. He appeals the sentence. (Silk Road is part of the dark web, areas on the internet that require special software or authorizations to access.)

2018 The US Justice Department indicted numerous Russian individuals and companies for conspiring to subvert the 2016 US presidential election, using techniques such as stealing the identities of American citizens and posting as political activists and distributing fake news online.

Ethical Dilemmas: Group Discussion and Debate

As a class or within an assigned group, discuss the following ethical dilemma.

The appeal of message boards is often the anonymity that the medium provides. In minutes, you can create a user name, log on to a board, and discuss the board topic with other a group of strangers—all without revealing your true identity. But what consequences would you face if the message board operator shared your account information with, say, your family, employer, or law enforcement authorities? What prevents the board operator from doing this? Would you have any recourse against the operator? Does thinking about this dilemma change your view on what you would and would not say on a message board?

Acronyms and Abbreviations

Acronym or Abbreviation	Meaning	Chapter No.	Learning Objective No.
3-D	three-dimensional	2	2.1
AAC	Advanced Audio Coding	5	5.7
AC	alternating current	2	2.3
ALU	arithmetic logic unit	2	2.4
app	application	4	4.1
ASCII	American Standard Code for Information Interchange	2	2.2
ATA	Advanced Technology Attachment	2	2.5
ATA	analog telephone adapter	5	5.7
ATM	automated teller machine	2	2.1
AVI	Audio Video Interleave	4	4.4
BD	Blu-ray disc	3	3.8
BD-R	Blu-ray recordable	3	3.8
BD-RE	Blu-ray recordable erasable	3	3.8
BD-ROM	Blu-ray read-only memory	3	3.8
BIOS	basic input/output system	3	3.1
bit	binary digit	2	2.2
blog	weblog	5	5.7
BMP	bitmap	4	4.4
CAD	computer-aided design	4	4.4
CBT	computer-based training	4	4.3
CCD	charge-coupled device	2	2.1
CCFL	cold cathode fluorescent lamps	2	2.6
CD	compact disc	3	3.8
CD-ROM	compact disc read-only memory	3	3.8
CD-RW	compact disc rewritable	3	3.8
CDA	CD audio	4	4.4
CES	Consumer Electronics Show	1	1.6
CF	CompactFlash	2	2.5
CGI	computer-generated imagery	4	4.4
CPU	central processing unit	2	2.4
CRT	cathode ray tube	2	2.6
DAS	direct-attached storage	3	3.9
DB15F	15-pin female D-sub connector	2	2.3
DBMS	database management system	4	4.2
DC	direct current	2	2.3
DDR	double data rate	2	2.5

Acronym or Abbreviation	Meaning	Chapter No.	Learning Objective No.
DDR SDRAM	double data rate SDRAM	2	2.5
DIMM	dual inline memory module	2	2.5
distro	distribution	3	3.4
DL	dual layer	3	3.8
DM	direct message	5	5.7
DMCA	Digital Millennium Copyright Act	5	5.8
DNA	deoxyribonucleic acid	1	1.2
DNS	Domain Name Service	5	5.3
dpi	dots per inch	2	2.1
DRAM	dynamic RAM	2	2.5
DSL	digital subscriber line	5	5.2
DSLAM	digital subscriber line access multiplexer	5	5.2
DVD	digital versatile disc (also called digital video disc)	3	3.8
DVD-DL	digital versatile disc double layer	3	3.8
DVD-R	digital versatile disc recordable	3	3.8
DVD-RAM	digital versatile disc random access memory	3	3.8
DVD-ROM	digital versatile disc read-only memory	3	3.8
DVD-RW	digital versatile disc rewriteable	3	3.8
DVD+R	digital versatile disc recordable	3	3.8
DVD+RW	digital versatile disc rewriteable	3	3.8
DVI	digital visual interface	2	2.3
EB	exabyte	2	2.2
ebook	electronic book	5	5.7
EEPROM	electrically erasable programmable ROM	2	2.5
email	electronic mail	1, 4	1.1, 4.6
EPROM	erasable programmable ROM	2	2.5
eSATA	external Serial ATA	2	2.5
ESnet	Energy Sciences Network	5	5.5
ESRB	Entertainment Software Ratings Board	4	4.5
EULA	end-user license agreement	4	4.1
FAA	Federal Aviation Administration	5	5.2
FAT	file allocation table	3	3.8
FAT16	16-bit file allocation table	3	3.8
FAT32	32-bit file allocation table	3	3.8
fax	facsimile	2	2.7
FCC	Federal Communications Commission	5	5.5
FLAC	Free Lossless Audio Codec	5	5.7
FLV	Flash Video	4	4.4

Acronyms and Abbreviations

Acronym or Abbreviation	Meaning	Chapter No.	Learning Objective No.
FTP	File Transfer Protocol	5	5.7
FTP-SSL	File Transfer Protocol over SSL	5	5.7
FTPES	File Transfer Protocol over SSL	5	5.7
FTPS	File Transfer Protocol over SSL	5	5.7
FTPS	FTP Secure	5	5.7
FTTdp	fiber-to-the-distribution-point	5	5.2
FTTH	fiber-to-the-home	5	5.2
FTTN	fiber-to-the-node service	5	5.2
Gb	gigabit	2	2.2
GB	gigabyte	2	2.2
Gbps	gigabit per second	5	5.2
GHz	gigahertz	2	2.2
GIF	Graphics Interchange Format	5	5.4
GIGO	garbage in, garbage out	1	1.2
GPS	global positioning system	2	2.1
GPU	graphics processing unit	2	2.6
GUI	graphical user interface	3	3.3
HD	high-definition	2	2.3
HDD	hard disk drive	3	3.8
HDMI	High-definition multimedia interface	2	2.3
HDTV	high-definition television	1	1.1
HGP	Human Genome Project	1	1.2
HTML	Hypertext Markup Language	5	5.4
Hz	hertz	2	2.2
IC	integrated circuit	1	1.1
ICANN	Internet Corporation for Assigned Names and Numbers	5	5.3
ICQ	"I seek you"	5	5.7
IDE	Integrated Drive Electronics	3	3.8
IDVL	Informedia Digital Video Library	5	5.7
IoT	Internet of Things	1	1.6
IP	Internet Protocol	5	5.3
IRC	Internet Relay Chat	5	5.7
ISDN	Integrated Services Digital Network	5	5.2
ISP	internet service provider	5	5.2
IT	information technology	1	1.1
IXP	internet exchange point	5	5.2
JPEG	Joint Photographic Experts Group	5	5.4
Kb	kilobit	2	2.2

Acronym or Abbreviation	Meaning	Chapter No.	Learning Objective No.
KB	kilobyte	2	2.2
Kbps	kilobit per second	5	5.2
KHz	kilohertz	4	4.4
LCD	liquid crystal display	2	2.6
LED	light-emitting diode	2	2.6
lm	lumen	2	2.6
LMS	learning management system	5	5.1
LPT	line printer	2	2.3
Mac	Macintosh	3	3.4
Mb	megabits	2	2.2
MB	megabytes	2	2.2
MBA	master's in business administration	5	5.1
MFD	multifunction device	2	2.7
MFT	master file table	3	3.8
MHz	megahertz	2	2.2
MIDI	Musical Instrument Digital Interface	4	4.4
MIT	Massachusetts institute of Technology	5	5.7
MKV	Matroska Multimedia Container	4	4.4
MMC	MultiMediaCard	2	2.5
MOOC	massive open online course	5	5.7
MP3	Moving Pictures Expert Group Layer III	5	5.7
MP4	Moving Pictures Expert Group Layer IV	5	5.7
MPEG	Moving Picture Experts Group	5	5.7
ms	millisecond	2	2.5
NAS	network-attached storage	3	3.9
NASA	National Aeronautics and Space Administration	1	1.2
NHGRI	National Human Genome Research Institute	1	1.2
nm	nanometer	2	2.5
ns	nanosecond	2	2.5
NSP	network service provider	5	5.2
NTFS	New Technology File System	3	3.8
OC	optical carrier	5	5.2
OCR	optical character recognition	2	2.1
OEM	Original Equipment Manufacturer device	2	2.1
OLE	object linking and embedding	4	4.2
OLED	organic light-emitting diode	2	2.6
OS	operating system	3	3.1
PATA	parallel ATA	3	3.8

Acronym or Abbreviation	Meaning	Chapter No.	Learning Objective No.
PB	petabyte	2	2.2
PC	personal computer	1	1.6
PCIe	Peripheral Component Interconnect Express	2	2.3
PDF	portable document format	4	4.2
PED	personal electronic device	5	5.2
piezo	piezoelectric	2	2.7
PIM	personal information management	4	4.2
PM	private message	5	5.7
PMPO	peak momentary performance output	2	2.6
PNG	Portable Network Graphics	5	5.4
POP	Post Office Protocol	4	4.5
POP	point of presence	5	5.2
POST	power-on self-test	3	3.3
POTS	Plain Old Telephone System	5	5.2
ppm	pages per minute	2	2.7
PRAM	phase-change RAM	2	2.5
ps	picosecond	2	2.5
RAID	redundant array of independent disks	3	3.9
RAM	random access memory	1, 2	1.5, 2.5
RCA	Radio Corporation of America	2	2.6
REM	remark	3	3.6
RF	radio frequency	2	2.2
RGB	red-green-blue	4	4.4
RIA	rich internet application	4	4.4
RJ	registered jack	2	2.3
RMS	root mean squared	2	2.6
ROM	read-only memory	2	2.5
rpm	revolutions per minute	3	3.8
SAN	storage area network	3	3.9
SATA	serial ATA	3	3.8
SCA	Stored Communications Act	3	3.8
SD	Secure Digital	2	2.5
SDR	single data rate	2	2.5
SDR SDRAM	single data rate SDRAM	2	2.5
SDRAM	synchronous dynamic RAM	2	2.5
SFTP	SSH File Transfer Protocol	5	5.7
SFTP	Secure File Transfer Protocol	5	5.7
SMS	Short Message Service	4	4.6

Acronym or Abbreviation	Meaning	Chapter No.	Learning Objective No.
SOC	system-on-chip	3	3.4
SOHO	small office/home office	4	4.2
SRAM	static RAM	2	2.5
SSD	solid-state drive	3	3.7
SSHD	solid-state hard drive	3	3.8
sync	synchronize	1	1.6
T1	tier 1	5	5.2
T3	tier 3	5	5.2
TB	terabyte	2	2.2
Tb	terabit	2	2.2
TIFF	tagged image file format	4	4.4
TLD	top-level domain	5	5.3
UEFI	Unified Extensible Firmware Interface	3	3.2
UN	United Nations	1	1.5
UPC	universal product code	2	2.1
URL	uniform resource locator	5	5.3
USB	universal serial bus	2	2.3
USGS	US Geological Survey	1	1.2
UW	University of Washington	1	1.6
VAN	value-added network	5	5.2
VGA	video graphics adapter	2	2.3
VoIP	Voice over Internet Protocol	5	5.7
VR	virtual reality	2, 5	2.6, 5.7
VTC	video teleconference	2	2.6
WAV	waveform file	4	4.4
wave	waveform file	4	4.4
WBT	web-based training	4	4.3
web	World Wide Web	5	5.1
WMA	Windows Media Audio	5	5.7
WMV	Windows Media Video	5	5.7
WWW	World Wide Web	1	1.2
WYSIWYG	what you see is what you get	4	4.4
XML	eXtensible Markup Language	5	5.4
YB	yottabyte	2	2.2
ZB	zettabyte	2	2.2
μs	microsecond	2	2.5

Glossary and Index

A

ABC News, 220

accelerometer An environmental sensor that reports how fast an object is moving., 54

accounting software Software that integrates all the financial activities of a company into a single interface., 155

activation A process that generates a unique code based on the hardware in your computer and the registration key you used when installing a software, locking that copy of the software to that computer., 140

Adobe Acrobat Reader, 205

Adobe DreamWeaver, 166

Adobe Flash, 166

Adobe Illustrator, 161

Adobe InDesign, 147

Adobe Photoshop, 160, 161

Adobe Premiere Elements, 165

Adobe Premiere Pro, 165

Adobe Shockwave, 166, 205, 206

Advanced Audio Coding (AAC) format A digital music file format used by iTunes that is thought to provide higher audio quality with comparable compression., 225

advanced search techniques, 211–212

advertising on websites
    banner ad, 206
    blind links, 207
    hijackers, 207
    overview, 206
    pop-ups, 207
    revenue from, 206

adware A form of malware that displays unwanted advertising on the infected computer., 114

Airplane mode, 194

Ajax, 205

Alexa, 212

all-in-one devices, 83

alphanumeric keyboard A keyboard that has both letters *(alpha-)* and numbers *(-numeric).*, 42–43

Amazon
    Kindle eBooks, 226

American Standard Code for Information Interchange (ASCII) An encoding scheme that assigns a unique 8-bit binary code to each number and letter in the English language plus many of the most common symbols such as punctuation and currency marks., 56–57

analog data Continuously variable data, such as a sound wave.
    vs. digital data, 3

analog telephone adapter (ATA), 223, 224

AND, 212

Android An open source operating system created by Google that is commonly used on smartphones and tablets, because it's simple and easy to use and a large number of apps are available for it., 20, 108
    intelligent personal assistant, 212

animation, viewing web pages and, 205–206

animation software, 163

antiplagiarism software, 158

antivirus software A utility program that monitors a system and guards against damage from computer viruses and other malware., 111
    function of, 111

AnyMeeting, 229

AOL Instant Messenger (AIM), 173

app An application designed for tablets and smartphones, and designed to be compact and to run quickly and efficiently on minimal hardware., 138
    app-ification of software industry, 139
    desktop application vs., 138
    desktop apps, 103
    for social media, 174
    for various mobile operating systems, 108–109
    Windows 8 apps, 103

Apple and Apple devices
    apps for, 108, 139
    FaceTime, 173
    iBooks, 227
    iCloud, 123
    iOS, 108
    iPad, 20, 108
    iPhone, 20, 108, 223–224
    iPhoto, 161
    iTunes, 224
    QuickTime, 205, 206
    Siri, 212

applet A small application. Many applets are written in Java or other platform-independent language so they can run on all types of computer operating systems.
    Java applets, 205

application software A program that enables users to do useful and fun things with a computer., 138
    app-ification of software industry, 139
    application download files, 142–143
    business productivity software, 143–155
    commercial software, 138–142
    desktop application vs mobile apps, 138
    desktop interface, 103
    freeware, 142
    individual, group and enterprise use, 138
    overview of, 20
    running, by operating system, 103
    sales and licensing of, 138–142
    shareware, 142
    tile-based interface, 103

archive files, 142–143

Arduino, 25

arithmetic logic unit (ALU) The part of a CPU that executes the instruction during the machine cycle; the ALU carries out the instructions and performs arithmetic and logical operations on the data., 64

ASKfm, 217

aspect ratio The ratio of width to height on a display screen., 74

AT&T
    DataConnect Pass plan, 194
    On the Spot, 194

attachments, adding to email, 215–216

Audacity, 165

audio card, 62

audio data Data consisting of sounds, such as recorded voice narration and music., 13
    downloading, 224
    input devices, 53–54
    output devices, 78
    speakers and headphones, 78
    streaming, 225–226

viewing web pages and, 205–206
audio-editing software, 163–165
audio input The speech, music, and sound effects that are entered into a computer., 53
audio port A small, circular hole into which a 3.5-millimeter connector is plugged for analog input devices (such as microphones), analog stereo equipment, and analog output devices (such as speakers and headphones)., 61
    color coding for, 61
augmented reality (AR)
    as visual output, 77
authentication The process of positively identifying a user. Forms of authentication include personal identification numbers (PINs), user IDs and passwords, smart cards, and biometrics., 123
AutoCAD, 163
Autodesk, 163
automation, creating your own, 25
avatar A virtual body that serves as a player's point of view in an online game world., 228
average access time The average amount of time between the operating system requesting a file and the storage device delivering it., 121

B

backup utility A utility that allows the user to make backup copies of important files., 113
banner ad An ad that invites the viewer to click it to display a new page or site selling a product or service., 206
bar code A set of black bars of varying widths and spacing that represent data and that can be read by a bar code reader., 50
bar code reader A scanner that captures the pattern in a bar code and sends it to an application that translates the pattern into meaningful data., 50
Barnes & Noble NOOK Books store, 226–227
BASIC Acronym for *Beginner's All-purpose Symbolic Instruction Code.* A high-level language that's friendlier and more natural than COBOL and FORTRAN. BASIC is still used professionally in an updated form (Visual Basic).
    as high-level language, 114

basic input/output system (BIOS) Software that is stored on a chip on a circuit board, such as the motherboard, and provides the initial startup instructions for the hardware. Also called *firmware.*, 96
    functions of, and replacing, 71
BD See **Blu-ray disc (BD)**, 125
binary digit, 55
binary number system analogy, 55
binary string A multidigit number consisting of only 1s and 0s., 55
Bing, 210
biometric security, 53
BIOS Setup utility A utility that allows the user to modify the BIOS settings. Sometimes called *CMOS Setup utility.*, 97–98
    changes made with, 98
bit Short for *binary digit.* Each individual 1 or 0 in the binary string., 55
    parity bit, 129
bit depth The number of bits used to describe a sample., 75, 164
bitmap A matrix of rows and columns of pixels., 48
bitmap (BMP) file, 162
bitmap images, 160
BitTorrent, 223
Blackberry 10, 109
Blackboard, 189
blind link A link that misrepresents its true destination, taking you to an unexpected page when you click on it. This type of deceptive device appears only on websites that aren't trustworthy, such as some free-host sites., 207
blog A frequently updated journal or log that contains chronological entries of personal thoughts and web links posted on a web page., 218
bloggers, 218
blogosphere The world of blogs., 218
Blu-ray disc (BD) An optical disc that can store up to 128 GB of data in up to four layers; used to distribute high-definition movies and to store and transfer large amounts of data., 125
    storage capacity and features of, 125
bookmark In some web browsers (such as Chrome) a web page that has been marked and organized to make it easier to find it again later. Also called *favorite* (e.g., in Internet Explorer)., 213

boot drive A drive that contains the files needed to start up the operating system., 99
booting computer
    cold or warm boot, 100
    defined, 100
    operating system and, 100
broadband connection A high-speed Internet connection. Broadband connection speeds are better suited to online activities such as viewing videos., 191
browser See **web browser**, 197
bus An electronic path along which data is transmitted., 59
    size of, 59
    types of, 59
business, online sources for, 220–221
Business Insider, 220
business productivity software, 143–155
    accounting software, 155
    database management software, 149–151
    desktop publishing software, 146–147
    note-taking software, 155
    overview of, 143–144
    personal information management software, 153–154
    presentation graphics software, 151–152
    project management software, 154
    software suites, 153
    spreadsheet software, 147–149
    word-processing software, 144–146
business-to-business (B2B) electronic commerce A form of e-business in which companies use the Internet to conduct a wide range of routine business activities with other companies, including ordering manufacturing parts and purchasing inventories from wholesalers.
    overview of, 186, 187
business-to-consumer (B2C) electronic commerce A form of e-business in which companies use the Internet to sell products and services to consumers and to receive payments from them. See also **online shopping**
    overview of, 186–187
Businesswweek, 220
bus width The number of bits a bus can transport simultaneously., 59

Glossary and Index

byte A group of 8 bits; an 8-digit binary number between 00000000 and 11111111., 56
    storage capacity and, 70–71

C

C++ A modern adaptation of the C programming language, with added features such as object-oriented programming. C++ is used for most commercial software development today.
    as compiled programming language, 114

cable Internet connection, advantages/disadvantages of, 193

cache A pool of extremely fast memory that's stored close to the CPU and connected to it by a very fast pathway; types include the L1, L2, and L3 caches., 65

CAD software, 162–163

capability-based OS, 107

capsule endoscopy, 27

career, online sources for, 220

CareerBuilder, 220

cascading style sheets (CSS) A separate formatting language that can be used along with HTML to make the design for a web page or site more consistent and easier to update., 202

cathode ray tube (CRT) monitor A nearly obsolete type of monitor that uses a vacuum tube to direct electrons to illuminate colored phosphors on the back of the screen to light up a display image., 72

CD audio (CDA) format A format used by audio CDs to store music files., 164

CD See **compact disc (CD)**, 123

cell The intersection of a row and column in a spreadsheet., 148–149

cell phone See also **smartphone**
    syncing with computer, 153

central processing unit (CPU) The device responsible for performing all the calculations for the computer system. Also called *microprocessor* or *processor.*, 63
    advances in design and manufacturing of, 66–67
    arithmetic logic unit, 64
    caches, 65
    control unit, 64
    functions of, 55, 63
    internal components of, 64–65
    looking inside, 64
    machine cycle, 64
    as main component of motherboard, 17
    number of transistors in, 66
    processing capability of, 66–67
    registers, 64
    speed of, 65–67
    system cooling, 67–68

charge-coupled device (CCD) The sensor that records the amount of light reflected from the scanned image in a scanner., 48

chart A visual representation of data that often makes the data easier to read and understand., 149

chat An online service that enables users to engage in real-time typed online conversations with one or more participants. Chats previously took place in forums called *chat rooms*. Many users now also call instant messaging interactions *chats.*, 216–217

chat rooms, 216

chat servers, software for, 173

child folder A folder that is located within another folder. Also called *subfolder.*, 116

chip, 4
    supersonic, 65
    transistors on, 65

chipset A controller chip (or set of chips) that controls and directs all the data traffic on a motherboard., 58

Chrome
    bookmarks, 213
    OS operating system, 20, 107

Chromebook, 107

Chromium OS, 107

Cisco Unified MeetingPlace, 173

Cisco WebEx, 173, 229

Citrix GoToMeeting, 173, 229

Close the Gap, 16

cloud The Internet, as used to deliver online services such as file storage and web-based apps.
    apps, 20
    photo storage on, 52
    services, 20

cloud apps
    overview of, 20

cloud drive A secure storage location on an Internet-accessible remote server. Examples include Microsoft's OneDrive and Apple's iCloud., 123

cluster On a storage disk, a group of sectors that are addressed as a single unit., 121

CMOS Setup utility, 97, 98

CNN, 220

COBOL Acronym for *COmmon Business-Oriented Language*. A programming language used chiefly for business applications by large institutions and companies. COBOL was designed to be an English-like language for handling database processing.
    as high-level language, 114

cold boot The act of booting (or starting) a computer after the power has been turned off., 100

cold cathode fluorescent lamps (CCFL), 73

collaboration, as function of Internet, 185–186

collaboration software An application that enables people at separate computers to work together on the same document or project. Also called **groupware.**, 138
    uses of, 173–174

color depth The number of bits of data required to describe the color of about each pixel in a particular display mode., 48, 75

command-line interface A text-based user interface in which users type commands to interact with the operating system., 101
    benefits of, 101
    folder structure in, 116

command prompt, 101

commercial software Software created by a company (or an individual, in rare cases) that takes on all the financial risk for its development and distribution upfront, and then recoups the costs by selling the software for a profit., 139
    overview of, 139–142

communications
    with computers, 9–12
    as function of Internet, 184–185

communications device A device that makes it possible for multiple computers to exchange instructions, data, and information., 18

communications software, 170–174
    email applications, 170–172
    groupware, 173–174

social media apps, 174
text-based messaging applications, 172–173
Voice over Internet Protocol (VoIP) software, 173
web browser applications, 172
web conferencing software, 173

compact disc (CD) An optical disc that can store up to 900 MB of data; commonly used for distributing music and small applications and for inexpensively storing and transferring data., 125
    storage capacity and features of, 125

compass An environmental sensor that reports the direction an object is pointing in relation to magnetic north., 54

compiler A computer programming tool that translates an entire program into machine language once and then saves it in a file that can be reused over and over., 114

component video port An older type of display connection consisting of color-coded red, green, and blue plugs. Also called an *RGB port*., 62

COM port, 62

compressed file A file created by using software to compress or "zip" large file attachments or multiple files into a single file that is smaller than the original file. Such a file takes less time to send and download. Also called **zipped file**., 216
    for downloading, 225

computer An electronic device with chips, memory, and storage that operates under a set of program instructions; a computer accepts user data, works with the data according to the program, produces results (information), and stores the results., 4
    accuracy of, 7
    advantages of, 5–12
    categories overview, 21
    characteristics of, 4–5
    communications with, 9–12
    compared to computerized devices, 4–5
    data processing by, 55–57
    embedded, 4, 22–24
    general-purpose, 4
    information processing cycle, 13–14
    mainframe, 32
    mobile devices, 27–28
    personal computer, 29–31
    servers, 31
    smart devices, 22–24
    speed advantage of, 6
    speed and processing capability, 65–67
    startup process, 100
    storage advantage of, 9
    syncing to phone, 153
    U.S. households with, 6
    versatility of, 8
    wearable, 25–26

computer-aided design (CAD) Technical object modeling with vector graphics, used to create architectural, engineering, product, and scientific designs with a high degree of precision and detail., 162–163

computer-based training (CBT) Training that is conducted using a computer., 159–160

computer-generated imagery (CGI) Images created using computer software., 163

computer system The system unit along with input devices, output devices, and external storage devices., 14
    components of, 15
    single-user vs. multiuser, 15

connected home, 24
Consumer Electronics Show (CES), 22
contact lens, smart, 26
controller, 47

control unit The part of a CPU that directs and coordinates the overall operation of the CPU; the control unit interprets instructions and initiates the action needed to carry them out., 64

convertible models, 30

cookie A very small file that a website places on a user's hard drive when he or she visits the site. Cookies can store login and activity information, and can be blocked via browser security settings., 205–206

copyright
    copyright infringement issues, 232
    Digital Millennium Copyright Act, 232
    streaming and, 225

core A set of the essential processor components (that is, the ALU, registers, and control unit) within a CPU; a single CPU may have multiple cores., 67

CorelDRAW Graphics Suite, 161
Cortana, 212–213
Coursera, 188, 229
CPU See **central processing unit (CPU)**, 63

cross-platform operating system An operating system that runs on computers of all kinds, from PCs to supercomputers., 109

crowdwork, 186
CyberLink PowerDirector, 165

D

dark web, 209

data Raw, unorganized facts and figures, or unprocessed information., 12
    combined to create information, 12–13
    computer processing, 55–57
    types of, 13

database A collection of data organized in one or more data tables stored on a computer, managed using database management system software.
    enterprise database, 149
    helper objects, 151
    relational, 150–151
    software for managing, 149–151
    uses of, 149

database management system (DBMS) An application used to manage a database., 149
    overview of, 149–151
    personal, 150

data processing A term that refers to the activities performed with a database, including various types of interactions and events., 13

data transfer rate The speed at which data can be moved from the storage device to the motherboard and then on to the CPU., 122

DB15F, 62
Debian, 106
decoding, in machine cycle, 64

desktop The background for a graphical user interface (GUI) environment., 102

desktop application An application designed to be run on full-featured desktop and notebook computers., 138

desktop computer A PC that's designed to allow the system unit, input devices, output devices, and other connected devices to fit on top of, beside, or under a desk or table., 29
    characteristics and uses of, 29

desktop interface, 102

desktop publishing software Software that allows users to create documents with complex page layouts that include text and graphics of various types., 146–147

deviantART, 217

diagnostic utility A utility that analyzes a computer's components and system software programs, and creates a report that identifies the system status and any problems found., 112

dial-up access A type of Internet access that allows connecting to the Internet over a standard land-based telephone line by using a computer and a modem to dial into an ISP., 192
    advantages/disadvantages of, 192–193

dial-up modem A device for connecting to the Internet or a remote network via a dial-up telephone connection., 62

Diccovery Channel, 227

Dice, 220

dictionary, 158

digital An electronic signal that's processed, sent, and stored in discrete (separate) parts called *bits.*, 3

digital camera A camera that captures images in a digital format that a computer can use and display., 51
    process of, 51–52
    resolution, 51–52

digital information, 3

Digital Millennium Copyright Act of 1998 (DMCA) A US law that, among other things, prohibits people from disabling software encryption programs and other safeguards that copyright holders have put in place to control access to their works., 232

digital subscriber line (DSL) Internet service A form of broadband delivered over standard telephone lines. DSL is as fast as cable modem service and provides simultaneous web access and telephone use, but has limited availability., 193
    advantages/disadvantages of, 193

digital versatile disc (DVD) An optical disc that can store up to 17 GB of data, although the most common type (single-sided, single layer) stores up to 4.7 GB of data; most often used to distribute large applications and standard-definition movies., 125
    storage capacity and features of, 125

digital video camera A digital camera designed primarily for capturing video; may be portable, like a regular camera, or may be tethered to a computer. See also **webcam**., 52

digital visual interface (DVI) port The most common type of port for display monitors in modern computer systems; also used for some consumer electronics devices and game consoles., 61

digitized information Data that has been converted to digital format (that is, stored as a string of numeric values)., 48

direct-attached storage (DAS) Local data storage; storage of data on the computer itself or on an external device directly plugged into it., 127

direct messages (DMs), 217

direct thermal printer A thermal printer that prints an image by burning dots into a sheet of coated paper when it passes over a line of heading elements; its quality isn't very good and it can't produce shades of gray or colors., 83

disc See **optical disc,** 123

discussion forum, 218

disk array A group of disks that work together, such as in a RAID system., 129

disk defragmenter A utility that scans the hard disk and reorganizes files and unused space in order to allow the operating system to locate and access files and data more quickly., 113

disk drive A mechanical device that reads data from and writes data to a disk, such as a hard disk or DVD., 115

disk scanner A utility that examines the hard disk and its contents to identify potential problems, such as physically bad spots and errors in the storage system., 112

display adapter The computer component that translates and processes operating system instructions to create the screen's display image., 75–76

displays
    adapters, 75–76
    performance and quality factors, 74–75
    types of, 72–74

distance learning A form of education consisting of online courses and programs offered by colleges and universities. It involves the electronic transfer of information, course materials, and testing materials between schools and students., 188–189
    platforms for, 228–229

distribution (distro) A collection of operating system files, utility files, and support programs sold as a package. Linux distros contain the Linux kernel and a variety of other applications., 106

dithering, 83

DMG files, 143

docking station A home base for a portable device, providing it extra capabilities, such as better speakers, more ports, and power recharging., 78

document formatting Formatting applied to the entire document, page, or section, such as changes to the paper size, margins, number of columns, and background color. Also called *page formatting* or *section formatting.*, 146

Dogpile, 211

Domain Name Service A system that maps the URL to the host's underlying IP address., 198

domain name The part of a URL that follows immediately after the protocol and is divided into various parts. First may come format information, such as *www* for World Wide Web pages., 199
    obtaining your own, 203

domain suffix The part of the domain name in the URL that follows after the second-level domain—for example, *.com* in Amazon.com. Also known as *top-level domain (TLD).* See also **domain name.**, 199
    commonly used, 199
    new, 199

DOS BASIC, 114

dot-com company A web-based enterprise, with the company's domain name ending with .com. The domain suffix *.com* also stands for *commercial organization.*, 199

dots per inch (dpi) On a scanner or printer, the number of pixels captured or printed per inch of the original page., 49

double data rate (DDR) SDRAM SDRAM in which data transfers twice as fast as with SDR SDRAM, because it is read or written at the rate of two words of per clock cycle., 70

download To receive information such as a document from a remote site to one's own computer via a network or the Internet., 191
    audio, video or podcasts, 224–225
    compared to streaming, 226
    music file format for, 225

drawing software Software that enables users to create and edit vector graphics., 160–161

driver A small program that translates commands between the operating system and the device., 103

drum A large cylinder, contained in a laser printer, that carries a high negative electric charge on which the page's image is written with a laser, neutralizing the areas that should pick up toner., 82

Drupal, 203

D-sub connector, 62

dual inline memory modules (DIMMs) A type of mounting for RAM chips to be installed in a desktop motherboard; the word *dual* indicates that both sides of the circuit board contain memory chips., 70

DVD See **digital versatile disc (DVD)**, 125

dye sublimation printer A thermal printer that produces an image by heating ribbons containing dye and then dispersing the dyes onto a specially coated paper or transparency. Also called *thermal dye transfer printer.*, 83

dynamic RAM (DRAM) RAM that requires a constant supply of electricity to keep its contents intact; used as the main memory in most computers. Also called *volatile RAM.*, 69

dynamic routing The capability for the Internet to reroute packets via different routes based on traffic loads. This helps make the Internet work well even with a heavy load of traffic., 202

E

ebook A downloadable electronic or digital book., 226–227
    publishing book on, 226–227
    retailers of, 226–227

EchoStar, 196

education
    distance learning, 188–189
    distance learning platforms, 228–229
    massive open online course (MOOC), 229
    online learning, 188–189

educational and reference software, 157–160

edX, 229

electrically erasable programmable ROM (EEPROM) A type of ROM chip that can be electrically reprogrammed using only the hardware that comes with the computer; allows upgrading the computer's BIOS without removing the BIOS chip from the motherboard., 71
    for file storage, 122

electronic commerce (e-commerce) The name for buying and selling products and services over the Internet.
    as function of Internet, 186–187
    overview of, 186–187

Electronic Communications Privacy Act (ECPA), 126

electronic mail (email) A means of online communication that involves the transmission and receipt of private electronic messages over a worldwide system of networks and email servers., 170
    adding attachments, 215–216
    common features of, 215
    firewall and, 112
    IMAP email, 170, 171
    managing email, 214
    number of people using, 214
    POP3 email, 170, 171
    privacy and, 232
    sending, 215
    SMTP, 171
    as store-and-forward system, 170
    tips for writing and sending, 231
    types of accounts for, 170–171
    web-based email, 170–171, 214–215

email client An application designed for sending, receiving, storing, and organizing email messages., 171

email See **electronic mail (email)**, 170

embedded computer A special-purpose computer, consisting of a chip that performs one or a few specific actions. Also called *embedded system.*, 4
    characteristics and uses of, 22–24

embedded system, 4

embed To insert data of one type into a file that stores its own data as a different file type (e.g., a picture into a Word document), so that if the embedded content is double-clicked, it opens for editing in its original application., 153

encoding scheme A system of binary codes that represent different letters, numbers, and symbols., 56–57

encyclopedia, 158

End User License Agreement (EULA), 141

enterprise application software, 138

enterprise database A large database that is stored on a server and managed using professional tools such as Oracle Database and Structured Query Language (SQL)., 149

Entertainment Software Ratings Board (ESRB) An organization that administers a game rating system, classifying each game according to the youngest suitable user., 169

Entrepreneur, 220

erasable programmable ROM (EPROM) An obsolete type of ROM chip that could be erased with a strong flash of ultraviolet light and then reprogrammed; enabled reuse but required removing the chip from the computer and placing it in a special machine., 71

eSATA port An external version of the Serial ATA standard used for connecting internal hard drives., 61

ESnet (US Department of Energy's Energy Sciences Network), 208, 209

ESPN, 220

Ethernet port A port that enables a computer to connect to networking device using a network cable, typically a UTP (unshielded twisted pair) cable with an RJ-45 connector on each end., 61

ethics The principles we use to determine the right and wrong things to do in our lives.
    citing Wikipedia in academic paper, 159
    keeping company communication clean, 220
    paying for shareware, 142
    privacy and law enforcement authority, 126
    radio frequency identification (RFID) chips, 54

exabyte, 56

Excite.com, 211

executable file A file that runs a program., 111

expansion board A circuit board that adds some specific capability to a computer system, such as sound, networking, or extra ports for connecting external devices. Also called *expansion card* or *adapter*., 59

expansion bus A bus on a motherboard that provides communication between the CPU and a peripheral device., 59

expansion card or adapter, 59

expansion slot A narrow slot in the motherboard that allows the insertion of an expansion board., 59

Extensible Markup Language (XML) A markup language used to organize and standardize the structure of data so computers can communicate with each other directly., 203
    website creation and, 203

F

Facebook, 216, 217
    apps for, 174
    Messenger, 172

FaceTime, 173, 223–224

Fast Company, 220

favorites, 213

fax machine A combination of a modem and a printer designed to send and receive copies of documents through a telephone line. The word *fax* is short for *facsimile,* which means "exact copy.", 83

fetching, in machine cycle, 64

fiber-optic cable A type of cable that uses a string of glass to transmit data using patterns of photons (beams of light).
    Internet connection, 194–196

fiber-to-the-distribution-point (FTTdp), 195

fiber-to-the-home (FTTH) service, 195

fiber-to-the-node (FTTN), 195

field A single type of information in a data table, such as a name, an address, or a dollar amount. A field may have multiple properties, such as data type, field name, and field size.
    in database, 149

file See also *file storage*
    disk defragmenter, 113
    file compression, 113
    graphics file formats, 162
    lossless and lossy compression, 162
    operating system and management of, 103
    path of, 117
    peer-to-peer (P2P) file sharing, 223
    transferring with file transfer protocol (FTP), 221–222

file allocation table (FAT) The disk's table of contents on a FAT or FAT32 volume., 121

file association A relationship set up in the operating system between a certain file extension and a certain application that can open that type of file., 117

file compression utility A utility that compresses (or shrinks) the size of a file so it occupies less disk space. A file compression utility can also combine multiple files into a single compressed file for easier transfer., 113

file extension A short code (usually three or four characters) that appears at the end of the file name, separated from the name by a period, to identify the file type., 117
    file types and extensions, 117–118

FileMaker, 150

file management, 118–119

file storage
    basic file management skills, 118–119
    devices for, 119–126
    file types and extensions, 117–118
    folder tree, 117
    on hard disk drives, 120–122
    on network and cloud dives, 122–123
    on optical storage devices and media, 123–126
    path of file, 117

    process of, 115–117
    shortcut, 118
    on USB flash drive and solid-state drives, 122

File Transfer Protocol (FTP) An Internet communications standard that enables users to transfer files to and from remote computers via the Internet., 221
    online collaboration with, 221–222
    software for, 221–222
    transferring files with, 221–222
    uploading web page with, 203, 204

FileZilla, 222

Firefox, bookmark, 213

firewall A security system that acts as a boundary to protect a computer or network from unauthorized access., 111–112
    network, 112
    personal, 112

FireWire port A general-purpose port often used to connect external drives and video cameras., 62

firmware See **basic input/output system (BIOS).**, 96

5G, 207

fixed storage Storage that is mounted inside the computer; to get it out, you would have to open up the computer's case. An internal hard disk drive is an example., 119

flaming Problem behavior that is the Internet equivalent of insulting someone face to face., 230

flashing the BIOS, 71

flash memory A type of EEPROM that is easily rewritten in small blocks, so it can be used for rewritable storage; commonly used on USB flash drives and solid-state hard drives., 72
    PRAM memory replacing, 71
    uses of, 72

Flash Video (FLV), 165

flat-bed scanner A scanner that has a large, flat, glass-covered surface, similar to a copy machine., 49–50

flat-screen monitor, 72–73

Flickr, 217

flux transition In magnetic storage, a transition point between positive and negative magnetic polarity on the disk surface., 121

folder A logical organizing unit for file storage on a volume., 116

child and parent folders, 116
structure of organization, 116–117
folder tree A conceptual graphic that helps illustrate the hierarchical structure of a file storage system., 117
Forbes, 220
formatting
in desktop publishing software, 147
in spreadsheet software, 149
in word-processing software, 145–146
formula An instruction to perform a math calculation in a spreadsheet., 148
forum, 218
frame A single still image from a video., 165
Free Lossless Audio Codec (FLAC) format An open source file format created using a method for compressing digital music files that provides efficient, lossless compression., 225
Freenet, 209
freeware Software that's made available at no charge to all., 142
Fuchsia OS, 107
function A named operation that performs one or more calculations on data in a spreadsheet., 148
function keys, 43
fuser The heating element in a laser printer that melts the plastic particles in the toner and causes the image to stick to the paper., 82

G

gambling, online, 228
game console A hardware platform designed specifically for running game software, 167–168
game controller pad A controller that offers the same button types and positions as dedicated gaming consoles. Also called *game pad* or *controller*., 47
game pad, 47
games and gaming
game consoles, 167–168
gaming-optimized PC, 169
gaming software, 167–169
online gaming, 186, 228
on PCs, 167
ratings for, 169
social gaming, 186
statistics on players, 167
system requirements for, 167

Gannett Company, 220
garbage in, garbage out (GIGO), 7
Gbps (gigabit per second) A data transfer rate equal to about a billion bits per second., 191
general-purpose computer, 4
Gfast, 195
GIF See *graphics interchange format (GIF)*, 162
gigabyte A data unit equal to 1,073,741,824 (approximately one billion) bytes., 56
gigahertz (GHz), 66
gig economy, 185–186
GlassDoor, 220
global positioning system (GPS) A device that provides location information by orienting the current location of the device to a signal from an orbiting satellite., 54
Gmail, 170, 214, 215, 217
GoDaddy, 203, 204
Google
advanced search page, 211
Gmail, 170, 214, 215, 217
Instant Apps technology, 174
Network+ Box, 195
search engine, 210, 211
Google Assistant, 212
Google Calendar, 154
Google Chrome, 20, 107, 172, 197
Google Earth, 7
Google Fiber, 195
Google Fuchsia OS, 107
Google Hangouts, 217
Google Home speakers, 212
Google Play, 226
Google Voice, 224
Google Web Designer, 203
GoToMeeting, 173, 229
government
portals, 221
websites for, 221
grammar checker An editing feature in a word-processing program that checks a document for common errors in grammar, usage, and mechanics., 145
graphical user interface (GUI) A user interface that enables the user to select commands by pointing and clicking with a mouse or other pointing device or by tapping with a finger or stylus., 102
desktop, 102

folder structure in, 116
interacting with, 46
graphic data Data consisting of still images, including photographs, charts, drawings, and maps., 13
graphics
bitmap images, 160
in desktop publishing software, 147
file formats for, 162
presentation graphics software, 151–152
vector graphics, 160
graphics and multimedia software
animation software, 163
audio-editing software, 163–165
painting and drawing software, 160–161
photo-editing software, 161
3-D modeling and CAD software, 161–163
video-editing software, 165
web authoring software, 166
graphics card, 75
graphics interchange format (GIF), 162
for website creation, 204
graphics processing unit (GPU) The processor built into the display adapter., 75
graphics tablet An input device that uses a grid of sensors to record drawing and handwriting using a stylus., 48
grids, in spreadsheet software, 148–149
groupware Software that allows people to share information and collaborate on a project, such as designing a new product or preparing an employee manual. Also called collaboration software., 138, 173–174
Guru, 220
gyroscope An environmental sensor that describes an object's orientation in space., 54

H

handheld computer, 27
characteristics and uses of, 27–28
hard copy, 18
hard disk drive (HDD) A mechanical hard drive consisting of one or more rigid metal platters (disks) mounted on a spindle in a metal box with a set of read/write heads—one read/write head for each side of each platter., 120
average access time, 121–122

data storage on, 120–122
data transfer rate, 122
hybrid, 122
inner workings of, 119–120
table of contents, 121
hardware All the physical components that make up the system unit plus the other devices connected to it, such as a keyboard or monitor; also includes other peripherals and devices., 16
communications devices, 18
input devices, 17
Internet connection requirements, 190–191
output devices, 17–18
overview of, 16–17
storage devices, 18
system unit, 17
headphones, 78
health information, on Internet, 227
helper objects, 151
hertz (Hz), 56
hexadecimal digit A number with a base of 16., 198
high-definition multimedia interface (HDMI) port A port used for connecting to HDMI-capable input and output devices, such as to high-definition TVs., 61
high-level language A programming language that uses coding that resembles human language, making the language easier for programmers to understand. Examples include Java, Basic, and COBOL., 114
translating into machine language, 114–115
hijacker A malware extension or plug-in that's installed with your web browser and functions by taking you to pages you didn't select—generally, pages filled with advertisements, 207
home page The first page that appears when you navigate to a website. The home page often provides an overview of the information and features provided by the website., 204
host computer, 31
hot keys, 119
hotspot An area of a graphic that has an associated hyperlink.
mobile, 196–197
Wi-Fi, 193–194
HowStuffWorks, 228

HTML See **Hypertext Markup Language (HTML),** 202
HughesNet, 196
Hulu, 226
Human Genome Project (HGP), 8
hyperlink An element on a web page that the user clicks to navigate to another web page or location. Hyperlinks most commonly appear as underlined text, buttons, photos or drawings, and navigation bars or menus. Also called *link* or *web link.*, 203
hypertext document A web document created with HTML that presents information usually enhanced with hyperlinks to other websites and pages., 203
Hypertext Markup Language (HTML) A tagging or markup language that has long been used to create web pages. The tags identify elements such as headings and formatting such as fonts and colors., 202
as interpreted programming language, 114–115
tags, 202–203
website creation and, 166, 202–203
hyper-threading Intel's version of multithreading. See also **multithreading.**, 67

I
i480, 62
IBM, supersonic transistors, 65
IBM-compatible platform Hardware that is based on the same standard as the original IBM PC back in the 1980s, which ran an operating system called *MS-DOS*. Also called **Intel platform.**, 99
iBooks, 227
icon A small picture that represents a file, folder, or program., 102
IEEE 1394, 62
IMAP, 170, 171
impact printer A type of printer that forms characters and images by physically striking an inked ribbon against the output medium, like an old-fashioned typewriter., 79
IM See **instant messaging (IM),** 216
individual application software An application that serves only one person at a time, such as a word processor, an accounting package, or a game., 138

information Data that's been processed (organized, arranged, or calculated) in a way that converts it into a useful form, such as a report., 12
data combined to create, 12–13
research and reference function of Internet, 187–188
information processing Using a computer to convert data into useful information. Also called **data processing.**, 13–14
information processing cycle A cycle in which the computer accepts input (data), performs processing on the data, displays or sends the resulting output, and then keeps the output in storage if required for future use., 13–14
information searches on Internet
advanced search techniques, 211–212
bookmarking and favorites, 213
default search engine, 211
online reference tools, 227–228
search engine choices, 210–211
information technology (IT) The use of computer, electronics, and telecommunications equipment and technologies to gather, process, store, and transfer data., 3
Informedia Digital Video Library (IDVL), 224
inkjet printer A nonimpact printer that forms characters and images by spraying thousands of tiny droplets of ink through a set of tiny nozzles and onto a sheet of paper as the sheet passes through it., 79
cost of operating, 78, 79, 81
features of, 80–81
printing photos, 80, 81
thermal and piezoelectric, 79–80
input The first step in the information processing cycle, or physically entering data into a computer or device., 14, 42
input device Any hardware component that enables you to perform input operations., 42
audio input devices, 53–54
bar code readers, 50
digital cameras and video devices, 51–52
graphics tables and styluses, 48
keyboards, 42–44
mice and other pointing devices, 46

overview of, 17
scanners, 48–50
sensory and location input devices, 54
touchscreens, 45
insertion point A blinking vertical line in a text entry area in an application, indicating where new text will appear when typed., 145
Instagram, 174, 216, 217
installation key, 140
instant messaging (IM) An online service, similar to chat, that enables users to hold typed online conversations. IM used to require separate messaging software, but today often has been rolled into or integrated with another web-based service., 216–217
software for, 172
integrated circuit (IC), 4
Integrated Services Digital Network (ISDN), 192–193
Intellicast, 220
intelligent personal assistant Software used by *smart speakers* that uses natural language processing to interpret verbal requests and translate them into response or actions., 24
searching Internet with, 212–213
Intel platform See **IBM-compatible platform**., 99
Internet A worldwide network made up of smaller networks linked by communications hardware, software, telephone, cable, fiber-optic, wireless, and satellite systems for communicating and sharing information. Also called *net.*, 10
antispyware and, 114
communications function of, 184–185
connecting to, 189–197
dark web, 209
distance learning function, 188–189
electronic commerce function, 186–187
entertainment and social connection function, 186
firewall, 112
guidelines for good behavior, 230–232
instant messaging and chat software, 172–173
Internet2, 224
IP address and URL, 198

navigating, 197–202
net neutrality, 231–232
number of devices connected to, 3, 11, 184
other resources and services, 213–229
overview of, 10–11
packets, 201–202
path of URL, 199–200
privacy and copyright infringement, 232
private internets, 208–209
purposes for using, 11
research and reference function of, 187–188
search engine choices, 210–211
searching for information on, 210–213
searching with intelligent personal assistant, 212–213
structure of, 189–190
surfing, 197
telecommuting and collaboration function, 185–186
text-based messaging applications, 172–173
viewing web pages, 204–209
web browser applications, 172
web browsers, 197–198
web conferencing software, 173
web page markup languages, 202–204
Internet2, 224
Internet connection, 189–197
cable, 193
dial-up access, 192–193
digital subscriber line (DSL), 193
download and upload speed, 191–192
fiber optic, 194–196
hardware and software requirements, 190–191
mobile hotspot, 196–197
overview of, 191–192
satellite, 196
structure of Internet, 189–190
types of, 191–197
wireless, 193–194
Internet Corporation for Assigned Names and Numbers (ICANN), 199
Internet exchange point (IXP) A physical connection through which an ISP may share data with other ISPs and networks., 189

Internet Explorer, 197
Internet of Things (IoT) Devices with embedded technology that can send and receive data wirelessly or via the internet., 22
Internet Protocol (IP) address A number used by web browsers to locate specific material or a specific location on the Internet. An IP address works like an Internet phone number. A URL is translated to an IP address during the connection process., 198
Internet Relay Chat (IRC), 173, 217
Internet service provider (ISP) An organization that provides user access to the Internet, usually charging a subscription fee., 189
account with, to connect with Internet, 190–191
choosing, 195
Internet service providers (ISPs), 10–11
Internet telephony See **Voice over Internet Protocol (VoIP)**., 223–224
interpreter A computer programming tool or an application that reads, translates, and executes one line of instruction at a time, and identifies errors as they are encountered., 114
Intuit QuickBooks, 155
InvestorVillage, 218
iOS The operating system used in Apple mobile devices, such as iPhones and iPads., 108
iPad, 20
iOS, 108
iPhone, 20
FaceTime, 223–224
iOS, 108
iPhoto, 161
IPv4 An IP address format that uses a four-group series of numbers separated by periods, such as 207.171.181.16. This format allows for only 4.3 billion addresses, and some of these addresses must be reserved for special purposes on the web., 198, 202
IPv6 An IP address format that uses eight groups of four hexadecimal digits, and the groups are separated by colons, as in 2001:0db8:85a3:0000:0000:8a2e:0370:7334. This format allows for 340 undecillion addresses. See also **128-bit address**., 198, 201, 202

ISO files, 143
iTunes, downloading from, 225
iTunes Radio, 225

J

Java A programming language that website designers frequently use to produce interactive web applications. Java was created for use on the Internet and is similar to the C and C++ programming languages., 205

JavaScript A popular scripting language used in building web pages; like Java, developed by Sun Microsystems., 205
    as interpreted programming language, 114–115

job postings, 220
Joint Photographic Experts Group (JPEG), 162
    for website creation, 204
Joomla, 203
joystick, 47
JPEG See *Joint Photographic Experts Group (JPEG)*, 162
junk mail, 113

K

Kbps (kilobit per second) A data transfer rate equal to about 1,000 bits per second., 191

kernel The portion of the operating system that manages computer components, peripheral devices, and memory. The kernel also maintains the system clock and loads other operating system and application programs as they are required., 100

keyboard A grid of keys used to enter text characters (letters, numbers, and symbols) and to issue commands., 42
    alphanumeric, 42–43
    keys not working, 43
    QWERTY layout, 42–43
    virtual, 44

keyword Search criteria that you type in a site's search text box or in your browser's address or location bar when performing a web search., 210
    advanced search techniques, 211–212
    in social networking service, 216

kilobyte A data unit equal to 1024 (approximately one thousand) bytes., 56
Kindle eBooks, 226

L

L1 cache, 65
L2 cache, 65
L3 cache, 65

label printer A small specialty printer that holds a roll of labels and feeds them continuously past a print head., 84

LAMP, 110

lamp A replaceable lamp (like a lightbulb) inside a projector that generates the brightness., 76

land Areas of greater reflectivity on the surface of an optical disc. Compare pit., 124

land grid array (LGA), 63

laptop computer A type of portable computer that can fit comfortably on your lap and has a flat screen, and whose screen and keyboard fold together like a clamshell for easy transport. Sometimes called **notebook computer**., 29

laser printer A nonimpact printer that uses a laser to write an image on a drum and then transfers the drum image to paper using powdered toner., 81
    costs and, 82
    process of, 81–82

learning
    distance, 188–189, 228–229
    massive open online course (MOOC), 229
    online, 188–189

learning management system (LMS) A type of online computing platform used to deliver distance learning courses (also called *online courses*). See also **distance learning**., 188–189

LED, 73–74

legacy port A computer port that is still in use but has been largely replaced by a port with more functionality., 62

legal software Software that helps users plan and prepare a variety of legal documents, including wills and trusts., 156

license, software, 141

license agreement The legal agreement that the user must consent to in order to complete the installation of a software product. Sometimes called *End User License Agreement (EULA).*, 141

lifestyle and hobby software, 156
light-emitting diodes (LEDs), 73
Lightning port, 61
Lightwave, 163
lightweight OS, 107
line printer, 62
LinkedIn, 216, 217

link To create a connection between embedded data and the file from which it was originally copied so that if the original file changes, the embedded data changes too., 153

Linux A UNIX-based operating system that runs on a number of platforms, including Intel-based PCs, servers, and handheld devices., 105
    advantages of, 110
    command-line interface, 101
    growing popularity of, 107
    operating system, 105–107
    as server operating system, 109–110

liquid cooling system A cooling system used for PCs that uses water in a closed system of plastic tubes to cool specific "hot spots"., 67–68

liquid crystal display (LCD) A screen that has two polarized filters, and between them liquid crystals that twist to allow light to pass through., 72–74
    display performance and quality factors, 74–75

local area network (LAN) A private network that serves the needs of a business, organization, school, or residence with computers located in a single building or group of nearby buildings, such as on a college campus.
    described, 9–10
    for Internet connection, 192

local data storage, 127

lossless compression A compression method in which none of the original sound or picture information is eliminated., 225

lossy compression, 162

low-level language See machine language., 114

LPT port, 62

lumens (lm) The measurement of a projector's brightness., 76

M

machine cycle A cycle consisting of the four basic operations performed by the internal components of a CPU: fetching an instruction, decoding the instruction, executing the instruction, and storing the result., 64

machine language A computer programming language consisting of only binary digits (1s and 0s). It is considered a **low-level language**., 114
    translating high-level language into, 114–115

macOS, 104
    application download files, 142–143
    desktop, 105
    Dock, 105
    file management, 103
    Finder, 104, 105
    graphical user interface, 102
    operating system, 104–105
    overview, 19–20
    System Preferences, 105

macro A recorded set of commands that can be played back to automate complex or repetitive actions., 149
    as helper object in database, 151
    in spreadsheet software, 149

magnetic storage A type of disk storage that stores data in patterns of transitions created by magnetizing areas of the disk with a positive or negative polarity., 121

mainframe computer A computer that is bigger, more powerful, and more expensive than a midrange server, and that can accommodate hundreds or thousands of network users performing different computing tasks., 32
    characteristics and uses of, 32
    compared to other computers, 21
    as early computers, 5

main memory Primary storage or random access memory (RAM) that consists of banks of electronic chips that provide temporary storage for instructions and data during processing., 17

malware Software that does damage to a computer, compromises security, or causes annoyance., 111

MarketWatch, 221

markup language A language that includes a set of specifications for describing elements that appear on a page—for example, headings, paragraphs, backgrounds, and lists., 202

massive open online course (MOOC) A free and open-access online course that frequently offers the best content from top schools and partners. MOOCs typically include online video lectures with accompanying project assignments and tests., 229

master file table (MFT) The table of contents on an NTFS volume., 121

Matroska Multimedia Container (MKV), 165

maximum resolution The highest display mode that a monitor can support. Sometimes called native resolution., 74

MayoClinic.org, 227

Mbps (megabit per second) A data transfer rate equal to about 1,000,000 bits per second., 191

McAfee, 112

McAfee AntiVirus Plus, 111

meetings, online, 229

megabyte A data unit equal to 1,048,576 (approximately one million) bytes., 56

megapixel One million pixels; a unit of measurement used to describe digital camera resolution., 51–52

memory A semiconductor chip that contains a grid of transistors that can be on (1) or off (0) to store data, either temporarily or permanently., 68
    flash, 72
    PRAM memory, 71
    RAM basics, 68–69
    random access memory, 68–71
    read-only memory, 68, 71–72
    storage capacity units, 56
    types of, 68

memory access time The time required for the processor to access (read) data and instructions from memory., 70

memory address The numeric address of a location in RAM., 69

memory card slots, for inkjet print, 81

memory resident The part of the operating system that remains in memory while the computer is in operation., 100

message board A type of online service that presents an electronically stored list of messages that anyone with access to the board can read and respond to. Often called *discussion forum* or simply *forum*., 218

messaging
    instant, 217
    private, 217

metalanguage A language for describing other languages., 203

metasearch engine, 211

microcomputer A computer that was designed for easy operation by an individual user and is capable of performing input, processing, output, and storage. Also called personal computer (PC)., 6

microprocessor, 4–5, 17, 63

Microsoft Access, 150, 153

Microsoft Edge, 213

Microsoft Excel, 147–148

Microsoft Exchange, 174

Microsoft Internet Explorer, 172

Microsoft Office, 153

Microsoft Office 365 A collaboration platform that gives small businesses the opportunity to combine the Microsoft Office suite applications or online apps with a SharePoint-based space in Microsoft's public cloud., 174

Microsoft OneNote, 155

Microsoft Outlook, 153, 172
    PIM components of, 154
    web-based email account, 170–171

Microsoft PowerPoint, 151–152
    online presentations, 229

Microsoft Project, 154

Microsoft SharePoint A collaboration platform popular in the corporate world that is typically used by large organizations to set up web- (cloud-) based team collaboration spaces.
    features and uses of, 173–174

Microsoft Windows A very popular operating system developed by Microsoft, used on personal computers and servers., 104
    application download files, 142–143
    Check Disk, 112
    command-line interface, 101
    Cortana, 212–213
    desktop, 104
    Disk Defragmenter, 113
    escaping from web page trap, 208
    File Explorer, 104, 105
    file extensions for, 117
    file management, 103
    graphical user interface, 101–102
    Internet Explorer, 197
    Media Player, 225
    mobile operating system for, 109
    New Technology File System (NTFS), 121

OneDrive, 123
operating system, 100, 104
Paint, 160
reinstalling OS, 104
Setting, 105
Start menu, 104, 105
Windows Defender, 114
Windows Server, 109
Microsoft Windows 8.1
apps, 103
as combination desktop and tablet operating system, 103
Microsoft Windows 10, 104
basic file management, 119
folder tree, 117
Microsoft Word, 146, 153
Print controls in, 146
micro tasks, 186
midrange server, 31
minicomputer, 31
mirroring Writing an identical backup of a drive's content to another drive simultaneously, so the data will continue to be available if the main disk fails., 128
mobile apps See **app**, 138
mobile computer, 29
mobile devices
characteristics and uses of, 27–28
compared to other computers, 21
email on, 214
operating systems for, 107–109
saving money for data plans, 185
trade-in and recycle programs for, 28
mobile-first design, 166
mobile hotspots, Internet connection with, 196–197
modem An electronic device that converts computer-readable information into a form that can be transmitted and received over a communications system., 18
historical perspective of, 18
Internet connection and, 192
moderated environments, 231
moderator An individual with the power to filter messages and ban people who break the rules for chat rooms, message boards, mailing lists, and other types of online communities., 231
monitor A display device that is separate from the computer and has its own power supply and plastic housing., 72

aspect ratio, 74–75
display adapters, 75–76
performance and quality factors, 74–75
pixels and, 73, 74
refresh rate, 75
types of, 72–74
monochrome display A type of display that shows output in only one color., 75
Monster, 220
Moodle, 189
motherboard The large circuit board that functions as the main controller board for a computer., 58
components of, 58
parts of, 17
system BIOS, 97–98
mouse A handheld pointing device that you move across a flat surface (like a desk) to move the onscreen pointer., 46
Moving Pictures Expert Group Layer III (MP3 format), 225
Mozilla Firefox, 172, 197
Mozilla Thunderbird, 172
MP3 format The Moving Pictures Expert Group Layer III digital music file format. MP3 compression reduces the size of CD-quality sound files by a factor of 10 to 14 by removing recorded sounds that the human ear can't perceive., 225
advantages and disadvantages of, 164
for website creation, 204
MP4 format A newer file format used for digital video files., 225
for website creation, 204
MPEG-4 (MP4), 165
MPEG (Moving Picture Experts Group) format A commonly used file format used for compressed digital video files., 225
MS-DOS, 99
command-line interface, 101
multicore processor A CPU that can process several instructions at once, as if the system physically contains more than one CPU, because it contains multiple cores., 67
multifunction device (MFD) A printer that also functions as a copier and a scanner (and sometimes a fax machine). Also called *all-in-one-device*., 83

multimedia software Software that allows users to combine content of multiple types, such as sound, video, and pictures., 160
animation software, 163
audio-editing software, 163–165
painting and drawing software, 160–161
photo-editing software, 161
3-D modeling and CAD software, 161–163
types of, 160–166
video-editing software, 165
web authoring software, 166
multithreading A processing technique that enables the operating system to address two or more virtual cores in a single-core CPU and share the workload between them., 67
multiuser computer system A computer system that can serve many users at a time; typically used by large organizations to enable multiple employees to simultaneously access, use, and update information stored in a central storage location., 15
multiuser operating system An operating system that allows multiple people to use one CPU from separate workstations., 109
Musical Instrument Digital Interface (MIDI) A type of music clip with no analog origin, created by digitally simulating the sounds of various musical instruments., 164–165
music See **audio data**, 13

N

nanobots, 68
nanotechnology, 68
Napster, 223, 226
NAS appliance Short for *network-attached storage appliance*. A specialized computer built specifically for network file storage and retrieval., 127
National Aeronautics and Space Administration (NASA), 7
National Geographic, 227
National Human Genome Research Institute (NHGRI), 8
native resolution See **maximum resolution**., 74
netbook, 30
Netflix, 226
netiquette A collection of guidelines that define good net behavior, based

on the idea that people should treat others as they would like others to treat them. The term *netiquette* combines the words *net* and *etiquette.*, 230
    core rules of, 231
    flaming and, 230

net neutrality Short for *network neutrality*. A doctrine or code of fairness that states that all Internet traffic will be treated with equal priority., 231–232

network Two or more computers or other devices, connected by one or more communications media, such as wires, telephone lines, and wireless signals.
    described, 9–10
    file storage, 122–123

network-attached storage (NAS) Storage that's made available over a network, such as on a centrally accessible file server., 127

network firewall A firewall that is a combination of hardware and software, and is designed for business network use., 112

network service provider (NSP) Providers of high-capacity Internet access. Other organizations, including ISPs, purchase bandwidth from NSPs., 189

network share A drive or folder that's been made available to users on computers other than the one on which the content physically resides., 122–123

Network Solutions, 203

news, online sources of, 220

New Technology File System (NTFS) The file system used in modern versions of Windows., 121

New York Time, 220

Nintendo Switch, 168

noncontiguous sectors Sectors that are not adjacent to one another on a hard disk drive's physical surface., 113

nonimpact printer A type of printer that forms the characters and images without actually striking the output medium; it creates print using electricity, heat, or photographic techniques., 79

nonresident The part of the operating system that remains on the hard disk until it is needed., 100

nonvolatile RAM RAM that does not lose its data when the power goes off; faster than dynamic RAM (DRAM), but more expensive, so not used as the main memory in PCs; also called *static RAM.*, 69

Norton AntiVirus, 111

NOT, 212

notebook computer A newer term that emerged to describe smaller, lighter laptop computers. Often used interchangeably with the term laptop computer., 29

note-taking software, 155

O

object The tables in a database plus the support items for working with the database such as forms, queries, reports, and macros., 151

object linking and embedding (OLE) The Windows and Office feature that allows content to be linked and/or embedded., 153

on die cache A cache that is stamped in the same piece of silicon wafer as the CPU, at the same time; an example is the L1 cache, which is quite small, so it can't hold everything the CPU has recently used or may need to use soon., 65

OneDrive, 222

128-bit address An address format or protocol that allows for 2128 addresses, or 340 undecillion addresses. IPv6 addresses are 128-bit addresses., 198

online courses, 188

online shopping Using a web-connected computer or mobile device to locate, examine, purchase, and pay for products. See also **electronic commerce (e-commerce)**, 219
    overview of, 186–187

OpenFirmware, 71

OpenOffice, 107

open source code, 142

open source software Software in which the developer retains ownership of the original programming code but makes the code available free to the general public., 106

Opera., 172

Opera browser, 197

operating system (OS) Software that provides the user interface, manages files, runs applications, and communicates with hardware., 96
    capability-based, 107
    Chrome, 107
    configuring and controlling devices, 103
    Fuchsia OS, 107
    lightweight, 107
    Linux, 105–107
    mac OS, 19–20, 104–105
    managing files, 103
    memory resident or nonresident, 100
    Microsoft Windows, 19–20, 104
    mobile, 107–109
    for mobile devices, 107–109
    overview of functions, 99–100
    for personal computers, 104–107
    providing user interface, 101–102
    running applications, 103
    for servers, 109–110
    simulators, 106
    starting up computer, 100
    universal, 107

optical carrier (OC) lines, 189, 190

optical character recognition (OCR) Specialized software that converts the bitmap images of text to actual text that you can work with in a word-processing or other text-editing program., 49

optical disc A disc that stores data in patterns of greater and lesser reflectivity on its surface. CDs and DVDs are optical discs., 123
    caring for, 126
    types of, and storage capacity, 125–126

optical drive A drive that contains a laser that shines light on the surface and a sensor that measures the amount of light that bounces back. CDs and DVDs are read and written in optical drives., 123

optical storage, process of, 124

OR, 212

Oracle Database, 149

O'Reilly, Tim, 208

organic light-emitting diode (OLED), 73–74

Outlook.com, 214, 215
    OneDrive, 222

output The information created as a result of computer processing; also the action of printing or displaying such information., 14

output device Hardware used to output data to a user., 72
    augmented reality (AR), 77
    overview of, 17–18
    projectors, 76
    speakers and headphones, 78
    virtual reality (VR), 76–77
overclock, 68

P

packet The unit that messaging software breaks a file into for Internet transmission. The various packets that hold the data for a single file travel over separate paths to a final Internet destination, where they are reassembled into a file., 201–202

packet switching The process of breaking a message into packets, directing the packets over available routers to their final Internet destination, and then reassembling them., 201–202

page formatting, 146

pages per minute (ppm) A measurement of printing speed., 80

painting software Software that enables users to create and edit raster images., 160

Pandora, 225

paragraph formatting Formatting applied to entire paragraphs of text at a time, such as changes to the line spacing, indentation, horizontal alignment between margins, and style of bullets and numbering., 146

parallel ATA (PATA) A hard disk interface that uses a 40-wire ribbon cable to transfer data in parallel fashion; limited to about 133 megabytes per second; mostly obsolete, having been superseded by serial ATA (SATA)., 122

parallel port A 25-pin female D-sub connector on a PC, which matches up with a 25-pin male connector on a cable; a popular port for connecting printers in the early days of computing but now obsolete. Also called *LPT port (short for line printer port)*., 62

parallel processing A processing technique that allows two or more processors (or cores within a single processor) in the same computer to work on different threads simultaneously., 67

parent folder A folder that contains a child folder., 116

parity bit An extra bit written to a disk array during each write operation that helps reconstruct the data in the event of the failure of one of the drives in the array., 129

path The complete description of the location of a file on a storage volume., 117

PC See **personal computer (PC)**, 29

peer-to-peer (P2P) file sharing An often-controversial type of service that enables users to download material directly from other users' hard drives, rather than from files located on web servers., 223

Peripheral Component Interconnect Express (PCIe) bus A popular standard for expansion slots in modern motherboards, available in several sizes including × 1, × 4, and × 16., 59–60

peripheral device A connected device located outside of (or peripheral to) the computer but typically controlled by it; for example, a keyboard, mouse, webcam, or printer., 16

per-seat site license A license agreement that grants use of a software for a certain number of computers, regardless of the number of users., 141

personal computer (PC) A self-contained computer that can perform input, processing, output, and storage functions. Also called **computer**. See also **microcomputer**., 6, 29
    characteristics and uses of, 29–31
    compared to other computers, 21
    desktop computers, 29
    operating systems for, 104–107
    portable computers, 29–30
    steps in startup process, 100
    workstations, 30–31

personal finance software Software that helps users pay bills, balance checkbooks, track income and expenses, maintain investment records, and perform other personal financial activities., 157

personal firewall A software-based firewall designed to protect a PC from unauthorized users attempting to access other computers through an Internet connection., 112

personal information management (PIM) software An application that provides an address book, a calendar, and a to-do list in one convenient interface., 153–154

personal software, 156–160

per-user site license A license agreement that grants use of a software by a certain number of users, regardless of the number of computers., 141

petabyte, 33

petaflop, 33

phablet, 28

phase-change RAM (PRAM), 71

photo-editing software Software designed for manipulating digital photos., 161

photos, storage on the cloud, 52

piezoelectric inkjet printer A printer that moves ink with electricity; each nozzle contains piezoelectric crystals, which change their shape when electricity is applied to them and then force out the ink. Also called *piezo printer.*, 80

pin grid array (PGA), 63

Pinterest, 174, 216, 217

pipelining A processing technique that enables the CPU to begin executing another instruction as soon as the previous one reaches the next phase of the machine cycle., 66

pit Area of lesser reflectivity on the surface of an optical disc. Compare **land**., 124

pixel An individual colored dot within a display or a graphic image., 48
    monitor screen and, 73, 74

plagiarism Using others' words and ideas without attributing the original creator(s) as the source.
    antiplagiarism software, 158

platform The hardware that a particular operating system will run on., 99

plotter A type of printer that produces large-size, high-quality precision documents, such as architectural drawings, charts, maps, and diagrams; used to create engineering drawings for machine parts and equipment., 84

plug-in A mini-program that extends the capabilities of web browsers in a variety of ways, usually by improving graphic, sound, and video elements., 206

PNG See *portable network graphics (PNG)*, 162

podcast Online content that is like a typical downloadable audio or video file (or even a PDF or ebook file), but different in the sense that each download is often part of an ongoing series., 225

downloading, 225
streaming, 225–226
pointing device An input device that moves an onscreen pointer (usually an arrow)., 46
actions of, 47
point of presence (POP) A connection through which an ISP provides Internet connections to customers., 189
POP3 email, 170, 171
Popular Science, 227
pop-up ad An online ad named for its tendency to appear unexpectedly in the middle or along the side of the screen. A pop-up ad typically hides a main part of the web page., 207
pop-up blocker A web browser security feature that you can activate to block pop-up windows., 207
port An external plug-in socket on a computing device, used to connect external devices such as printers, monitors, and speakers to a computer. Sometimes called *interface*., 60
audio, 61–62
component video, 62
dial-up modem, 62
digital visual interface (DVI), 61
eSATA, 61
Ethernet, 61
FireWire port, 62
high-definition multimedia interface (HDMI), 61
legacy, 62
parallel, 62
port 110, 112
PS/2, 62
serial, 62
S-Video, 62
Thunderbold port, 61
types of, 60–62
universal serial bus (USB) port, 60–61
video graphics adapter (VGA), 62
portable computer, 29–30
portable document format (PDF), 152
portable network graphics (PNG), 162
for website creation, 204
portable printer A lightweight, battery-powered printer that can be easily transported., 84
portal A special type of website that acts as a gateway for accessing a variety of information and serves as a "launching pad" for users to navigate categorized web pages within the same website or across multiple websites., 221
postage printer A printer that is similar to a label printer but may include a scale for weighing letters and packages, and may interface with postage-printing software., 84
power-on self-test (POST) A test of the essential hardware devices at startup to make sure they are operational., 100
PowerPoint, 151–152
power supply The component that converts the 110-volt or 220-volt alternating current (AC) from your wall outlet to the much lower voltage direct current (DC) that computer components require., 57–58
presentation, online, 229
presentation graphics software An application that allows users to create computerized slide shows that combine text, numbers, animation, graphics, audio, and video., 151–152
primary storage Storage where data is placed immediately after it's input or processed. Primary storage is by nature temporary. Generally, the term *primary storage* refers to dynamic RAM., 17, 119
printer A device used to produce hard-copy output on paper or another physical medium, such as transparency film., 79
impact vs. nonimpact, 79
inkjet, 79–81
laser, 81–82
multifunction device, 83
plotter, 84
special-purpose, 84
thermal, 83
3-D, 84
print preview An onscreen preview of a document as it will appear on the printed page., 146
privacy
email and, 232
law enforcement authority and, 126
private internets, 208–209
private message (PM) A form of online messaging that is visible only to you and others invited to the conversation. PM is usually offered as part of a social media service such as Facebook., 217

processing The operations carried out in the computer's electrical circuits that transforms input (data) into information or output., 14
processor, 17, 63
productivity software See *business productivity software*, 143–155
program A set of instructions executed by a computer. Types of programs include operating systems, utilities, and applications. See also **software**, 4
programming language A set of coding specifications used by programmers to create a program.
for compiling or interpreting, 114–115
project management software A type of software that helps manage complex projects by keeping track of schedules, constraints, and budgets., 154
projectors, 76
proprietary software Software that an individual or company holds the exclusive rights to develop and sell., 106
proxy server A server that intercepts and processes network requests for a variety of purposes, including security, content filtering, or network performance optimization., 112
PS/2 port A small, round port that was the standard for connecting keyboards and mice in the years before USB; named for a very old IBM computer, the PS/2, which was the first to use this connector type., 62
public domain Software that's not copyright protected, so anyone may use or modify it., 142

Q

QR code, 50
quantified self, 25
Quantum computers, 33
Quark XPress, 147
Quicken, 157
QWERTY layout The standard English keyboard layout, named for the first five letters in the top row of letters on the keyboard., 42–43

R

radio frequency identification (RFID) chips, 54
RAID0 A type of RAID that improves performance by striping the data., 128

RAID1 A type of RAID that improves data safety by using mirroring., 128

RAID5 A type of RAID that combines striping (for performance) with distributed parity (for data protection)., 129

random access memory (RAM) A type of memory that can be written and rewritten easily as the computer operates; the primary type of memory used in almost all desktop and notebook computers., 17, 68
- basics of, 68–69
- functions of, 69
- memory access time, 70
- speed of and performance of, 70
- static vs. dynamic, 69–70
- storage capacity of, 70–71
- upgrading, 69

raster image An image made up of a grid of colored pixels (dots). Also called *bitmap image.*, 160

RCA jacks, 78

read-only disc (ROM), 126

read-only memory (ROM) A type of memory that stores data permanently, even when the computer is not powered on; individual bytes of ROM cannot be easily rewritten., 68
- upgrading, 71
- uses of, 68, 71

read/write head, 121

recordable disc (R), 126

Reddit, 216, 217

Red-Green-Blue port, 62

Red Hat Linux, 106

redundant array of independent disks (RAID) A technology that attempts to improve performance, data safety, or both by combining multiple physical hard disk drives (HDDs) into a single logical volume., **128**

Refdesk.com, 227–228

reference
- function of Internet, 187–188
- online reference tools, 227–228

refresh rate The number of times per second that each pixel on a monitor is re-energized; measured in hertz (Hz)., 75

register An area inside of a CPU that functions as a "workbench," holding the data that the ALU is processing., 64

registration key A string of characters that uniquely identifies the user's purchase of a software. Sometimes called *installation key.*, 140

relational database A model used by most modern databases in which fields with similar data can be connected between data tables or files in the database, making it possible to retrieve data from the related tables.
- overview of, 150–151

removable storage Storage media that can easily be separated from the computer. CDs, DVDs, external hard drives, and USB flash drives are all examples., 119

resolution On a scanner or printer, the number of pixels recorded or produced per inch; in a display, the number of pixels horizontally and vertically that comprise the display mode., 49
- digital cameras, 51–52
- monitors, 74
- printers, 80
- scanners, 49

rewriteable disc (RW), 126

RGB port, 62

Rhapsody, 223

rip To copy songs from a CD., 225

RJ-11 connector A telephone cable connector with a single-line cable with two wires., 62

RJ-14 connector A telephone cable connector with a dual-line cable with four wires., 62

RJ-45 port The official name of the connector on an unshielded twisted pair (UTP) Ethernet cable. RJ stands for *registered jack.*, 61

root directory The top level of the storage hierarchy within a volume., 116

router An enhanced type of switch that directs network traffic, like a switch does, but is able to do so on a larger scale, sending data out to other networks.
- wireless, for Internet connection, 192

Roxio, Inc., 223

Roxio Creator NXT, 165

S

Safari, 172, 197
- favorites, 213

sampling A process in which sound waves are recorded thousands of times per second to create a digitized version., 164

satellite, 7
- Internet connection with, 196

Scanner A light-sensing device that detects and captures text and images from a printed page., 48
- resolution, 49
- scanning process, 48–50
- **3-D scanner**, 50

Science, 227

Science Daily, 227

science information, on Internet, 227

Sciencemag.org, 227

screen projector See **video projector**., 76

screen The viewable portion of a display device., 72

Search.com, 211

search engine A website or service you use to locate information on the web., 210
- choosing, 210–211
- default, 210, 211
- metasearch engine, 211

search operator An operator used in a logic statement when performing an advanced web search. Three common search operators are AND, OR, and NOT., 212

secondary storage Storage that holds data until the data is removed from it intentionally. Generally, the term *secondary storage* refers to storage volumes such as hard drives., 119

second-level domain The part of the domain name in the URL that follows the format information and identifies the person, organization, server, or topic (such as Amazon) responsible for the web page. See also **domain name**., 199

section formatting, 146

sector A numbered section of a track on a hard disk., 121

security measures Methods used in a database management system to protect and safeguard data.
- biometric, 53

semiconductor material A material that is neither a good conductor of electricity (like copper would be) nor a good insulator against electricity (like rubber would be); used for the internal circuits in CPUs and memory., 63

seminar, web-based, 229
sensory input device, 54
serial ATA (SATA) A hard disk interface that transfers data in serial fashion at up to 308 megabytes per second; used on most desktop and notebook systems today., 122
serial port A 9-pin male D-sub connector on the PC; a popular port for connecting external devices in the early days of computing but now obsolete. Also called *COM port (short for communications port).*, 62
server A powerful computer that's capable of accommodating numerous client computers at the same time. Formerly known as a *midrange server* or *minicomputer.*, 31
    characteristics and uses of, 31
    compared to other computers, 21
    operating systems for, 109–110
SharePoint, 173–174
shareware A commercial software product that is released on a try-before-you-buy basis; purchasing the product unlocks additional features or removes time restrictions on the trial version., 142
    paying for, 142
shell See **user interface.**, 101
shopping, online, 219
shortcut, 118
Short Message Service (SMS), 172
Simply Hired, 220
single data rate (SDR) SDRAM SDRAM in which data moves into or out of RAM at the rate of one word per system clock cycle., 70
single-user computer system A computer system that can accommodate one user at a time; found in homes and small businesses and organizations., 15
Siri, 212
site license A license agreement that grants permission to make multiple copies of a software and install it on multiple computers., 141
Skype, 173, 217
Slack, 217
Slide An individual page or screen that's created in presentation graphics software., 152
slide show A collection of slides in a single data file in a presentation graphics program., 152

small office/home office (SOHO), 143–144
smart device Any type of device or appliance that has embedded technology., 22
    characteristics and uses of, 22–24
    compared to other computers, 21
    examples of, 23
smart home A concept where numerous networked home devices provide specific, usually automated functions, and share data and media., 24
smart home hub, 25
smartphone A cell phone that can make and receive calls and text messages on a cellular network, and that can also connect to the Internet via a cellular or wireless network and perform numerous computing functions, such as email and Web browsing., 27
    characteristics and uses of, 27–28
    compared to other computers, 21
    mobile-first design, 166
    tethering, 197
    trade-in and recycle programs for, 28
smart speaker, 24–25
smartwatch, 25
SMTP, 171
Snapchat, 174
snapshot printer, 83
social connection function of Internet, 186
social gaming, 186
social media
    popular social media services, 215–217
    sharing and networking with, 215–217
    worldwide revenue from, 186
social networking service A new genre of web services that enables people to create personal online spaces, share content, and interact socially. Examples include Facebook, Twitter, Pinterest, LinkedIn, and Reddit., 216
soft copy, 18
software Programs that consist of the instructions that direct the operation of the computer system and enable users to perform specific tasks, such as word processing. Types of software include system software and application software. See also application software, 19
    accounting software, 155

antiplagiarism software, 158
app-ification of software industry, 139
application download files, 142–143
application software, 20
commercial software, 139–142
communications software, 170–174
database management software, 149–151
desktop application vs mobile apps, 138
desktop publishing software, 146–147
distribution of, 139–140
educational and reference software, 157–160
gaming software, 167–169
graphics and multimedia software, 160–166
individual, group and enterprise use, 138
Internet connection requirements, 190–191
legal software, 156
licensing, 141
lifestyle and hobby software, 156
multimedia software, 160–166
note-taking software, 156
open source vs. proprietary, 106
personal finance software, 157
personal information management software, 153–154
presentation graphics software, 151–152
project management software, 154
purchasing and downloading online, 141
sales and licensing of, 138–142
shareware and freeware, 142
software suites, 153
spreadsheet software, 147–149
system software, 19–20
tax preparation software, 157
word-processing software, 144–146
Software as a Service (SaaS) Software delivered on demand from the cloud (Internet), usually via a web browser.
    advantages of, 139–140
software piracy Stealing commercial software by using it without paying when payment is required., 140
software suite A group of applications bundled and sold as a single package., 153

solid-state drive (SSD) A storage device that stores data in nonvolatile memory rather than on a disk platter., 115, 122

solid-state hybrid drive, 122

solution stack A set of complementary applications that work together to customize an operating system for a specific purpose., 110

Sony PlayStation, 168

Sony Vegas Pro, 165

sound card The adapter in a computer that translates between the analog sounds that humans hear and the digital recordings of sound that computers store and play back., 53, 61–62

sound See **audio data**, 13

source code The uncompiled programming code for an application., 142

spam Unwanted commercial email, or junk mail., 113

spam blocker A utility program often used to filter incoming spam messages., 113–114

spam trap A folder—usually named *Spam* or *Junk*—designated to automatically collect incoming email messages that are known or suspected spam., 216

speaker-dependent systems, 54

speaker-independent systems, 53

speakers, connecting to computer, 78

Spectrum Internet, 193

spelling checker An editing feature in a word-processing program that checks each word in a document against a word list or dictionary and identifies possible errors., 145

Spotify, 225

spreadsheet software Software that provides a means of organizing, calculating, and presenting financial, statistical, and other numerical information., 147–149

    features of, 148–149

spyware A form of malware that secretly tracks the activities of an Internet user and relays the gathered information back to a third party., 114

startup process, computer, 100

static RAM (SRAM) RAM that does not lose its data when the power goes off; faster than dynamic RAM (DRAM), but more expensive, so not used as the main memory in PCs., 69–70

steering wheel A specialized input device used to simulate a vehicle's steering wheel., 47

storage Recording processed output (information) on a medium such as a hard disk, flash drive, external hard disk, or cloud storage area for future use; also the media used to store the information., 14

    in data processing cycle, 14

    direct-attached storage, 127

    file storage, 115–126

    fixed, 119

    large-scale, 127–129

    local vs. network, 127

    magnetic, 121

    network-attached storage, 127

    optical, 123–126

    primary, 119

    RAID, 128–129

    removable, 119

    secondary, 119

    storage area network, 127

storage area network (SAN) A network storage technology that enables users to interact with a large pool of storage (including multiple devices and media) as if it were a single local volume., 127

storage bay A space in the system unit where a storage device (such as a hard drive or a DVD drive) can be installed. Also called *bay.*, 63

storage device A hardware component that houses a storage medium that provides more permanent storage of programs, data, and information; you can later retrieve the stored data to work with it. Also called secondary storage., 18

store-and-forward system A message delivery system in which messages are stored on a server until they are picked up by the recipient., 170

Stored Communications Act (SCA), 126

streaming An alternative method to downloading that sends a continuous stream of data to the receiving computer's web browser, which plays the audio or video., 225

    audio, video and podcasts, 225–226

    compared to downloading, 226

striping Spreading out data to be stored across multiple discs, and writing to all the discs simultaneously, to improve the speed at which data can be written and read., 128

style A named set of formatting that can be consistently applied to multiple paragraphs throughout a document., 146

stylus A handheld wand that resembles an inkless pen and that is used with a graphics tablet to record handwriting and drawing., 48

subwoofer, 78

Summit, 33

supercomputer An extremely fast, powerful, and expensive computer capable of performing quadrillions of calculations in a single second., 33

    characteristics and uses of, 33

    compared to other computers, 21

superscalar architecture A type of CPU design that enables the operating system to send instruction to multiple components inside the CPU during a single clock cycle., 67

surfing Navigating between pages and locations on the web using the browser to follow links on various pages., 197

SUSE Linux Enterprise, 106

S-video port A port used for video output for standard-definition TVs. Also known as *i480.*, 62

Symantec, 112

synchronous dynamic RAM (SDRAM) RAM in which the speed is synchronized with the system clock., 70

system BIOS The basic input/output system (BIOS) for the motherboard., 97

system bus The bus that connects the CPU to the main memory., 59

system clock A small oscillator crystal that synchronizes the timing of all operations on the motherboard. Also called *system crystal.*, 65–66

system crystal, 65

system requirements The minimum configuration needed to install and run an application or operating system., 168

system software Software that runs application software, starts up the computer, keeps it running smoothly, and translates human-language instructions into computer-language instructions., 96

    category overview, 96–97

    operating system, 99–110

    overview of, 19–20

    system BIOS, 97–98

translators, 114–115
utility program, 110–114
system unit The main part of the computer that houses the major components used in processing data., 57
   buses, 59
   components of, 57–63
   expansion slots and boards, 59–60
   motherboard, 58
   overview of, 17
   ports, 60–62
   power supply, 57–58
   storage bay, 63

T

tablet A computing device that is larger than a smartphone and generally used on your lap or on a table, and that can connect wirelessly to the Internet and perform computing activities. Also called *tablet computer* or *tablet PC.*, 27
   characteristics and uses of, 27–28
   compared to other computers, 21
tablet computer, 27
tablet PC, 27
tag Code inserted within an HTML file to define various page elements, such as language type, body, headings, and paragraph text. A tag is enclosed in angle brackets. Tags generally must be used in pairs—an opening tag and a closing tag., 202
tagged image file format (TIFF), 162
tax preparation software Software designed to aid in analyzing federal and state tax status, as well as preparing and filing tax returns., 157
TED Talks, 188
telecommuting A work arrangement that enables millions of workers to perform their work activities at home using a computer and an Internet connection. Also called *teleworking.*, 185–186
telephone
   dial-up modem, 62
   using Internet to make telephone calls, 223–224
television, streaming shows, 226
template An application designed to help users create a variety of text-based documents., 145
terabyte, 56

tethering A setup where you use your smartphone as a hotspot when you need only a single Internet connection while on the go., 197
text-based messaging applications, 172–173
text data Data consisting of alphabet letters, numbers, and special characters, such as in spreadsheet formulas, and that typically enable the computer to produce output such as letters, email messages, reports, and sales and profit projections., 13
text formatting Formatting applied to individual characters of text, such as changes to the font, size, and color of the text., 145–146
text message, 172
texture A pattern added to the wireframe surface of a 3-D– modeled object to make it look more realistic., 162
thermal dye transfer printer, 83
thermal inkjet printer An inkjet printer that heats the ink to about 400 degrees Fahrenheit, which creates a vapor bubble that forces the ink out of its cartridge and through a nozzle, which in turn creates a vacuum inside the cartridge, which then draws more ink into the nozzle., 79
thermal printer A printer that uses heat to transfer an impression onto paper., 83
thermal wax transfer printer A thermal printer that adheres a wax-based ink onto paper, using a thermal print head that melts the ink from a ribbon onto the paper; images are printed as dots, so they must be dithered to produce shades of colors., 83
thin client A computer that has only minimal capabilities and was designed to be small, inexpensive, and lightweight, to provide Internet access, and to interface with other computers (such as on the Internet)., 107
thread A part of a program that can execute independently of other parts., 66–67
3-D modeling program An application used to generate the graphics for blueprints and technical drawings for items that will be built, as well as 3-D representations of people and objects in games., 161–163
3D printer A printer that uses a special kind of plastic, metal, or other material to create a three-dimensional model of just about any object you can design in a 3-D modeling program on a computer., 84
3-D scanner A scanner that captures the complete size and shape information of an object which can be used in a 3-D modeling application to print the 3-D object., 50
Thunderbolt port A high-speed, multipurpose port found in Macs and some PCs used to delivery multiple signals such as PCIe, DisplayPort, and power., 61
tile A rectangular block on a graphical user interface screen (such as the Windows 8.1 Start screen) that represents an application, a folder, or a file. The user taps a tile to select it., 102
tile-based interface, 102
T1 line A T line that carries data at 1.5 Mbps over its 24 internal lines, with each line running at 64 Kbps.
   speed and cost of, 189, 190
T3 line A T line that contains a bundle of 672 individual lines. Working collectively, these lines can transfer data at up to 43 Mbps.
   speed and cost of, 189, 190
toner A powdered combination of iron and colored plastic particles that is contained within a reservoir inside a printer., 82
top-level domain (TLD), 199
Tor Browser, 208–209
Tor Project, 208–209
Torvalds, Linus, 105
touchpad A small, rectangular pad that's sensitive to pressure and motion, used as a pointing device. Also called *track pad.*, 46
touchscreen A touch-sensitive display that is produced by laying a transparent grid of sensors over the screen of a monitor., 45
   gestures for, 45
   user interface for, 102
trace A conductive pathway on a circuit board., 58
track A numbered, concentric ring or circle on a hard disk., 121
trackball A stationary pointing device with a ball that you roll with your fingers to move the pointer., 46
track pad, 46
transistor
   number of, inside CPU, 66
   supersonic, 65

translation software, 114–115
　　compiler, 114
　　function of, 114–115
　　interpreter, 114
translator Software that translates programming code into instructions that can be run as programs., 96
trunk lines, 189
Tumblr, 217
Turnitin and SafeAssign, 158
tutorial A form of instruction in which the student is guided step by step through the learning process., 159–160
Twitter, 174, 216

U

Ubuntu Linux, 106
Udacity, 229
Ultrabook, 30
Unicode An encoding system that assigns a unique 16-bit binary code to each number and letter, and can represent up to 65,536 characters including many different languages., 57
Unified Extensible Firmware Interface (UEFI), 97
uniform resource locator (URL) A web address that corresponds to a website or page. URLs generally use descriptive names rather than numbers, so they are easier to remember than IP addresses., 198
　　parts of, 199
　　path of, 199–200
uninstaller A utility program for removing software and its associated entries in the system files., 112
universal OS, 107
universal product code (UPC) A bar code on a product for sale, used to provide information about the product at the point of sale and in the store's inventory., 50
universal serial bus (USB) port A general-purpose port used by many different devices, including input devices (such as keyboards and mice) and external storage devices (such as external hard drives, USB flash drives, and digital cameras)., 60–61
UNIX A powerful multiuser command-line operating system designed for servers., 109
　　advantages of, 109
　　command-line interface, 101

file management, 104
　　as server operating system, 109
upload To send information such as a document from one's own computer to a remote computer or location via a network or the Internet., 191
Upwork, 220
URL See **uniform resource locator (URL)**, 198
USA.gov, 221
USB flash drives, solid-state storage, 122
USB See **universal serial bus (USB) port**, 60–61
user interface The interface that allows communication between the software and the user. Also called shell., 101
　　graphical user interface, 101–102
　　operating system providing, 101–102
　　text-based, 101
　　for touchscreen, 102
utility program Software that performs troubleshooting or maintenance tasks that keep the computer running well and protect the computer from security and privacy violations., 96
　　antispyware, 114
　　antivirus software, 111
　　backup utility, 113
　　diagnostic utility, 112
　　disk defragmenter, 113
　　disk scanner, 112
　　file compression utility, 113
　　firewalls, 111–112
　　overview of types and functions, 110–111
　　spam blocker, 113–114
　　uninstaller, 112

V

value A numeric entry in a spreadsheet cell or formula., 148
value-added network (VAN) A large ISP that provides specialized service or content, such as reports from a news service or access to a legal database., 190–191
vector graphic A graphic created using a mathematical equation, resulting in a line drawing that can be scaled to any size without loss of quality., 160
Verizon
　　mobile hotspot, 196
ViaSat Internet, 196

video adapter, 75
video card, 75
video data Data consisting of moving pictures and images, such as webcam data from a videoconference, a video clip, or a full-length movie., 13
　　downloading, 225
　　streaming, 225–226
　　viewing web pages and, 205–206
video-editing software Software that enables you to import digital video clips and then modify them in various ways., 165
video graphics adapter (VGA) port An older type of monitor port still used in many systems; on a PC, a 15-pin female D-sub connector (often abbreviated *DB15F*)., 62
video graphics array (VGA) port, 62
video projector A device that captures the text and images displayed on a computer screen and projects them on to a large screen so an audience can see them clearly. Also called **screen projector**., 76
virtual keyboard A software-generated, simulated keyboard on devices with a touchscreen, such as a tablet PC or smartphone., 44
virtual reality (VR) A computer simulation of an imagined but convincing environment or set of surroundings.
　　gaming and, 228
　　as visual output, 76–77
virus Maliciously designed code that infects executable files and spreads the code to other executable files, as well as performing harmful or annoying actions on the computer., 111
virus checker, 111
Visual Basic (VB) A Microsoft-owned language that is commonly used for developing software prototypes and custom interfaces for Windows platforms.
　　as compiled programming language, 114
voice input Technology that enables users to enter data by talking into a microphone connected to the computer., 53
Voice over Internet Protocol (VoIP) A technology that enables two or more people with good-quality connections to use the Internet to make telephone-style audio and video calls around the world. Also called **Internet telephony**., 223–224

software for, 173
    using, 223–224
volatile RAM, 69
volume A storage unit with a letter assigned to it, such as *C* or *D*., 116
Vonage, 173

W

Wall Street Journal, 220
warm boot The act of restarting a computer without powering it off., 100
wave file format An uncompressed sound file type identified with a .wav extension., 164
waveform A sound file that has an analog origin—that is, an origin outside a computer system., 164
wearable computer A computer device that's worn somewhere on the body—most often, on the head or wrist—and that typically provides mobile computing capabilities and Internet access., 25
    characteristics and uses of, 25–26
    compared to other computers, 21
weather, online sources of, 220
Weather Channel, 220
Web 3.0, 208
web applications, 139–140
web authoring software Software that helps users develop web pages without having to learn web programming., 166
    mobile-first design, 166
web-based email, 170–171, 214–215
web-based training (WBT), 159
web browser A program that displays web pages on the screen of a computer or mobile device and enables the user to move between pages., 197
    ads, 206–208
    apps, 172
    audio, video and animation elements, 205–206
    basic browsing actions, 204
    common features of, 198
webcam A digital video camera that must be connected to a computer to take pictures; commonly used to capture live video that's streamed via a website., 52
web conferencing software, 173
WebEx, 229
web hosting, 203

web links, 203
weblog, 218
webmail A type of email you can work with and manage via a smartphone or tablet in addition to a computer. Webmail enables you to use your web browser to navigate to your email account on your ISP's website., 214
WebMD, 227
web page A single document stored on the World Wide Web, created using HTML and containing text, graphics, links, and more. See also **website**, 204
    HTML and CSS format, 202–203
    publishing basics, 203–204
    viewing, 204–209
    web page markup language, 202–204
    web page traps, 207–208
    XML format, 203
web page trap A situation where a website changes your browser's settings permanently or attempts to prevent viewers from leaving by continually popping up more windows and disabling the Back button., 207–208
Webroot SecureAnywhere, 114
website One or more web pages devoted to a particular topic, organization, person, or the like, and generally stored on a single domain. See also **web page**, 204
    ads on, 206–208
    basic browsing actions, 204
    cookies, 205–206
    home page, 204
    IP address, 198
    plug-ins, 206
    steps for creating, 203–204
    URL and, 199–200
    web page markup language, 202–203
WhatsApp, 217
Whisper, 217
wide-format inkjet printer, 84
Wi-Fi hotspot A location with a wireless access point for connecting to the Internet via a mobile computer or device, often found in public places such as coffee shops and airports., 193–194
wiki A website that works like an online encyclopedia, but allows anyone to contribute information., 228
Wikipedia, 159

window A well-defined rectangular area on the computer screen, in which file listings and applications appear., 102
Window Defender Firewall, 112
Windows See **Microsoft Windows**, 104
Windows Defender utility, 111
Windows Mail, 172
Windows Media Audio (WMA) format A digital music file format used to compress audio file size in Windows., 225
Windows Media Player, 225
Windows operating system, 19–20
Windows Phone OS, 109
Windows Server A variant of Microsoft Windows designed for servers., 110
Wired, 220
wireframe A surfaceless outline of an object composed of lines. A 3-D vector graphic is first drawn as a wireframe., 161–162
wireless Internet connection, advantages/disadvantages of, 193–194
Wireless ISPs (Wi-Fi ISPs), 194
WMV (Windows Media Video) format The format commonly used when creating digital video files in Windows, or viewing them with Windows Media Player., 225
WordPress, 203, 204
word-processing software An application designed to help users create a variety of text-based documents., 144–146
word size The number of bits that a computer can manipulate or process as a unit; for example, a 32-bit CPU can handle 32-bit blocks of data at a time., 66
workstation A high-performance, single-user computer with advanced input, output, and storage components., 30
    characteristics and uses of, 30–31
    compared to other computers, 21
World of Warcraft, 228
World Wide Web (web) The part of the Internet that enables users to browse and search for information using a web browser application. Users can follow links and read and download information. See also Internet, 187
    overview of, 11–12
worm A form of malware that actively attempts to move and copy itself across a network., 111

wrist-based devices, 25–26
WYSIWYG Acronym for *what you see is what you get.* An approach to content creation that allows you to see, during the development process, the layout and content as they will appear when distributed to recipients (such as on the web or in print)., 166

X

x64 The 64-bit version of the Intel platform., 99

x86 The 32-bit version of the Intel platform. The name *x86* is a reference to the numbering system of Intel CPUs used in early computers: 286, 386, and 486., 99

Xbox 360, 168

Xfinity Internet, 193

XML See **Extensible Markup Language (XML)**, 203

Y

Yahoo!
    Finance, 218
    Mail, 170, 214, 215
    search engine, 210
yottabyte, 56
YouTube
    video and music streaming, 225–226

Z

zettabyte, 56
zip file, 113, 143
zipped file See **compressed file**., 216

Image Credits

Microsoft images used with permission from Microsoft.

Chapter 1
Page 1 © Creativa/Shutterstock.com, © AdobeStock/sdecoret; 2 Baramee Thaweesombat/123rf.com, Courtesy of Whirlpool Corporation; 4 Used with permission from iDevices LLC., 5 © Bettmann/Corbis, 5 Used with permission from Intel Corporation; 7 Landsat imagery courtesy of NASA Goddard Space Flight Center and U.S. Geological Survey or Landsat imagery courtesy of USGS/NASA Landsat; 8 © Shutterstock/wavebreakmedia, Courtesy of the National Human Genome Research Institute; 9 Courtesy of Sandisk Corporation; 11 Courtesy of ATT&T Intellectual Property. Used with permission.; 12 Google and the Google logo are registered trademarks of Google Inc., used with permission; 14 Courtesy of Sandisk Corporation; 15 pryzmat/Shutterstock.com (if using); 17 © romvo/Shutterstock.com, © tele52/Shutterstock.com, © Popcic/Shutterstock.com, © Nata-Lia/Shutterstock.com, © AlinaMD/Shutterstock.com; 18 © dny3d/Shutterstock.com, © iStock.com/greg801; 19 Courtesy of Apple Inc.; 20 Courtesy of Apple Inc.; 23 Courtesy of Garageio., Courtesy of Honeywell International, Inc., Courtesy of iSmart Alarm, Inc., Courtesy of Spectrum Brands Holdings, Inc., Courtesy of TCP International Holdings., Courtesy of Whirlpool Corporation, Used with permission from Belkin, International, Inc.; 24 Zapp2Photo/Shutterstock; 25 Courtesy of Apple Inc.; 26 Used with permission from Vandrico Solutions.; 27 © TATSIANAMA/Shutterstock.com; 29 © Oleksiy Mark/Shutterstock.com; 30 © iStock.com/MichaelDeLeon; 31 © Can Stock Photo Inc. / AndreyPopov, © iStock.com/GodfriedEdelman; 32 Courtesy of International Business Machines Corporation. Jon Simon/Feature Photo Service for IBM.; 33 Image courtesy of Oak Ridge National Laboratory, U.S. Department of Energy

Chapter 2
Page 41 © iStock/gorodenkoff; 44 Pressmaster/Shutterstock.com; 45 ivelly/Shutterstock.com, ivelly/Shutterstock.com, ivelly/Shutterstock.com, ivelly/Shutterstock.com, ivelly/Shutterstock.com; 46 goldyg/Shutterstock.com, Mau Horng/Shutterstock.com, Olga Popova/Shutterstock.com; 47 Emanuele Ravecca/Shutterstock, Isil Akdede/Shutterstock; 48 Valeriy Lebedev/Shutterstock.com; 49 Mile Atanasov/Shutterstock.com; 50 mrkob/Shutterstock.com; 52 Maridav / Shutterstock, Oleksiy Mark/Shutterstock.com, Portfolio/Shutterstock.com, Volodymyr Krasyuk/Shutterstock.com; 53 chungking/Shutterstock.com, wavebreakmedia/Shutterstock.com; 55 Courtesy of Fred the Oyster/Wikipedia Commons; 58 Courtesy of ASUSTek Computers, Inc.; 59 © ASUSTek Computers, © Bastiaan va den Berg/Creative Commons Wikipedia; 60 Courtesy of Tosaka/Wikipedia Commons, Courtesy of Andreas Pietzowski/Wikipedia Commons, Courtesy of Fred the Oyster/Wikipedia Commons, Courtesy of IngenieroLoco/Wikipedia Commons, Dragana Gerasimoski/Shutterstock.com, 63 jelome/Shutterstock.com, John Kasawa/Shutterstock.com; 67 KsanderShutterstock, THANAN KONGDOUNG/Shutterstock; 68 Andrea Danti/Shutterstock.com; 70 Julio Embun/Shutterstock.com; 71 Jianghaistudio/Shutterstock.com; 72 cobalt88/Shutterstock.com, pryzmat/Shutterstock.com; 74 © Shutterstock/Sorbis; 75 Olena Zaskochenko/Shutterstock.com; 77 © Shutterstock/Monkey Business Images, Zapp2Photo/Shutterstock; 78 QiLux/Shutterstock.com; 80 Dja65/Shutterstock.com, photosync/Shutterstock.com; 81 Tomislav Pinter/Shutterstock.com; 82 Piotr Adamowicz/Shutterstock.com; 83 Singkham/Shutterstock.com; 84 © Hewlett-Packard Company, © Newell Rubbermaid Inc.

Chapter 3
Page 95 © AdobeStock/Tomasz Zajda; 96 kaczor58/Shutterstock.com; 102 Courtesy of Apple Inc.; 105 iMac®, iTunes®, and iPhoto® are registered trademarks of Apple Inc.; 106 Used with permission from the Linux Foundation; 108 Courtesy of Google.; iPad®, Facetime®, and iTunes® are registered trademarks of Apple Inc.; 109 © 2018 BlackBerry Limited. All Rights Reserved. Used with permission.; 110 Courtesy of International Business Machines Corporation.; 127 © Can Stock Photo Inc. / Scanrail, © Can Stock Photo Inc. / Vlucadp

Chapter 4
Page 137 © AdobeStock/georgejmclittle; 147 Adobe product screenshot(s) reprinted with permission from Adobe Systems Incorporated; 151 Peter Bernik/Shutterstock.com; 155 Reprinted with permission © Intuit Inc. All rights reserved.; 156 Used with permission from Rocket Lawyer.; 157 Reprinted with permission © Intuit Inc. All rights reserved.; 158 Couresy of Turnitin®; 159 Used with permission from Dictionary.com; 161 Adobe product screenshot(s) reprinted with permission from Adobe Systems Incorporated.; 162 © iStock.com/mevans; 163 Ragma Images/Shutterstock.com; 166 Pinnacle Systems, Inc.; 168 Courtesy of Blizzard Entertainment®, R-O-M-A/Shutterstock.com, Rvlsoft/Shutterstock.com, Supertrooper/Shutterstock.com; 169 Courtesy of Dell, Inc.

Image Credits

Chapter 5
Page 183 © AdobeStock/Rawpixel.com; *187* naqiewei/Shutterstock.com; *188* Used with permission from Blackboard, Inc.; *191* Adisa/Shutterstock.com, Alexey D. Vedernikov/Shutterstock.com, Andrew Donehue/Shutterstock.com, © iStock.com/-Oxford-, Johnny Lye/Shutterstock.com, zentilia/Shutterstock.com; *194* iPhone® is a registered trademark of Apple Inc.; *195* Courtesy of Google; *196* Courtesy of Verizon Wireless; *197* Courtesy of Paradigm Education Solutions.; *202* © iStock.com/gunnargren; *205* Courtesy of Nasa.gov.; *207* Courtesy of Today Financial; *208* ALMAGAMI/Shutterstock.com; *210* Google and the Google logo are registered trademarks of Google Inc., used with permission., Reproduced with permission of Yahoo.; *211* Google and the Google logo are registered trademarks of Google Inc., used with permission.; *214* Reproduced with permission of Yahoo.; *215* Google and the Google logo are registered trademarks of Google Inc., used with permission.; *217* Courtesy of deviantART., Courtesy of Instagram. All rights reserved., Courtesy of LinkedIn Corporation, Courtesy of Pinterest. All rights reserved., Courtesy of Reddit, Inc., Courtesy of Tumblr, Inc., Google and the Google logo are registered trademarks of Google Inc., used with permission., Reproduced with permission of Yahoo. © 2018 Yahoo. FLICKR, the FLICKR logo, YAHOO and the YAHOO logo are registered trademarks of Yahoo.; *218* Courtesy of Investor Village; *219* Courtesy of Southwest Airlines/southwest.com; *221* Courtesy of USA.gov; *222* FileZilla is a registered trademark of its respective owners.; *224* Gladskikh Tatiana/Shutterstock.com; *225* issumbosi/Shutterstock.com; *227* Courtesy of 1x1 Media; *229* Tashatuvango/Shutterstock.com; *230* © iStock.com/onivelsper.